Beta maritima

Enrico Biancardi · Leonard W. Panella ·
J. Mitchell McGrath
Editors

Beta maritima

The Origin of Beets

Second Edition

 Springer

Editors
Enrico Biancardi
Formerly Stazione Sperimentale
di Bieticoltura
Rovigo, Italy

J. Mitchell McGrath
USDA-ARS, Sugar Beet and Bean Research
Michigan State University
East Lansing, MI, USA

Leonard W. Panella
Crop Research Laboratory
Department of Soil and Crop Sciences
Colorado State University
(Formerly USDA-ARS)
Fort Collins, CO, USA

ISBN 978-3-030-28750-4   ISBN 978-3-030-28748-1   (eBook)
https://doi.org/10.1007/978-3-030-28748-1

This Springer imprint is published by the registered company Springer Nature Switzerland AG
The registered company address is: Gewerbestrasse 11, 6330 Cham, Switzerland

The only success achieved, but without doubt petty and insignificant if compared to the juvenile hopes, dates back to the beginning of the Century, when seed of Beta maritima collected along the Adriatic coast was crossed with sugar beet varieties. It was possible to identify some genealogies endowed with an actual resistance to cercospora leaf spot
Ottavio Munerati
Rovigo, 1949

Foreword

It might be tempting to ask "why a book about sea beet?": a wild plant with no immediately obvious attraction or significance, a somewhat limited geographical distribution, and for a scientist an underlying genetics that doesn't lend itself to easy experimentation. This book provides counterarguments to allay such misapprehensions, detailing its journey through pre-history, its contribution to one of the world's most recently evolved crop plants, and its significance in terms of modern biodiversity conservation. To emphasize its significance, aside from a book on teosinte written in the 1960s, there are probably no other books that focus specifically on a single crop wild relative.

While sea beet is commonly thought to be an inhabitant of Europe, North Africa, and the Near East, closely related leaf forms of beet were undoubtedly used as a medicinal plant and as a herb or vegetable in Chinese cuisine as far back as the first millennium BC. In 1976, I received correspondence from William Gardener, who was an obsessive collector of plant data and who lived part of his life in China, fluent in both the spoken and written languages. He had recorded that the leaves of "t'ien ts'ai" or cultivated beet, along with some fish, could be used in the preparation of a preserve called "cha". Cha is a preparation originating from the Yangtze valley and Gardener's research led him to believe that t'ien ts'ai, when brought into culinary use, was a coastal plant from anywhere south of Shantung, and perhaps a riparian plant from along the lower Yangtze. However, there are now no records of wild beets growing anywhere in China, so Gardener's assumption that wild as well as cultivated beets existed in China in these times represents one of the enigmas surrounding this crop wild relative.

Considering geographical range and moving to a different continent, it has long intrigued me as to how wild forms of beet, closely related to *Beta maritima*, come to exist in California. The fact that genetic evidence suggests that there are two distinct forms living in the Imperial Valley, both having European origins, only partly clarifies the situation. One form is likely to be a naturalized or de-domesticated cultivated beet, while the other closely resembles the wild *Beta macrocarpa* (a sister species to *maritima*). So a second enigma exists as to precisely how both forms of wild beets reached California.

What else is intriguing about *Beta maritima*? For me, it is its place in the history of genetic resources conservation. I believe that it could comprise one of the first crop genetic resources to have been actively conserved. As a postgraduate student, I was first introduced to the needs of "genetic conservation" by my mentor Prof. Jack Hawkes in Birmingham who, along with Prof. Trevor Williams, my supervisor, collected beet germplasm with me in Turkey in 1972. Other key figures who passed through Birmingham at the time such as Jack Harlan, Erna Bennett, and Otto Frankel were also key to my education. Jack Hawkes, in particular, had met the great Russian geneticist Nikolai Ivanovič Vavilov in the Soviet Union and acknowledged him to be the "father" of plant genetic resources. Vavilov had proposed in the 1920s that crop improvement should draw from wide genetic variation and on this premise collected cultivated plants and their wild relatives from most parts of the world. The germplasm that he collected was for immediate use for the development of new crop varieties in the Soviet Union, and not specifically for conservation. George H. Coons on the other hand was a US scientist, sugar beet breeder, and germplasm collector, who also influenced my early thoughts and activities ahead of my germplasm collecting missions to Turkey back in the 1970s. Remarkable for me, some of Coons's material was actually conserved, allowing me to use some of it in my research, and indeed still survives within the USDA-ARS system in Salinas, California. In many ways, Coons was no different to Vavilov; expeditions to Europe in 1925 and 1935 allowed him to collect and then evaluate diverse germplasm and put it to good use in sugar beet improvement programs and so Coons should be placed alongside Vavilov in the promotion of germplasm conservation.

Maybe as a plant scientist one could easily be put off working on beet. But really its basic genetics is what makes it fascinating. *Beta maritima* and its relatives range from being short-lived annuals where flowering and seed set can be as short as 6 to 8 weeks, to long-lived perennials that are known to survive for as long as 8 years. They can be strongly inbreeding on the one hand but exhibit genetic incompatibility and obligate out-crossing on the other. In light of the most recent taxonomy where *Beta maritima* is actually a subspecies of *Beta vulgaris*, then this wide range of habits and genetic tendencies is all to be found within a single species and may make it much less vulnerable to climate change, unlike other crop wild relatives. Again, because the wild and cultivated are so close genetically, this is a benefit if genes from wild populations need to be used in crop improvement. In contrast, this represents a serious problem in terms of breeding strategies where hybrids can easily occur and contaminate sugar beet seed crops. This may also leave wild beets vulnerable to contamination from GM sugar beet crops.

These features of beet, particularly related to the life cycle are what has made it worthwhile to sequence its genome along with that of sugar beet, something that has happened between now and the first edition of the book.

With a reference genome in place, and the sequence availability of closely related species such as spinach, we will rapidly be able to answer some of the intriguing questions, particularly regarding genes conferring diverse genetic adaptation exhibited by this enigmatic species, many of which are covered in this valuable book. Finally, I strongly believe the value of the book lies in its contribution to avoiding "reinventing the wheel." Combining historical perspective with sound taxonomy, plant breeding, and molecular genetics, it will provide an important overview of the current state of crop wild relative and sugar beet research. It will also provide access to knowledge for new researchers who may wish to revisit the enigmas that wild beet represents.

Birmingham, UK   Brian Ford-Lloyd
2012

Preface

The publication of a book dealing only with a plant without any direct commercial interest is a task requiring some explanation. Given that *Beta maritima* is believed to be the common ancestor of all cultivated beets, the collection in a single publication of the more relevant references concerning the species is useful for biologists, agronomists, and researchers who have the task of preserving, studying, and utilizing the *Beta* gene pool. Indeed, *Beta maritima* is necessary to ensure a sustainable future for the beet crops. This very important reason is the easiest to present, but not fully satisfactory to explain a book dedicated to a single wild plant.

Among other reasons, increasing attention must be paid to wild germplasm as source of potentially useful traits in cultivated crops. Indeed, genetic resistances are a crucial argument due to the urgent need to minimize both production costs and the use of chemical. The need is especially apparent for sugar beet, which is cultivated on about 5.2 Mha in 38 countries, supplying around 20% of the sugar consumed worldwide.

Editing this book, particular attention was paid to the history of the use, recognition, and knowledge of *Beta maritima*. This was done because little has been collectively recorded, and because science evolves also on the foundations of the past. This interpretation of the flow, distillation, and accumulation of knowledge that points forward is another task of the book. The information was collected from the literature dealing in medicinal and food plants in general, and, to a lesser extent, with cultivated beets. This part required reading reprinted manuscripts written over almost two millennia, but the search gleaned information sometimes unknown even to insiders.

Recently, an increasing number of scientific papers related to *Beta maritima* have been published, based on the developments and applications of molecular biology. Several doctoral theses concerning particular aspects of the species have been authored as well. In fact, sea beet germplasm currently is used as a model for gene flow experiments, owing to the frequent coexistence in the same area of different and interfertile genotypes belonging to the genus *Beta*. Being a littoral species and consequently distributed in populations more extended in length (along the beach) than in width, *Beta maritima* fits very well into research concerning population genetics, natural selection, colonization, speciation, etc. In these fields of research, *Beta maritima* is surely one of the more interesting and studied wild plants.

Part of historical information was collected through digital libraries listed in the Appendices. The traditional system of bibliographic research has retained its importance not only for the large amount of yet to be digitized books (and therefore named "analogic" by some) but also for old collections of journals no longer in print or with limited distribution, such as the "Österreiche-Ungarische Zeitung für Zuckerindustrie und Landwirtschaft", where the first important experiences on *Beta maritima* were published at the end of 1800s.

The large number of researches concerning molecular genetics recently undertaken and the download of more than 4200 single chapters of the first edition have led the publisher to propose the second edition. The request surprised the editors, absolutely unready for this occurrence. It seemed impossible that there was any interest in a book with such a limited and specific target, moreover, concerning mainly a noncommercial plant. In the end, the proposal was accepted, despite the need to rewrite at least half of the first edition. The rapid evolution of the matter required the involvement of other experienced researchers and the remaking *ex novo* of the chapters regarding molecular genetics.

The modern breeding techniques have moved mainly to glasshouses and laboratories. This evolution resulted in researchers having less and less contact with the real crop and its background. A further task of the book is to try to provide them an updated, comprehensive summary on everything that involves the species. The outlook should be appreciated, given the future difficulties to put together the variety of skills that allowed the publication of this book.

Owing to the huge amount of recent papers, the editors apologize for possible omissions.

Rovigo, Italy   Enrico Biancardi
Fort Collins, USA   Leonard W. Panella
East Lansing, USA   J. Mitchell McGrath

Acknowledgements

First of all, the editors would like to thank the friend Bob Lewellen for his fundamental contribution to the first edition of this book. He justified his resignation in this way: "I have been retired for a few years and I am not enough updated on the latest literature, so I do not feel able to participate in the role of editor." Notwithstanding his huge modesty, it can be recalled that he will be considered the most prolific (in terms of released genotypes) and meritorious breeder in the history of sugar beet.

It must be recalled with gratitude the collaboration given by Luciano Soldano, Hsing-Yeh Liu, Gudrun Kadereit, and Nigel Maxted in some basic matter. Thanks are also addressed to Marco Bertaggia, Gianpaolo Fama, and Davide Drago for their contribution in the preliminary bibliographic searches. Special appreciation is given to Donatella Ferraresi for her original watercolor paintings and to Mauro Colombo for his support in the graphics and organization of the second edition of the book.

Thanks must be given to the entire private beet seed, production, and processing companies that have been so generous in their funding of the public research efforts, and in North America to the Beet Sugar Development Foundation, which functions as the beet sugar industry's research support organization.

Finally, for their help in historical researches, the authors wish to express their appreciation to the staff of the public and private libraries listed in Appendix E.

Rovigo, Italy   Enrico Biancardi
Fort Collins, USA   Leonard W. Panella
East Lansing, USA   J. Mitchell McGrath

About This Book

Along the undisturbed shores of the Mediterranean Sea and the European North Atlantic Ocean, the plant called *Beta maritima* or more commonly "sea beet" is quite widespread. The species has had and will continue to have invaluable economic and scientific value. Indeed, according to Linnaeus, it is considered the progenitor of the cultivated beet crops, which has been confirmed by recent molecular research. Something similar to mass selection applied after domestication has created many cultivated types with different uses (fresh vegetable, fodder, sugar, ethanol, etc.). Also, the wild plant has been harvested since antiquity and used both for food and for its medicinal properties. Sea beet hybridizes easily with the cultivated types. This facilitates the transmission of genetic traits partly lost during the domestication processes aimed at increasing the features useful to farmers, consumers, and sugar industry. In the last decades, modern breeding techniques have moved mainly to the laboratory. As for other crops, this evolution has resulted in researchers having less contact with the real crop and its cultivation practice. Also for this reason, one of the objectives of the book is to provide an updated summary of everything that involves sea beet, including history, distribution, physiology, breeding, and taxonomy.

Beta maritima has been successfully used to improve the genetic resistances against diseases and pests of the crop, allowing some of the more important results in plant breeding. In fact, without the recovery of traits of resistance preserved in the wild germplasm, the cultivation of sugar beet would be today impossible in almost all countries.

Enrico Biancardi
Leonard W. Panella
J. Mitchell McGrath

Note to the Reader

To make more comprehensible the rare and fragmentary references, the knowledge regarding *Beta maritima* and synonyms was ordered chronologically and placed in its historical framework. In fact, it has been necessary to briefly review information on the evolution of scientific though. Because of the direct parentage with *Beta maritima*, the similarity of the two taxa and their continuous interrelationships after domestication, some information involving the beet crops has been required. Actually, without molecular analyses, differences in morphology are frequently not sufficient for the correct classification inside the section *Beta*, leading to some uncertainties in current *in situ* and *ex situ* collections.

In this book, *Beta maritima*, now classified *Beta vulgaris* L. subsp. *maritima* (L.) Arcang., is called for the sake of brevity "*Beta maritima*" or "sea beet". To avoid confusion and other complications, these names will be utilized also before the taxonomy of Linnaeus. The term "wild beet" is used to indicate the species (spp.) and subspecies (subsp.) belonging to the genus *Beta* excluding *Beta vulgaris* L. subsp. *vulgaris* (cultivated beets). In order to avoid confusion, *Beta maritima* is considered species (spp.) or subspecies (subsp.) according to the respective reference.

In the references of old books, and manuscripts, after the anglicized name of the author, year of publication and title, the printer or publisher (when available), are listed the anglicized name of the printing location and the country. The printer or publisher is typed in Roman fonts. References of more recent reprintings are indicated as well, where applicable.

For uniformity, the initial of the word *Beta* is always capitalized, even though this was not compulsory before Linnaeus. Latin phrases, words, and botanical names (genus, sections, species, and subspecies) are written in Italic (APG II 2003). According to the same classification, subfamily, family, and superior categories are written in Roman. Latin or Latinized names of the authors are typed in Italic or in Roman if Anglicized. The common or vulgar names of plants are also typed in Roman. The initials of the common names of diseases are written in lowercase, as are the acronyms of viruses. Words and phrases in other languages are written in Italic between brackets, whereas the English translation is written in Roman between parentheses. The captions of figures (Fig.) and tables (Tab.) without indication of the source are intended as supplied by the editors. The references cited in captions, notes, and in Appendix A are included in the first chapter.

Contents

Affiliation of Editors and Collaborators

Detlef Bartsch: University of Aachen, Germany

Enrico Biancardi (ed.): Formerly Stazione Sperimentale di Bieticoltura, Rovigo, Italy

Henry Darmency: Formerly INRA, Dijon, France

Marco de Biaggi: Sugar Beet Breeding Consultant, Massalombarda, Italy

Brian Ford-Lloyd: Formerly University of Birmingham

Lothar Frese: Formerly Julius KuhnInstitute, Kuedlingsburg Germany

Nina Hautekeete: University of Lille, France

Robert T. Lewellen: Formerly USDA-ARS, Salinas CA, USA

J. Mitchell McGrath (ed.): USDA-ARS, East Lansing MI, USA

Leonard W. Panella (ed.): Formerly USDA-ARS, Fort Collins CO, USA

Rania Pavli: Agricultural University of Athens, Greece

George N. Skaracis: Formerly Agricultural University of Athens, Greece

Piergiorgio Stevanato: DAFNAE, University of Padua, Italy

Pascal Touzet: University of Lille, France

Henk van Dijk: University of Lille, France

Abbreviations

ARS	Agricultural Research Service (USA)
ASSBT	American Society Sugar Beet Technologists
BMV	Beet Mosaic Virus
BNYVV	Beet Necrotic Yellow Vein Virus
BYV	Beet Yellowing Virus
CMS	Cytoplasmic Male Sterility
CRA	Consiglio per la Ricerca e Sperimentazione in Agricoltura
CRISPR	Clustered Regularly Interspaced Short Palindromic Repeat
CWR	Crop Wild Relatives
DNA	Deoxyribonucleic Acid
FAO	Food and Agriculture Organization
GABI	German Agricultural Biotechnology Initiative
GP1, GP2, GP3	Gene pool 1, Gene Pool 2, Gene Pool 3
GRIN	Germplasm Resource Information Network (USA)
GWP	Gathered Wild Plant
IBPGRI	International Plant Genetic Resources Institute
IIRB	International Institute of Sugar Beet Research
IPK	Leibniz Institute of Plant Genetics and Crop Plant Research (Germany)
IRBAB	Institut Royal Belge pour l'Amelioration de la Betterave
ISCI	Istituto Sperimentale perle Colture Industriali (Italy)
LDC	Least Developed Countries
mM	milliMolar
NMS	Nuclear (or Mendelian) Male Sterility
OECD	Organisation for Economic Cooperation and Development
OT	O-type
QTL	Quantitative Trait Locus (Loci)
R-gene	Resistance gene
RI	Recombinant Inherited
RNA	Ribonucleic Acid

SNP	Single-Nucleotide Polymorphism
SSR	Simple Sequence Repeat
USDA	United States Department of Agriculture
WBN	World Beta Network

Chapter 1
History and Current Importance

Enrico Biancardi and Robert T. Lewellen

Abstract The ancestors of *Beta maritima were* known from prehistory. After domestication, beet became more important not only for food and drug source, but also as sugar (sucrose) producer. The cultivation for leaves and root to be used as vegetable or cattle feed retains its economic value. *Beta maritima* was described by several authors, becoming in the last century crucial as source of traits disappeared in the beet crops after domestication. The research has led to important results, especially in the field of resistance to severe diseases. An increasing numbers of publications are dedicated to *Beta maritima* because it fits well into studies concerning breeding in general, population genetics, natural selection, colonization, speciation, gene flow, transgenes pollution, and so on. The discovery of new useful qualities in the wild germplasm is expected by the application of molecular biology.

Keywords *Beta maritima* · Origin · Domestication · History · Crop evolution · Breeding

Beta maritima,[1] commonly named "sea beet", is a very hardy plant that tolerates both high concentrations of salt in the soil and severe drought conditions (Shaw et al. 2002). Thus, it can also grow in extreme situations such as along the seashores almost in contact with saltwater "frequently between the high tide zone and the start of the vegetation, or where the wastage of the sea is deposited" (Figs. 1.1 and 1.2). On the contrary, sea beet is sensitive to competition with weeds especially under water and nutritional deficiency (Fig. 1.3) (Coons 1954; de Bock 1986). Sea beet seems to take advantage of its salt and drought tolerance to reduce the presence of competitor plants in the neighborhood (Coons 1954; Biancardi and de Biaggi

---

[1]*Beta maritima*, now classified *Beta vulgaris* L. subsp. *maritima* (L.) Arcang, is called for the sake of brevity "*Beta maritima*" or "sea beet".

E. Biancardi (✉)
Stazione Sperimentale di Bieticoltura, viale Amendola 82, 45100 Rovigo, Italy
e-mail: enrico.biancardi@alice.it

R. T. Lewellen
USDA-ARS, Salinas, CA, USA

This is a U.S. government work and not under copyright protection in the U.S.; foreign copyright protection may apply 2020
E. Biancardi et al. (eds.), *Beta maritima*,
https://doi.org/10.1007/978-3-030-28748-1_1

1

Fig. 1.1 Sea beet on a stone bank at the mouth of Po di Levante River, Italy. The plant grew on a few grams of sea debris and was able to flower and set seeds notwithstanding being surrounded by salty water. Any other superior plant can survive in these conditions, thus demonstrating the very high environmental adaptability and stress tolerance of the species. Due to the uneven distribution of rains and the limited water supply, *Beta maritima* can be observed in this site only after rainy season, that is, once in about a decade. Therefore, the survival of the populations, at least in the mentioned location, implies also a longlasting germination ability under high salt concentration and unknown interactions with the seed dormancy (Biancardi, unpublished)

Fig. 1.2 Site with optimal growing conditions for *Beta maritima*: vicinity to the seawater; sandy/stony soil; low presence of competing weeds; tourism connected activities; grazing cattle; etc. Baja California USA (Courtesy, Bartsch)

Fig. 1.3 *Beta maritima* competing against weeds (Torcello, Italy)

1979). Salty soils, frequently caused by seawater spray, tidal flows, storms, and so on, also induce relatively low pathogen pressure, thus may be helpful for the survival of the species. von Proskowetz (1910) referred to having never seen cysts of nematodes on sea beet roots, likely due to their very high woodiness. Conversely, Munerati et al. (1913) observed severe attacks of *Cercospora beticola*; *Uromyces betae*; *Peronospora schachtii*; and *Lixus junci* along the Italian-Adriatic seashores. Bartsch and Brand (1998) referred to the absence of beet necrotic yellow vein virus (BNYVV), the causal agent of rhizomania, as likely related to the high salt content in soils.

Saltwater plays an important role in the dispersal of the species. Less frequently, also for this reason, sea beet populations are localized in interior areas, in the presence or absence of beet crops in the vicinity. In the first case, the wild populations are likely to be feral or ruderal beets[2] that are more or less aged offspring of beet cultivation (Ford-Lloyd and Hawkes 1986; Bartsch et al. 2003).

1.1 Predomestication

The first use of sea beet (or one of its earlier relatives) goes back to prehistory, when the leaves were gathered and used as raw vegetable or pot herb (von Boguslawski 1984). The leaves, shiny and emerald green even in winter (Fuchs 1551), were unlikely confused with those of other plants, a feature that was very important for the first harvesters. The separation of the sub-family Betoideae (to which the genus *Beta* belongs) from the ancestral family Chenopodiaceae is estimated to have occurred between 38 and 27 million years ago (Hohmann et al. 2006). Therefore, it is possible that sea beet already was known to our ancestors in their remote African dawn.

Further confirmation of sea beet's ancient and widespread use are the remains of desiccated seed stalks, carbonized seeds, and fragments of root parenchyma found in the sites of Tybrind Vig and Hallskow, Denmark, dated from the late Mesolithic (5600–4000 BC) (Kubiak-Martens 1999, 2002; Robinson and Harild 2002). Pals (1984) reported on the discovery of similar remains in the Neolithic site (around 3000 BC) at Aartswoud, Holland. In agreement with Kubiak-Martens (1999), evidence of harvest and use of sea beet also are present at the Neolithic site at Dabki, Poland. Pollen of *Beta* wild plants was recognized in sediments sampled at Lake Urmia (Iran), Lake Jues (Germany), and Adabag (Turkey) dated around 10,000 years BC (Voigt et al. 2008; Bottema 2010).

The presence of fragments of root in the sites suggests that this part was used as frequently as the leaves. It is important to remember that in northern regions, the roots of sea beet are much more regular and developed than in southern environments. Therefore, the root better lends itself to harvest (Fig. 1.4) most likely beginning in

---

[2]Feral beets originate by a "dedomestication" of the crop. The process starts with the early flowering (bolting) of some cultivated beets before harvest.

Fig. 1.4 Atlantic *Beta maritima* with regular and swollen root (Smith 1803)

August, whereas the leaves were collected mainly in winter through spring (Kubiak-Martens 1999). After the discovery of fire, leaves and roots were eaten after cooking (Turner 1995). The frequent presence of remains of other wild plant species in these sites suggests the key role that vegetables played in the hunter–gatherer's diet even in pre-agrarian times (Kubiak-Martens 2002).

Charred remains of sea beet seeds were identified in late Mesolithic sites located in the northern region of the Netherlands, demonstrating the ancient presence of the species along the North-Atlantic seashores (Perry 1999), as it was further confirmed by the remains of sea beet found at the site of Peins, the Netherlands, dated to the first century BC (Nieuwhof 2006). Collecting data from 61 archeological sites in different parts of Egypt dated from predynastic to Greco-Roman times, Fahmy (1997) recognized 112 weed species including sea beet. Macro remains of the plant (seeds, leaves, stalks, etc.) were preserved by desiccation in sites dated from 3100 BC until the middle of the Pharaonic period (2400 BC).

As to the area of origin of the species, de Candolle (1885) wrote: "beets originated from Central Europe or from nearby regions, due to the large amount of wild species of the genus *Beta* present throughout the area". Some years later, de Candolle (1884) asserted that the beet crop, "which is the more easily [plant] to be improved by selection", was derived from the species now classified *Beta cicla* (or *Beta vulgaris* L. subsp. *vulgaris* Leaf Beet Group), very similar to sea beet. He also affirmed that *Beta cicla* expanded from the Canary Islands along the North-Atlantic coasts to the Mediterranean areas, up to the countries around the Caspian Sea, Persia, and Mesopotamia. The hypothesis of de Candolle, perhaps reasonable because of the numerous *Beta* species present today on Canary Islands, has not been confirmed by later authors (Meyer 1849; Pitard and Proust 1909; Francisco-Ortega et al. 2000). According to Coons (1954), the origin of sea beet could be located to the areas delimited by Ulbrich (1934) some decades before (Fig. 1.5).

Southwest Asia could be the area of origin, not only of sea beet and many other important crops (wheat, barley, etc.), but also of the family Chenopodiaceae (now Amaranthaceae), in which the genus *Beta* is included. Avagyan (2008) suggested that the species could have originated in Armenia. A number of authors: Honaker, Koch, Boissier, Bunge, Radde, and others reviewed by von Lippmann (1925), agree in locating the origin of the genus *Beta* in the area comprising the shores of the Caspian Sea, Transcaucasia, the East and South coasts of the Black Sea, Armenia, Asia Minor, the shores of the Red Sea, Persia, and India. Analyses of cytoplasmic diversity confirmed that the area of origin of sea beet should be the Mediterranean countries, where it is widely diffused even today (de Bock 1986; Cheng et al. 2011).

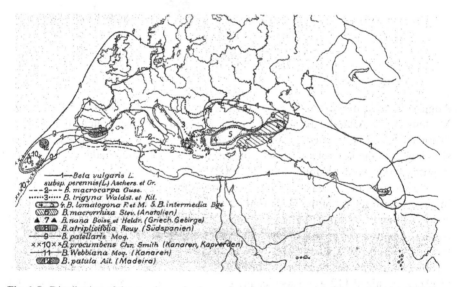

Fig. 1.5 Distribution of the species and sub-species of genus *Beta* according to Ulbrich (1934)

1.2 Domestication

Domestication can be described as the changes necessary to adapt plants to habitats especially prepared by man (van Raamsdonk 1993). Based on the rudimentary tools found in settlements of Neolithic age, the first farming of wheat (*Triticum* spp.) and barley (*Hordeum* spp.) is thought to have arisen in the Near East, perhaps earlier than 8500 BC (Zohary and Hopf 2000). The agricultural practices then would have spread into the Mediterranean areas through the ship routes of that time, and more slowly toward Central Europe. At least three millennia were necessary for agriculture to arrive in the British Islands, Scandinavia, and Portugal (Zohary and Hopf 1973, 2000): that is spreading at a rate of about 1 km per year (Cavalli-Sforza and Edwards 1967).

Beet cultivation may have begun, perhaps more than once, in Mesopotamia around 8000 BC (Simmonds 1976; McGrath et al. 2011). According to Krasochkin (1959), the first beet cultivation occurred in Asia Minor, mostly in localities at relatively high altitude with a cool growing season. Subsequently, the practice spread to Mediterranean areas, developing a great diversity of primitive forms of beet still existing today. The wild ancestor may have resembled types currently present in western Anatolia and Afghanistan, characterized by short life span, large seed-balls, elongated and fangy roots, tendency to flower very early, and so on (Krasochkin 1959, 1960). Using analyses of mitochondrial DNA, Santoni and Bervillè (1992) confirmed the hypothesis that cultivated beets likely originated from a unique ancestor quite different from the one currently known. After domestication, sea beet has continued to be harvested in wild sites and to be used as a vegetable, a custom still widespread in many coastal areas (Thornton 1812). According to Magnol (1636) *"Nihil in culinis Beta frequentius est"* (nothing is more used in the kitchen than beet). Rivera et al. (2006) consider the sea beet among the most gathered wild plants for food (GWP) in the Mediterranean and Caucasian regions. In the mentioned paper, the local names of sea beet are listed in 25 languages.

van Zeist and de Roller (1993) argued that beet farming had spread throughout much of Egypt by the time of construction of the pyramids of Giza (around 2700 BC). This hypothesis is supported by Herodotus (von Lippmann 1925). Because of the large quantity of beet that would have been required, the vegetable must have been domesticated. According to Buschan (1895), some wall paintings (Fig. 1.6) inside the tombs of Beni Hassan near Thebes, and dating to the 12th Dynasty (2000–1788 BC), represent beet and not horseradish (*Cochlearia armoracia*), as speculated by others. In a second painting inside the same tomb (Fig. 1.7) the farmer seems to have a beet in his hand, while the plants on the ground most likely are garlic (*Allium sativum*) (Woenig 1866). In both paintings, the regular shape of the root suggested that should be a cultivated variety of beet. Given the extensive spread of sea beet along the northern Egyptian coasts, Buschan (1895) speculated that its cultivation in the region had begun much earlier. In Fig. 1.8, the word meaning "beet" is written in ancient Egyptian (Kircher 1643; Veyssiere de la Croze 1755). Other findings dating from the third Dynasty (2700–2680 BC) have been made at Memphis, Egypt (Zohary

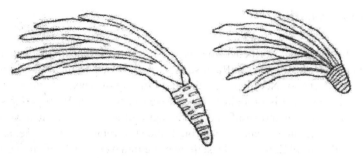

Fig. 1.6 Sea beet (or something similar) drawing at Beni Hassan, Egypt (Buschan 1895)

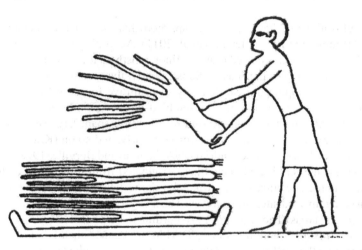

Fig. 1.7 Sea beet (likely) in the hands of the farmer. Painting at Beni Hassan, Egypt (Woenig 1866)

Fig. 1.8 The word meaning
"beet" written in old
Egyptian alphabet (Veyssiere
de la Croze 1755)

and Hopf 2000). The lack of morphological differentiation often does not allow the establishment of whether remains are from wild or cultivated beets. In general, if the beet plant remains are found far from the sea and after the spread of agriculture in the area, it may be assumed that they are derived from cultivated beets. This is the case of beet seeds found in central Germany in sites dating to the Roman Empire (Zohary and Hopf 2000). A very original hypothesis was proposed by Stokes (1812).

He restored the old name *Beta sylvestris* and the likewise old name *Pyrola major*, establishing that it is "native of North America and Europe".

The cultivated beets have been adapted in response to selective pressures imposed by growers, who instinctively selected for reproduction the plants with the best expression of the traits of interest. The domestication process was hastened by utilizing plants showing mutations as well, but only if the new trait enhanced the qualities required by the farmers (Fehr 1987). This early selection, according to Ford-Lloyd et al. (1975), gave rise to a taxon classified as *Beta vulgaris* subsp. *provulgaris*, an ancestral form selected both for root and leaf production. The inherited offspring of this plant is believed still existent in Turkey (Ford-Lloyd et al. 1975).

Some traits necessary for survival in the wild became superfluous in cultivated field (Zohary 2004). For example, cultivation by the farmer reduced the beet's already poor competitive ability against weeds, a trait which is not necessary or of reduced in artificial monoculture. The annual cycle, necessary for increasing seed production and thus essential for the survival in the wild (Biancardi et al. 2005, 2010), slowly became biennial. In this way, as with other vegetables, was increased the duration over which leaves and roots remained edible (Harlan 1992). As a consequence of the selection process, genetic diversity decreased rapidly (Bartsch et al. 1999). Santoni and Bervillè (1995) observed in cultivated beets the lack of the rDNA unit V-10.4-3.3, common vice versa in wild beets. Because *Beta maritima* has been used in the last century as a source of resistances, the authors suspected the elimination of this DNA unit occurred through the selection processes. Recently, Li et al. (2010) confirmed the key role of genetic variation for the traits of interest in the first phase of sugar beet breeding (Ober and Luterbacher 2002).

The first written mention of beet farming goes back to an Assyrian text of the eighth century BC, which described the hanging gardens of Babylon (Meissner 1926; Ulbrich 1934; Körber-Grohne 1987; Mabberley 1997; Zohary and Hopf 2000). As has happened with the most important crops, the cultivated beet left its first domestication sites (Kleiner and Hacker 2010). Whereas Cheng et al. (2011) speculated that *Beta* has been domesticated in the Mediterranean area. Some centuries BC, the leaf beet was called "selga" or "silga", words that, according to Winner (1993), would have the same origin as the Latin adjective "*sicula*" (Sicilian). Around 400 BC, the cultivated leaf beet returned to Asia Minor (whence the sea beet had spread some millennia earlier) from Sicily, whose population of Greek origin had extensive trade relations with Mycenae and the eastern Mediterranean harbors (Becker-Dillingen 1928; Ulbrich 1934). Older European peoples, such as the Arians, did not cultivate beet (de Candolle 1885; Geschwind and Sellier 1902).

1.3   Athens and Rome

The first unambiguous written reference to beet cultivation dates back to Aristophanes, who mentions beet, at the time called τευτλον (*seutlon* or *teutlon*), in the plays "The Acharners", "The Frogs", and "Friends" (Winner 1993). According to von Lippmann (1925), in an old edition of "*War between frogs and mice*", a comedy written by Homer, there are some words resembling τευτλον, but their meaning is still uncertain. Again, according to von Lippmann (1925), the first written reference positively alluding to beet dates back to Diocles from Carystos (end of fourth century BC), who included its dried leaves in a medicinal mixture with other herbs. Diocles stated that the wild beet (τευτλον άγςια or άγριον) was very common along the coasts of Greece and its islands. The wild plant was rather different when compared to the cropped *Beta* (Jaeger 1952). The cultivated beet is of two types: white (λευχόυ) and black (μελαν). For sea beet, Diocles used also the terms "βλιτος (*blitos*)" and "λειμωνιον (*leimonium*)", which certainly can be attributed to the plant. Diocles is believed to be the author of the first illustrated herbal considered the prototype of several later authors (Collins 2000).

In "*Historia plantarum*" (295 BC?), the philosopher Theophrastus confirmed the existence of two varieties of cultivated beets: the black "τευτλον μελαν (*nigra*)" and the white "τευτλον λευχόυ (*candida*)" also called "*cicla*". Both display a long and narrow root similar to horseradish and have a sweet and satisfying taste. This description coincides with the shape of the plants painted at Beni Hassan. Both Diocles and Theophrastus described a beet, like the black one, and grown at the time for its roots. According to Sturtevant (1919), Aristotle himself cited the existence of a third cultivated type: the red beet. Theophrastus also listed the medicinal properties of sea beet. Since that time, the plant has taken on the dual nature of food and of medicinal herb against some diseases.

As for other types of beet, with rare exceptions, the therapeutic use was the most prevalent in books written until the end of the twelfth century (Jackson 1881; Lamarck 1810). The medicinal properties of sea beet were best described by the physician Hippocrates, who is recognized as the founder of medicine based on proto-scientific basis (Dalby 2003). von Lippmann (1925) argued that the dark-leaved variety (*nigra*) was cultivated extensively in the Grecian world also for the root.

In "*De Re Rustica*" (274 BC), the Roman writer Cato used the word "*Beta*" for the first time without giving indication of its source (Schneider 1794). The term appeared in the following phrase regarding the composition of a laxative mixture:

"*Si ungulam non habebis, adde betae coliculos cum radice sua*"
(If the nail of jam is not available, use the beet stalk and its root).

According to Columella and several later writers, the name seems to derive from the second letter of the Grecian alphabet, that is, the letter whose form looks like the embryo of the seed in the early stages of germination (Berti-Pichat 1866). de Lobel (1576) confirmed:

"*Betam etenim a litera graeca β sic dictam vocant*"
(It is believed that *Beta* is so-called from the Greek letter β).

Whitering, cited by Baxter (1837), approved that the name is derived from the form of its seed vessel, which, when swollen with seed, resembles the letter "β". The hypothesis that "*Beta*" was derived from the Celtic "bett" (red), or from the Irish "biatas" (red beet) (Kirby 1906; Baxter 1837) does not seem to be supported due to the infrequent contacts that Rome had at the time with the British Islands (Poiret 1827; von Lippmann 1925). Moreover, according to Geschwind and Sellier (1902), people of Celtic origin began to grow beets in Central Europe only around the fourth century AD. According to Strabo (cited by von Lippmann 1925), the use in North Sea area of "wildwachsene Gemüse" (wild vegetables) including beet, was dated earlier. An original hypothesis was given by Pabst (1887): in his opinion the word "*beta*" derived from the Latin "*meta*", which means, among other things, "conic heap of stones", similar to the spindle form of the beet root. Because the germinating seed resembles α (alpha) more than to β (Fig. 1.9), the assonance of the Greek word "βλιτος" cannot be missed. The etymological evolution of the word may be as follows: βλιτος → *Blitos* → *Blitum* → *Bleta* → *Beta* (Becker-Dillingen 1928).

The beet crop was mentioned several times by Latin writers including Plautus, Cicero, Catullus, Virgil and Varro. Martial (80 AD?) listed the beet "among the abundance of the rich countries", and defines it as "unserviceable to a sluggish stomach" (Feemster-Jashemsky and Meyer 2002). Beet was cited in two epigrams:

"*Pigroque ventris non inutiles betas*" (Beet is useful for lazy bowel).

"*Ut sapiant fatua fabrorum prandia betae, o quam saepe petet vina, piperque cocuus*"

(Insipid beet may bid a tradesman dine, but asks abundant pepper and wine)

Suetonius wrote that the emperor Caesar Augustus invented the verb "*betizare*" to indicate man showing effeminate behavior (Tanara 1674). Pliny the Elder (75 AD?) provided important information on the crop in "*Historia Naturalis*", mentioning both agricultural methods of cultivation and medicinal properties. Like Hippocrates and Theophrastus, Pliny mentioned the existence of varieties with white roots (*candida*) and dark green leaves (*nigra*). The plant could be sown either in spring or autumn; the seed took 6 days to germinate in summer and 10 days in winter. Germination of some seeds also occurred after two or more years. Among the uses of beet as food, Pliny also mentioned the root. This seems to confirm the hypothesis that in Roman times some new varieties (*Beta rubra*) appeared whose root, tender and sweet, was eaten after cooking. The use of the root, perhaps only of sea beet, was already common for medicinal uses in Greece, as reported by Hippocrates. For Pliny, the wild beet, named "*Beta silvestris*" (see footnote 4) corresponded to the plant called "*limonium*" or "*neuroides*", words dating back to Hippocrates:

"*Est et Beta silvestris quam limonium vocant, alii neuroidem, multo minor tenuoribusque ac densioribus*"

(Sea beet is called "*limonium*" by some and "*neuroidem*" by others, it has smaller and shallower leaves than the cultivated one)

Fig. 1.9 Painting of *Beta vulgaris* showing some particulars of flower and seed (www.bodley.ox.ac.uk)

Pliny also mentioned the existence of illustrated herbaria drawn up by a physician of the Aristotelian school (likely Diocles by Carystos), which described the medicinal properties of plant, mineral, and animal substances (Collins 2000).

The word "*Beta*" was written in some mural graffiti found at Pompeii. The wall inscription in Fig. 1.10, dating before 79 AD, is abbreviated or partially removed and is probably, together with the following, the oldest original writing of the name *Beta*. In another graffiti (Fig. 1.11) was written:

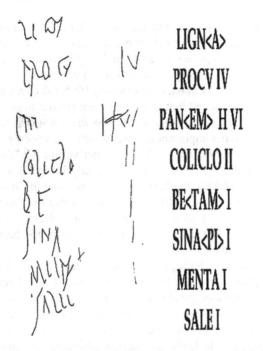

LIGN‹A›
PROCV IV
PAN‹EM› H VI
COLICLO II
BE‹TAM› I
SINA‹PI› I
MENTA I
SALE I

Fig. 1.10 List of prices written near a food shop at Pompeii: *procu* (pig) 4; *panem* (bread) 6; *coliclo* (small cabbage) 2; *betam* (beet) 1; *sinapi* (mustard) 1; *menta* (mint) 1; *sale* (salt) 1. Prices are in *axa* (around half dollar at the time) pro *libra* (around 350 g) (Ciarallo 2004)

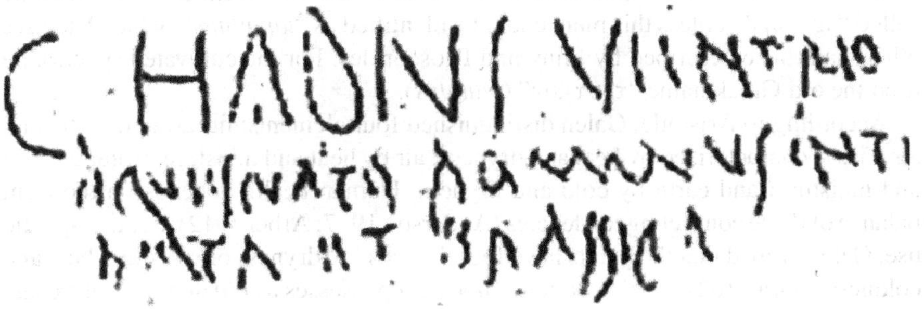

Fig. 1.11 Wall inscription at Pompeii (Ciarallo 2004)

"*C. Hadius Ventrio equus natus romanus intra beta et brassica*"
(*C. Hadius Ventrio*, knight, born Roman citizen among beets and cabbages)[3]

---

[3]The graffiti refers to the vulgar origin of the man, likely "nouveau riche", alluding to the digestive consequences of consuming the mentioned vegetables Funari (1998).

After Pliny, beet was referred to by the name *"Beta"*, at least in books written in Latin, but an incredible amount of synonyms of *"silvestris"*[4] was created (Appendix D). Dioscorides, a contemporary of Pliny and physician of the emperor Claudius Nero, described in *"De materia medica"* (89 AD?) the various medicinal properties of *Beta silvestris*. About *limonium*, mentioned by Pliny (also called *lonchitis, sinapi aselli, etc.*), Dioscorides stated that the leaves were similar to beets, but were more slender, long, and numerous. In other words, *limonium* was a different species with other uses. Attached to the important treatise, which was widespread and influential during the Middle Ages was believed to be a herbarium probably dating back to Crateuas, which included a color drawing certainly referring to beet (Fig. 1.12). According to Collins (2000), the herbarium seems to be attributed to Diocles by Carystos. The caption written in old Greek indicated that the illustration represents the "wild beet" called *"sylbatica"* (synonymous of *"silvestris"*) by the Romans" (Biancardi et al. 2002). But the plant resembles a cultivated beet more than wild because of the regular shape of the root.

As used by astrologers, 18 chapters of *"De materia medica"* described the influence of stars and planets over the herbs and their medicinal effects. Indeed, it was believed that successful therapy always was linked with the astral influence (Riva 2010). Magical properties, such as keeping away the devil, curing the plague, and stimulating sexual attraction, often were attributed to some herbs until a couple of centuries ago. As regards, the herbarium sample attributed to Crateuas, it seems quite unlikely that it was appended to the original *"De Materia Medica"*, because the text makes no references to enclosed drawings (Ventura 1998).

Galen opened a sort of pharmacy in downtown Rome. In *"De alimentorum facultatibus"* he (190 AD?) claimed to be unaware of the wild form of beet, which he called *"agrestis"*, unless this plant could be identified as *"lapathum"*, which had uses other than those described by Pliny and Dioscorides. For the cultivated species, he used the old Greek name "τευτλον" (*teuthlus*).

According to Aristotle, Galen distinguished four elements: fire, water, earth, and air. Fire is characterized by heat and dryness; air by heat and moisture; water by cold and moisture; and earth by cold and dryness. Human health depends on the right balance of these conflicting tendencies (Anderson 1977; Arber 1912). For therapeutic use, Galen argued that the plants have four degrees of "dryness or moisture, heat and coldness" (Gray 1821). Galen believed that beet possesses a cold and wet nature and must be used accordingly. As a Christian, Galen believed in a unique divinity, for this reason his theory was well accepted also by Jews and Arabs (Jackson 1881; Pezzella 2007).

---

[4]The correct Latin adjective first used by Pliny is *"silvestris"*, and not *"sylvestris"* as was written by later authors.

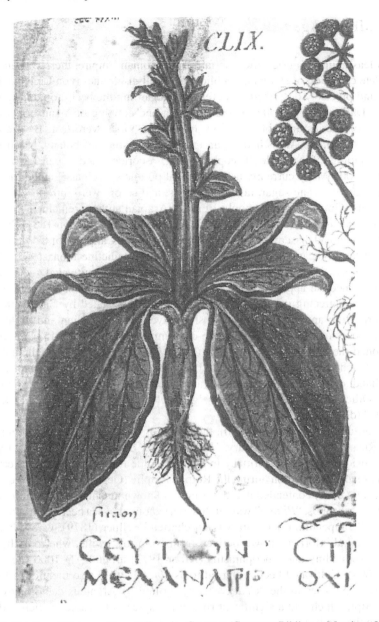

Fig. 1.12 Painting of *Beta maritima* attributed to Crateuas (Courtesy: Biblioteca Marciana, Venice. Reproduction is prohibited)

1.4 Middle Age

For at least eight centuries after the fall of the Roman Empire, there was an almost complete cessation of study and publication in all disciplines. von Lippman (1925) listed and precisely described the references regarding the beet crop during the so-called "Dark ages". Despite the conservation and copying of manuscripts carried out in monasteries and abbeys, many invaluable books were lost. By the end of the millennium, the Arabs had begun to pursue the study of botany, based mainly on translations of Aristotle, Theophrastus, Dioscorides, and Galen (Arber 1912; Collins 2000). Many currently used botanical products, such as camphor, lavender, rhubarb, opium, cane sugar, tamarinds, hops, and so on, were introduced by Arab physicians (Gray 1821). The books of many Arabian authors remained confined to libraries because of the difficulty of writing and reading (Jackson 1881), but some found widespread dissemination in Europe through the Caliphate of Cordoba, Spain and translations, particularly those made in the Benedictine monastery of Monte Cassino, Italy. Some Arabian books mentioned wild and cultivated beets together with their medical applications. Ibn Sina (Avicenna) recommended the use of sea beet leaves, agreeing on their wet and cold nature (as stated by Galen), in different therapeutic applications. Aven Roshdi (Averroes), physician and philosopher, used sea beet named "decka" in some drug mixtures (Bruhnfels 1534). Ibn Beith mentioned the existence of wild beets (likely weed beets) alongside the cultivated fields, which were characterized by a different shape and color. Avicenna, on the other hand, called "selq" the more isolated beets likely *Beta maritima* (Sontheimer 1845). Other Arabic names such as "selg" and "silg" resemble the old Greek name, "sevkle" (de Candolle 1884, 1904).

According to Krasochkin (1960), the beet crop likely spread from Byzantium to Kiev, Russia, in the tenth century. Hildegard von Bingen (Throop 1998) reported this diffusion throughout Germany in the same time frame, but surely the crop had already reached the region during the Roman Empire (Geschwind and Sellier 1902). Shun et al. (2000) contended that the beet was known in China around 500 BC.

In the early 800s, *"Blitum"* was quoted as a synonym of sea beet in the anonymous treatise "Compendium der Naturwissenschaften" Fellner (1879), whose botanical and medicinal information was derived from Isidor of Seville, who took the information from Pliny and Theophrastus (Arber 1912) around AD 1000 (de Divitiis et al. 2004). Cultivated beets, referred to as *"bleta"*, were also mentioned for several medicinal uses in the *"Codex"*, likely written by Arnaldus de Villanova. The manuscript, which had a significant role in the spread of Arabic medicine, did not mention the sea beet.

Albertus Magnus, Bishop of Ratisbona (Regensburg, Germany), reported some recipes based on *blitum* and parsley (*Petroselinum* spp.) (Kennedy 1913). He held the theory that species are mutable, in fact, cultivated plants might run wild and degenerate, and the wild plants could be domesticated. Matteo Silvatico cited *Bleta silvestris* for some therapeutic applications taken from the Arabic literature (Silvatico 1523).

As stated by von Lippmann (1925), identification of *sea beet* in *herbaria*[5], books, descriptions, and indexes of botanical gardens, all written with increasing share after the invention of the printing press, is often difficult. Moreover, confusion exists, not only among the various synonyms and varieties obtained by selection, but also in the identification problem (which still exists) between beets and turnip (*Brassica* spp.) in the case of roots, and between beets and spinach (*Spinacia oleracea*) in the case of leaves (Fischer-Benzon 1894). One must also remember the multitude of local names given to various types of cultivated beet. Because the wild and cultivated beets easily cross with one another, one also must take into account wild populations derived from spontaneous crosses.

Among the herbals of Greek origin recalled previously, we also must mention the "*Herbarium Apuleii Platonici*" and translated into several European languages around 1480. The manuscript "*Tractatus de virtutibus herbarum*" written by Arnoldus de Villanova is illustrated with very simple drawings of various plants, including the beet, here called "*bleta*" (Fig. 1.13). The drawing is accompanied by a short description of the medicinal properties taken mainly from Theophrastus. As regards the sea beet, the hypothesis of Pliny, that identified the *Beta silvestris*

---

[5]The books describing the medicinal applications of plants are named "*herbaria*" or "*dynamidia*" whether they include or not drawings of the plants (Piccoli 2000). The use of *dynamidia* seems to date back to the Chinese, Assyrian-Babylonian, and Egyptian medicine. The "Pents'ao" was written in China around sixteenth century BC (Pezzella 1993). The "Papyrus of Luxor" dated 1550 BC was essentially a list of medical properties of plants (Pezzella 2007). Further examples are given by the *Herbaria* attributed to Crateuas and Apuleius: at least one copy of the latter was employed in the abbeys. In the Middle Age, the herbaria become banal reproductions of ancient manuscripts (Lazzarini et al. 2004). Many transcriptions made by copyists not involved in botany lead to a considerable increase of mistakes in texts and illustrations (Weitzmann 1979). The drawings became very formal and simple, sometimes with complete bilateral symmetry and often included only for embellish the manuscript (Arber 1912). Therefore, the identification of the represented or described plants became quite impossible. The language was a mixture of Latin, vulgar, common, and foreign terms frequently difficult to translate, as the names given to the plants. In the manuscripts, the name of the author was often omitted as the references regarding the hand written book (Gasparrini-Leporace et al. 1952). Only toward the end of the thirteenth century, when it was necessary to print the most important manuscripts, they began to check the names and the correspondence with the reality of descriptions and illustrations. The first printed herbaria were also named "Book of Nature" from the "Půch der Natur" written likely by Konrad von Megenberg (1348?) and published around 1470.

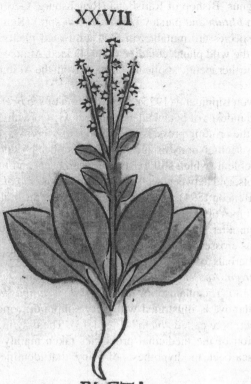

Fig. 1.13 *Beta maritima* here named "*Bleta*" (Courtesy: Orto Botanico, Padua. Reproduction is prohibited)

with *limonium*, was partially confirmed in the *"Liber simplicibus"*[6] written by Roc-cabonella (1457) (de Toni 1925; Pitacco 2002; Teza 1898). The manuscript,[7] pro-viding the illustration of sea beet (Fig. 1.14), is accompanied by its names in Greek, Latin, Italian, German, French, and so on (Fig. 1.15). Roccabonella explained that the drawing of Andrea Amodio represents the *bleta silvestris*, corresponding to *limolion* or *limonion*, the Greek names of wild beet according to Pliny. The small and fangy root seems confirm that it could be a sea beet, likely widespread in the lagoon around Venice at the time. The realism of the hand drawing can be noted, especially when compared with other contemporary illustrations (Fig. 1.16). Signs of the changing times also can be seen in the work by Hermolao Barbaro (1494). In his treatise *"Castigationes Plinianae"*, the author erased from the text of Pliny's *"Historia Naturalis"* the mistakes collected during the frequent recopying that took during the Middle Age.

The thinking of Aristotle, who was, among other things, the author of two lost treatises on botany, dominated all scientific disciplines for a long time, delaying and, in many cases, preventing the development of modern science. The books of Aristotle, Theophrastus, and Hippocrates were transcribed by hand many times, losing in part, as it has been said, their relationship with the originals. Only around the fourteenth

---

[6]*"Simplices"* were called medical substances extracted from various sources and used without any further processing. Those mixed or treated were called *"compositae"*. The first category of drugs is currently called "Galenic" as well; the second "Hippocratic" in agreement on the respective authors. A very useful list of the *simplices* at the time available in the pharmacies of Ferrara, Italy, is given by Musa Brassavola (1537). The medical substances are divided into herbs (including *Beta nigra* and *alba*), seeds, fruits, roots, barks, gums, metals, soils, salts, oils from flowers, oils from mine, and so on. The last ones are named *"petroleum et asphaltum"* as well. At the end of the treatise, as for the modern drugs, are written the applications and the warnings which can be paid before using. The *"Hortus simpliciorum"* or *"Hortus sanitatis"*, *and so on* (Garden of simple drugs or Garden of health) were the ancestors of the current *"Hortus botanicus"* (Botanical garden), where a number of plants are grown and studied. According to Schulters (1817), the first *Hortus* arose in Padua, Italy (1533)

[7]The manuscripts are books written by hand on different substrates (papyrus, animal skin, parchment, handmade paper, etc.). Given the reproduction system and the very high costs, the spread was limited to the libraries of monasteries, universities, royal courts, etc. *Incunabula* are called the books produced by the invention of printing (1455) until around the middle fourteenth century. These printed books distinguished by preserving the setting of the old manuscripts, which were often loose-pages, with any title, page number, index, and with any indication about the author or subsequently of the printer. Thanks to the increased share and the lowering costs, the printed books took gradually a set-up similar to the modern publications. The first *incunabulum* was the Latin version of the Holy Bible printed around 1455 by Gutenberg. The Pliny's *"Historia Naturalis"* was printed in 1478, whereas the Dioscorides *"Materia medica"* was the first printed book regarding medicine and botany (Gray 1821). *Tacuina sanitatis* were illustrated books containing popular therapeutic remedies, taken in part from the Arabic literature, at the time considered most effective and innovative than the traditional Greek-Roman medicine. The term *"tacuinum"* derives from the Arabic "Taqwin al sihha" (Tables of health). Reworked and translated into Latin around 1200, these booklets began to spread in Tuscany and Lombardy, Italy. Because this sort of manuals was intended mainly to the aristocracy, the manuscripts were embellished with precious decorations and miniatures. In addition to plant drawings, scenes from daily life were illustrated with great richness of details. Unlike *herbaria*, descriptions of the plants were summarized in few lines each illustration (Fig. 5.2).

Fig. 1.14 Painting of *Beta sylvestris* attributed to Andrea Amodio (Courtesy: Biblioteca Marciana, Venice. Reproduction is prohibited)

Century, did scientific thinking begin struggling to rid itself of the ancient classical approach. Schultes (1817) in "Grundniss einer Geschichte und Literatur der Botanik", terminates with Lorenzo de' Medici (1449–1492) the first period of the history of botany, which began with Theophrastus. In this case, the Florentine is seen as the initiator of the new course, firstly in the arts (Renaissance) and then in the sciences. Jackson (1881) agrees with Schultes, but he finishes the first period with Bruhnfels (1488–1534) (Figs. 1.17 and 1.18). A new era of botanical illustration also began, clearly anticipated by Roccabonella. Incidentally, should be remembered the relative independence that Venice had in the relations with the Roman Catholic Church that had supported the thinking of Aristotle until recent times (1492).

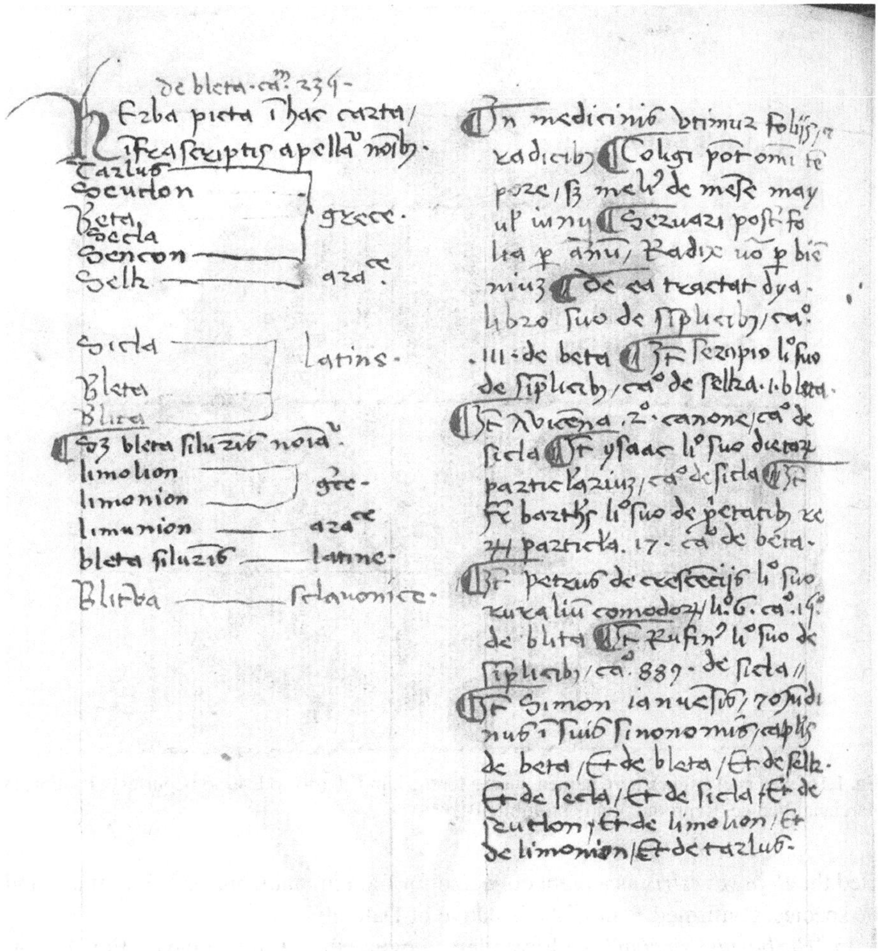

Fig. 1.15 The verso of the former page with translations of *Beta sylvestris* in some languages (Courtesy: Biblioteca Marciana, Venice. Reproduction is prohibited)

1.5 Renaissance

The study of plant physiology and pathology began to develop during this time, though not without difficulties. New plants and herbal drugs coming from the Americas became commonplace in European pharmacies, with applications taken mainly from the native people (Ximenez 1615). Otto Bruhnfels published the treatise *"Herbarium vivae iconae etc."* in 1534 which contained illustrations that clearly were free from the old tradition (Bruhnfels 1534). Previously, often the differences between the actual observation of the plants (Fig. 1.14) and the description given by ancient authors are very evident (Fig. 1.16). The *Herbarium* of Bruhnfels (Fig. 1.18)

Fig. 1.16 Sea beet (right), here named "herba ferella", in Erbario cod. 4936 (Courtesy: Biblioteca Marciana, Venice. Reproduction is prohibited)

cited the *Beta silvestris* as a plant collected for food in many places in Germany, and the species, confirmed Bruhnfels, is native of Dalmatia.

In "*Herbarium siccum*", Aldrovandi catalogued the dried sample on Fig. 1.19 as "*Beta carota*" (carrot beet)*, Rapum sativum etc.* (Soldano 2003). But in the explanations reported in the manuscripts, written by Aldrovandi himself, the plant is named "*Beta silvestris marina*" (Fig. 1.20). Here the word "*marina*" appears for the first time related to the sea beet. A third manuscript reported that "*Beta silvestris marina nascit in Lio prope mare*" (sea beet grows on the seashore near the Lido of Venice) (Soldano 2003). In another page of *Herbarium,* the stalk, surely of *Beta maritima,* is classified as *Spinacium silvestre* (wild spinach), which is described as "growing between Ancona and Senigallia" on the Italian coast of Adriatic Sea, where *Beta maritima* is still very common on the undisturbed seashores.[8]

The drawings of plants began to become very accurate in "*De historia stirpium commentarii insignes*" edited by Fuchs (1551) who catalogued *Beta sylvestris* as *limonium* and cited other names given to the plant: *tintinabulum terrae* (Latin), pyrola (vulgar Italian), Wintergrün, Holtzmangold or Waldmangold (German), and so on.

---

[8]In Fig. 1.19, it is still possible to see salt crystals on every part of the plant.

Fig. 1.17 Mengelwurtz
(fodder beet) in a drawing of
Bruhnfels (1534)

The illustration of *limonium* (Fig. 1.21) does not correspond to the characteristics of *Beta maritima*. Other mistakes arise through the author's willingness to apply the names taken from Dioscorides to the plant from Northern Europe. The majority of these mistakes were made because the real functions of the different parts of the plants were not understood yet. It was not until 1682 that the sexual and reproductive functions of flowers were explained by Grew (Arber 1912). Another source of errors was the absence of a common terminology. According to Arber (1912), Fuchs and later Dodoens, were the first botanists who attempted to introduce common botanical terms. Fuchs wrote that the *limonium* grew in shady places and flowered in June. The white and red beets (*Beta candida* and *Beta nigra*) were described and illustrated (Fig. 1.22) in a section of the book, where the heading is the ancient Greek word "teutlo".

Fig. 1.18 Front page of "Kreuterbuch etc." written by Bruhnfels (1534)

Fig. 1.19 In "*Herbarium siccum*" (Soldano 2003), Aldrovandi collected a plant classified wrongly *Spinachium sylvestre* (spinat). In reality, the plant is surely a sea beet, named *Beta marina* by Aldrovandi (Courtesy: Museo Aldrovandi, Bologna. Reproduction is prohibited)

The book "*De plantis*" written by the physician Cesalpino was published in 1583. According to Geschwind and Sellier (1902), he was among the first to describe the plants using a rather scientific approach that took into account the flower and the seed traits and, therefore, was the first attempt at plant classification using modern standards. He might be considered as the last representative of Aristotelian botany (Gray 1821). Dodoens (1553) described a drawing representing the *blitum* (Fig. 1.23) as "*Beta sylvestris ac terrae tintinabulum*, also named Wintergruen, Holtzmanngoldt in German, and Wintergruen, Officinis Pyrola in Brabantis". In the 1554 edition, Dodoens changed completely the illustrations representing the *Beta nigra* and *Beta candida* (Fig. 1.24) with drawings taken from Fuchs (1551).

A few years later, Luigi Squalermo (named also Anguillara) wrote in the work "*De simplicibus*" (1561) to be aware that *limonium* is sea beet, then known in Italy as "piantaggine acquatica" or "giegola silvestre" or "helleboro bianco". This opinion was not confirmed by later writers. Squalermo, mentioning the books of Pliny and Dioscorides, stated that the cultivated beets are black or white. Moreover, there exists

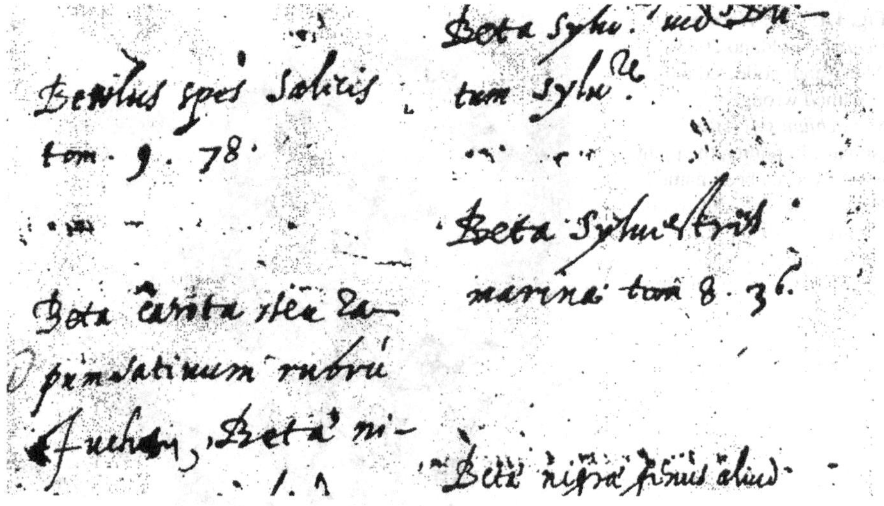

Fig. 1.20 In the explanations of the former page, Aldrovandi used for the first time the term "*marina*" (marine) (Courtesy: Museo Aldrovandi, Bologna. Reproduction is prohibited)

a third variety in Greece called "cochinoguglia", whose roots are bright red and round like the turnip (*Brassica rapa* L.).

Mattioli (1557) cited the opinion of Galen, who claimed not to know of any kind of wild beets, unless it was the plant named "*rombice*" or "*lapatio*". The same observation appeared in the treatise "Il Dioscoride" (Mattioli 1565), a translation and commentary on the work of the ancient physician. The book was among the most popular until the time of *Linnaeus*, and was printed in 60 editions and more than 32,000 copies (Gray 1821). In Fig. 1.24, it is possible to see the limitation imposed on drawings by the small size of the carved wooden blocks used in the first printed books. *Beta sylvestris* also is called "*pyrola*" by Mattioli (1586). Antonio Michiel (1510), who, after quoting several names in various languages, wrote that *Beta sylvestris* probably corresponded to the *limonium* mentioned by Dioscorides. The plant grows "in forests and shady places, along the river Reno, Italy, and around the Castle of Sambuca, Italy".

Hieronymus Bock in "Krauter Buch" described the characteristics of the cultivated *Beta nigra* and *Beta agrestis* (Fig. 1.25) (Bock 1560). The name "*agrestis*", was commonly used as synonymous to "*silvestris* or *sylvestris*". Sea beet here is called "Wald Mangold", "Winter grün", "Winter grün Pyrola", "*Betula Theophrasti*", and so on (Bock 1552). Bock confirmed the correspondence of the name, "*Beta agrestis*", with the "*limonium*" mentioned by Pliny (75 AD). The name "Winter grün" (winter green) derives from the ability of the sea beet leaves to remain green and alive throughout the winter. Anonymous (1852) better explained: "How dark and rich is the green tint of those leaves, which, on their long stalks, lie about the root of sea beet, and how well does the deep green hue contrast with the pale sea-green tint of the perfoliate yellow wort, and of many other plants of the rock".

Fig. 1.21 Drawing of
limonium or Bette saulvage
(Fuchs 1551)

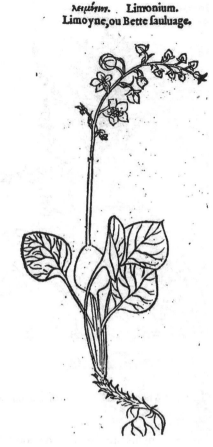

The invention of microscope introduced another revolution in the seventeenth century. This instrument enhanced exponentially the knowledge of anatomy, histology, and physiology of living organisms (Malpighi 1688), exactly as the telescope had in astronomy. The new discovery, developed at least in part thanks to progress in glass processing in the Netherlands and at Murano, Venice (Italy), revealed the real structure of plant and animals. Malpighi published the results of the first observations on plants in his "*Anatome plantarum*" printed by the London Royal Society Malpighi (1675). Five drawings of germinating seed of *Beta* are included (Fig. 1.26).

Johann Günther von Andernach, when commenting on the work of Paulus Aegineta, used the ancient Greek name "*teutlon*" as did Fuchs. The wild beet was described by Castore Durante in "Herbario Nuovo" (1635) with the name "piombagine" (*plumbago*) and "bietola salvatica". The author reported that leaves and stalk are similar to *limonium*, and consequently, it is called "false *limonium*". The plant grows "along streets and hedges, and also in wild places" The medical properties and

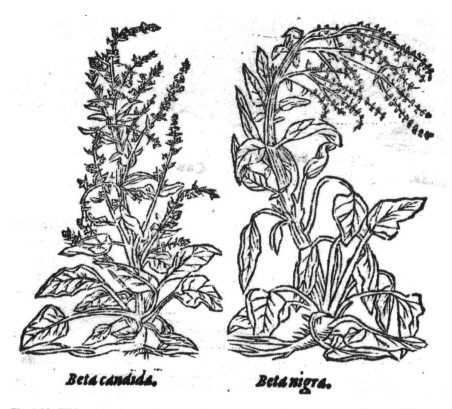

Beta candida.   *Beta nigra.*

Fig. 1.22 White and red beets (*Beta candida* and *Beta nigra*) represented by Fuchs (1551)

some synonyms are listed in the book, together with drawings of *Beta alba*, *nigra*, *rubra*, and *plumbago*.

In "*Historia generalis plantarum*", Dalechamps (1587) included drawings and descriptions of the known types of beets (*alba*, *nigra*, *rubra vulgatior*, *rubra Matthioli*), and those of *Beta erythrorhiza* (with red root) and *Beta platicaulis* (with flattened seed stalk). The first name was taken from Dodoens (1553), the second was given by Dalechamps himself. In describing *Beta platicaulis*, he considered it as a different species, although we know today that the plants were suffering from a rather common anomaly known in Italy as "fasciazione" (Munerati and Zapparoli 1915).

A very accurate description of growing and harvesting techniques was given in the "*Ruralium commodorum*", written by Pietro de Crescenzi (1605), which also described an important feature of the beet crop, the bienniality (namely, that beets had been selected for flowering in the second year, i.e. after overwintering), which made the crop more nutritional and suitable for cultivation. The "herbalist" William Coles (1657) included in the list of all sorts of beets both "sea beet" and "prickly beet of Candy; the former is surely *Beta maritima* and the latter is the species named *Beta agrestis or Beta cretica semine spinoso* (Fig. 1.24) by some later authors. Coles

Fig. 1.23 *Blitum* represented by Dodoens (1553)

seems to be the first who used the English term "sea beet". Tanara (1674) reported that when the leaves of beets were cut in the fall of the moon, they grow back with greater vigor and speed. He also reported some interesting observations on contamination of varieties caused by foreign pollen. Indeed, for the red beets, it was necessary to use seed coming yearly from France to get uniform color in the roots. The seed produced in Italy likely was contaminated often by pollen spread from other types of beets, which gave rise to hybrids with different color and shape.

BIETOLA NERA.   BIETOLA BIANCA.

Fig. 1.24 *Beta sylvestris* black (left) and white (Mattioli 1557)

Chabray (1666), in "*Stirpium sciagraphia et icones, etc.*", together with the drawings, described various types of cultivated and wild beets. In the appendix of the book, he cited several synonyms of *Beta sylvestris* (*limonium, trifolium palustre, lampsana, pyrola, mysotis, potamogaton, carduus pratense, plantago aquatica, lapathum*) mostly of unknown origin. Chabray reported that the name "*blitum*", while in use, was attributed to a plant different from the *Beta sylvestris* described by Theophrastus; an example of this confusion is seen by Dorsten (1540), who confused *Beta* with *Brassica* spp. Among the wild beets, only the drawing of *Beta cretica* is reported by Chabray.

Pena and de Lobel (1576) began grouping the plants by their characteristics (grass, grass-like plants etc.) in "*Adversaria nova*" (1576). In "*Plantarum seu stirpium historia etc.*," de Lobel (1576) mentioned the sea beet under the name "*Beta sylvestris spontanea marina*" surely derived from the adjective used by Aldrovandi some decades before. In the second letter of the word *silvestris*, de Lobel used the letter "Y" not existing in the Latin alphabet.

Fig. 1.25 *Beta cretica*
(Chabray 1666)

Gaspard Bauhin, in *"Pinax theatri botanici"* (1623), assembled a number of synonyms for *Beta* partly adopted by Linnaeus. Bauhin began grouping species according to their botanical affinities, thus pioneering the binomial classification. The book reported a complete reference of the authors involved in botany and medicine. Bauhin is thought to be the author of the name *"maritima"* given to the sea beet. In *"Paradisi in sole, etc."*, Parkinson (1629) sought to clarify the uncertainties about the correct identification of the ancient term *"Beta nigra"*. Sea beet was called "common green beete" found in "salt marshes near Rochester". Parkinson also hinted at a "great red bete" recently imported to London "by Master Lete and given unto Master Gerard for his herbal". The plant is similar to the Italian beet (*Beta romana*), but larger and with red petioles. The latter, also called *Beta raposa* for its resemblance to the turnip could be used for both the leaves and roots. *Beta maritima* was called *"blitum,"* and eaten cooked together with other herbs. In the revised edition of de Lobel's (1591) *"Stirpium illustrationes etc.,"* Parkinson (1655) described two types of sea beets, *Beta maritima syl(vestris) spontanea* and *Beta maritima syl(vestris) minor*, the roots of the former were much more developed. Both were grouped with *Beta maxima* i.e. the cultivated type.

Gerard and Poggi (1636) wrote in "The herball or general histoire of plants" "… the ordinary white beet (*Beta alba*) growes wilde upon the sea-coast of Tenet and diners others places by the sea." In reference to the confusion caused by the different names given to *Beta sylvestris*, he added: "For the barbarous names we can say nothing: now it is said to be called *limonium* because it growes in wet or overflown meadows: it is called *neuroides* because the leaf is composed of divers strings or fibers running from one end thereof to the other, as in plantaine (*Plantago* spp.) … In addition, it may be das fitly termed *lonchtis* for the similitude that the leafe hath

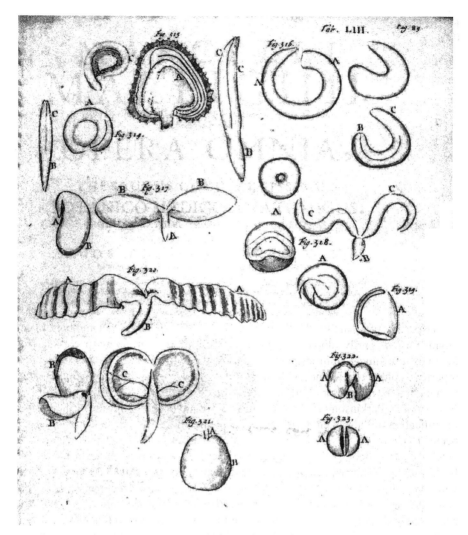

Fig. 1.26 Above left are the germination phases of beet seed: "… the flattened seed cavity contains the embryo (**a**); the rootlet is pushed by the elongated plantula (**b**), which bears two thick and equal cotyledons (**c**) (Malpighi 1675)

to the top of head of a lance … And for *potamogaton*, which signifies a neighbor to the riner or water, I thinke it loves the water as well, and is as neere a neighbour to it as that which takes its name from thence and is described by Dioscorides. Now to come to Pliny, he calls it *Beta sylvestris*, *limonium* and *neuroides*. The two later names are out of Dioscorides, and I shall shew you where also you shall finde the former in him. Thus much I thinke might serve for vindication of my assertion, for I dare boldly affirm, that no late writer can fit all these names to any other plant; and that makes me more to wonder that all our late herbarilists, as Mattioli, Dodoens,

Fuchs, Cesalpino, Dalechamps, but above all Pena and de Lobel, should not allow this plant to be *limonium*, especially seeing that Anguillara had before or in their time asserted it so to be: but whether he gave any reason or no for his assertion, I cannot tell because I could never by any means get his opinion, but onely find by Bauhin *Pinax*, that such was his opinion hereof".

John Ray extended and corrected the intuitions of Cesalpino (Gray 1821). In *"Historia plantarum"* (Ray 1693, 1724), he described *Beta sylvestris maritima*, named that by Bauhin and Parkinson. Parkinson also named it *Beta sylvestre spontanea maritima* and *Beta commune viridis* (Parkinson 1655). The species, as suggested by Johnson (1636), resembled *Beta alba*, but it grew in marshy places and, more frequently, along beaches. Ray wrote that the *Beta sylvestris maritima* differed from all other beets because it was perennial, a statement sustained also by Coakley (1787) and Koch (1858). According to Ray, *Beta sylvestris maritima* was similar to *Beta communis viridis*, however, in disagreement with Gerard and Poggi (1636), he found it rather different from *Beta alba*. In *"Methodus plantarum"*, Ray (1703) pointed out that the single beet flowers developed seeds with a single embryo (monogerm), whereas multiple flowers developed glomerules containing the same number of embryos as there were flowers (multigerm). Johnson (1636), author of *"The Herbal etc."*, identified *Beta maritima* in the coastal area of Tenet and other locations near the sea, as had been reported by Gerard and Poggi (1636).

The images of *Beta alba*, *nigra,* and *rubra* appeared in the book (Bauhin 1731) with the title "Kräuter Buch" but here the figures were accompanied by a more precise explanation. *Beta sylvestris* was drawn under the heading "Wintergrün" (also called "Holz Mangold", "Wald Mangold", and "Waldkohl"). The book cited several synonyms in various languages (Appendix C): *"Pyrola"*, *"Beta sylvestris"*, *"Pyrola rotundifolia mayor,"* *"Limonium"* (Latin), "Wintergreen" (English); Pyrol (French), and Pirola (Italian). Under the heading "Wald Mangold", there were the drawings of Gross Limonium, Wald Mangold (*Limonium*, pyrola), and Klein Limonium mit Olivenblätter (little *Limonium* with leaves as olive tree). The figures referred to Wintergrün and Wald Mangold bore no resemblance to *Beta maritima* or *sylvestris*. The same is true for *limonium*, which was repeatedly mentioned in the text.

Beta sylvestri maritima was mentioned briefly by Blackwell (1765), in "Sammlung der Gewächse", the German edition of *"Herbarium Blackwellianum"*. The author clearly distinguished *Beta maritima* from *Pyrola* (Fig. 1.27), instead of treating them the same, as was done by Bock and Fuchs. The names *Beta sylvestris maritima* and *Beta sylvestris spontanea marina* are attributed to Bauhin and de Lobel, respectively. In the treatise were included color illustrations and the description of the characteristics of *Beta rubra vel nigra* (Fig. 1.28).

Zanichelli (1735) reported the "presence of *Beta maritima* in various parts of the lagoon around Venice and in particular around the harbor of Malamocco". This location is near the Lido cited by Soldano (2003). The similarity between the cultivated and wild forms was confirmed, excluding the shape and smaller size of the *Beta maritima* root and its annual life cycle. The observations of Zanichelli were shared by Naccari (1826), who defined the plant as "biennial and bearing sessile flowers,

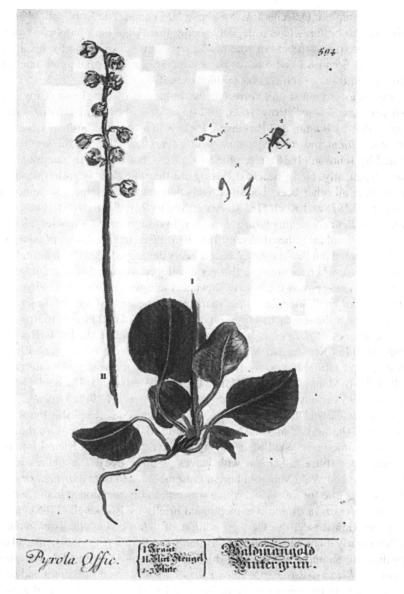

Fig. 1.27 Pyrola (Blackwell 1765)

Fig. 1.28 *Beta rubra vel nigra* (red or black beet) according to Blackwell (1765)

often lonely". Note that "lonely" could be synonym of "monogerm" in the Savitsky meaning.

The species were ranked under a new grouping called "*genus*" (pl. *genera*) in "*Institutiones rei herbariae*" written by de Tournefort (1700). About 10,000 names of *genera*, including *Beta*, have survived, not only in the Linnaean system (Schultes 1817), but also in the current taxonomy. He cited two species of sea beet: *Beta sylvestris maritima* (also named *sylvestris*, *spontanea*, and *marina*) and *Beta sylvestris* (also named *cretica*, *maritima*, and *foliis crispis*). Seed and flower of beets were shown in the third volume of the cited book.

Smith (1803), after an accurate description on some quite original illustrations of *Beta alba* and *Beta rubra* represented without the root (Figs. 1.29 and 1.30), included *Beta sylvestris* under the heading *bistorta*, which "Andere nennen sie *Lappam minorem*, andere *Bardanam minorem*, andere *Limonium*, andere *Britannicam*. Bei dem Plinis heisst sie *Beta sylvestris* (by some called *lappam*, by others *bardanam* or *limonium*. By Pliny, it was named *Beta sylvestris*").

The first published work of Linnaeus (1735) was *Systema naturae*. Every plant was identified by the name of the species preceded by the corresponding genus as was done by Cesalpino and Bauhin. The main part of botanists and zoologists rapidly adopted this system. Linnaeus (1735) observed that beet, if returned to the wild environment, never took the original form of sea beet. Therefore, the two types were

Fig. 1.29 *Beta rubra* (Weinmann 1737)

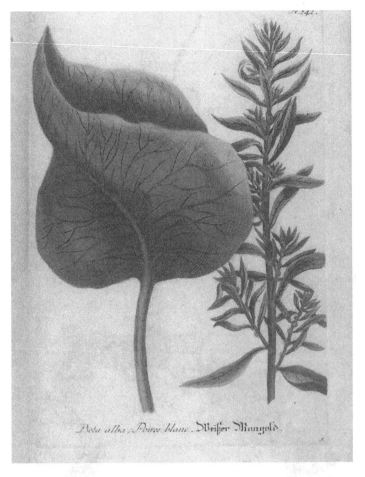

Beta alba, Poire blanc. Weißer Mangold.

Fig. 1.30 Drawing of sea beet showing the development of seed on the stalk, the shiny leaf blades of different shape/dimension and some red-veined parts

classified as distinct species: *Beta vulgaris* and *Beta maritima* (Figs. 1.31 and 1.32). The first included all cultivated varieties, the second derived directly from "the native original unknown species, probably extinct in the prehistory" (Ford-Lloyd et al. 1975; Greene 1909a, b).

The Gardner's Dictionary (Miller 1768) declared that sea beet "is probably the parent of all garden beets". Hill (1775) described the drawings of three types of beet: "common", "ciclane", and "sea beet". The first had the leaves more or less colored in red, it is biennial, and native of the coasts of Italy. The second one had light green leaves and corresponded to *Beta cicla*. The third one also was biennial and native to the English coasts. Hill reported, "It has been said that the first two species were produced by culture from this. Tis soon said, but will not bear enquiry; at least, experience here at Bayswater, perfectly contradicts it". Hill's posthumous edition of

Fig. 1.31 Stalk and seed of *Beta maritima* (Linnaeus 1735)

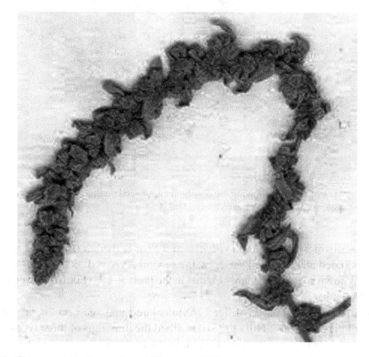

Fig. 1.32 Close up of the former figure (Linnaeus 1735)

Fig. 1.33 Flowers and germinating seeds of *Beta maritima* on the "Encyclopedie" (Lamarck 1810)

the book "*Synopsis plantarum*" (written by Ray) was among the first to adopt the new taxonomic system of Linnaeus.

Smith (1803) gave us, along with a colored drawing of *Beta maritima*, a precise description of its morphology and physiology. The stem "bears in the axils clusters of small leaves and flowers solitary or in pairs". Smith argued that sea beet is certainly distinct from *Beta vulgaris*, as described by Linnaeus, since it flowered during the first year. He stated that "With us it appears to be perennial, flowering in August and September. The stigmas are very frequently three in number". Also Hardwicke (1887) confirmed never having seen beet flowers with more or less than three stigmas.

Lamarck (1810) briefly described *Beta maritima* in the Encyclopèdie edited by Diderot and D'Alambert (1751). The drawing (Fig. 1.33) illustrates the characteristics of the seed stalk and flowers. He cited Oliver de Serres, who, describing some red beets "just arrived from Italy", referred to the sugar syrup extracted from the roots. This observation likely addressed Margraaf's (1907) research in obtaining crystals of sucrose from beet juice. The adventure of sugar beet crop began at the end of the same century (Achard 1907; von Lippmann 1929).

A very original description of *Beta maritima* was given by Gray (1821) "Stem prostrate at bottom; lower leaves triangular, petiolate; flowers solitary or in pairs, lobes of the perigonium quite entire. Root: black, internally white, stems many, much branched at the top; flowers racemose". By the end of 1700, countless reports on the local flora had been published. These sorts of surveys, which gradually ceased in the subsequent century, are still useful for locating the ranges of wild species and detecting any changes in their geographical distribution and botanical characteristics (Jackson 1881).

1.6 Age of Science

After the rediscovery of the experiments of Gregor Mendel (Tschermak-Seysenegg 1951), botany gradually evolved from the mere description, localization, collection, and classification of plants, primarily toward studies aimed at physiology and scientific improvement of the production traits. Mendel (1865) established the fundamental "laws of inheritance", which became the basic rules of the modern plant breeding (Allard 1960; Fehr 1987; Poehlman 1987). Initially plants were evaluated by investigating their behavior in homogeneous environments, and then they were selected, crossed, and reproduced using appropriate systems (Bateson 1902).

By the beginning of the 1800s, beet varieties adapted to sugar production were being selected in Germany. In the course of only a few years, sugar production quickly became the most important use of cultivated beet. In France and Germany, private seed companies began breeding programs that were very successful in improvement of sugar production, mainly through mass selection. Genetics, breeding, plant pathology, and other disciplines took advantage of the advances in analytical instrumentation (i.e. the polarimeter), primarily developed for rapid analyses in sugar factory (de Vilmorin 1850, 1856). de Vilmorin (1923) successfully developed the first methods of family selection.

Brotero (1804) identified populations of sea beet "*ad Tagi ripas, et alibi in maritimis*" (along the Tagus River, Portugal, and in other marine sites). The plants exhibited the following traits: "*caulis ex decumbenti erectus; flores saepius gemini, axillares, sessiles, in spicam foliaceam tenuem digesti*" (the stalks are prostrate or erect; flowers are often twin sessile flowers located in the bract axils). They are not distributed closely on the leafy inflorescence. Another detailed description of some sea beet traits was given by Baxter (1837): "Roots: large, thick, and fleshy, blackish on the outside, white within. Stems: procumbent at the base, from 6 inches to 2 feet long, angular and furrowed, alternatively branched, leafy, often reddish. Root-leaves: large, spreading, slightly succulent, stalked, egg-shaped, veiny, and more or less wavy at the edges. Stem-leaves: nearly sessile, alternate, and, in consequence of the position of the stem, oblique or vertical. Flowers: greenish, usually in pairs, rarely solitary, sessile, in the axils of the leaves, of which the uppermost are diminished almost to bracteas". A similar description was given by Hooker (1835).

Reichenbach and Reichenbach (1909) confirmed that:

"*Est planta silvestris a qua omnes betarum stirpes culti originem trahunt*"

(There is a wild plant from which all the cultivated beets originated)

The plant also may be annual, and it grows "*in omnibus terris mediterranei*". Another brief description of *Beta maritima* was given by Bois (1927):

"*C'est une plante vivace ou bisannuelle, à racine dure et grêle, à feuilles un peau charnues, les radicales ovales ou rhomboïdales, les caulinaires ovales ou lancéolées*"

(Sea beet is a vivace or biennial plant, with hard and skinny roots, the leaves a little fleshy, oval and rhombic if developed from the root, oval and pointed if attached to the stem)

With the theory of inheritance allowing the basis for plant breeding, statistics provided a tool to maximize the gain that plant breeders could make with their selections. Much of this work was begun on crops such as maize, wheat, barley, and so on (East 1912; East and Jones 1919). Statistics and field plot design were valuable tools in the improvement of sugar beet as well (Harris 1917); it was immediately recognized as a powerful mean for reducing error when evaluating the results of replicated field trials. Rimpau, Schindler, and Munerati were among the first researchers who focused their research primarily on sugar beet. But von Proskowetz was certainly the first to understand the importance of *Beta maritima* not only as a donor of useful traits but also to perform the first crosses with the commercial varieties (Kajanus 1910).

The book written by von Lippmann (1925) provided an excellent summary of these early researches. But the basic contribution by Mendel was at all ignored by the book. At this time *Beta maritima* began to be regarded as a potential source of useful traits for beet crop. The same was tried for other species in the genus *Beta*, but there were problems still existing in obtaining viable hybrids (Rimpau 1891; Campbell and Russell 1964).

Publications concerning medicine and botany were written or printed primarily as books until the end of seventeenth century. These books rarely included sea beet. Since then, journals, reviews, and proceedings of scientific societies have become prevalent. Although the number of papers reporting on sea beet has increased almost exponential over the last decades (www.newcrops.uq.edu.au), only a few book chapters and dissertations on sea beet have been edited. No book has been published until the first edition of this book.

Historically, publications have been written in the dominant scientific language of the time: Greek and Latin until the Imperial Period of Rome, Latin until Linnaeus and beyond, and German until World War II. Since then, English has gradually become the dominant language of sciences. Unlike other sciences, botany retained the traditional use of Latin until the early eighteenth century. Also for this reason, it was customary for botanists to adopt Latin names (pseudonym or pen name) until around the end of sixteenth century. The German language dominance lasted longer in botany than in other sciences, especially in studies related to sugar beet, in part because the crop and its technology were born and developed in Germany. Many of the fundamental books on botany were written in German in the seventeenth century. As medical plant, sea beet was mentioned primarily in books printed in Latin and German; the species was almost ignored in the English literature until the beginning of the last century. The literature on botany and medicine written in Arabic from the ninth to twelfth century was also important. From fourteenth century onward, important works were published in many other languages (English, Italian, French, Spanish, etc.). In the last few decades, English has become dominant because, among other things, the important journals are published in USA and UK. Almost all papers on sea beet published by international journals, certainly in the last three decades, have been edited in English.

1.7   State-of-the-Art and Prospects

For the presence in the same environment of different types of beets perfectly interfertile (*Beta maritima*, *Beta vulgaris* crops, seed production fields, ruderal, weed, and feral beets), condition rarely possible for other crops, *Beta maritima* is becoming a reference plant for other species and disciplines (Auer 2003). This is proved by the growing number of publications dealing with the plant. This is true especially for the research regarding molecular biology.

The major risk for survival of sea beet germplasm in natural condition is represented by the climatic changes, which, at least in Europe, seems reducing significantly the amount and the frequencies of summer rains. Taking the Fig. 1.1 as an example, it is easily foreseen that the life of the represented plant is closely linked with the rain water supply. In slightly better situation, lives the great part of *Beta maritima*, that is on sandy or stony soils, without water capacity and not or rarely provided by water table. In these conditions, the minimum shortage or delay of rain can represent the death of the plants. This is the situation observed recently in some sites of West Adriatic Sea, once composed by hundreds of individuals, where the plants are completely disappeared.

The in situ organization and conservation of sea beet populations worldwide will be an important mean to follow and at due time utilize the reactions induced by the new climatic conditions. As regards, the tolerance to abiotic stresses, until now without significant results, the in situ selection of *Beta maritima* in environments, where selection pressures modified the population originating adaptive ecotypes, will enable to identify potential hotspot of genetic diversity in order to enhance the tolerances to abiotic stresses (Monteiro et al. 2018).

Thanks to the expected development of molecular analyses and the still quite unexplored germplasm of garden and leaf beet (Cheng et al. 2011), many progresses are still possible against diseases and stress.

References

Achard FC (1803) Anleitung zum Anbau der zur Zuckerfabrication anwendbaren Runkelrüben und zur vortheilhaften Gewinnung des Zuckers aus denselben. Reprinted in: Ostwald's Klassiker der exacten Wissenschaft (1907) Engelmann, Lipsia, Germany

Allard RW (1960) Principles of plant breeding. Wiley, Hoboken NJ, USA

Anderson FJ (1977) An illustrated history of the herbals. Columbia Univ. Press, New York, USA

Anonimous (1852) Book for the sea-side. Religion Treet Society, London UK

Arber AR (1912) Herbals, their origin and evolution. History of botany. Cambridge Univ. Press, Cambridge, UK, pp 1470–1670

Auer CA (2003) Tracking genes from seed to supermarket. Trends Plant Sci 8:531–538

Avagyan A (2008) Crop wild relatives in Armenia: diversity, legislation, and conservation issues. In: Maxed M, Ford-Lloyd BV, Kell SP, Iriondo JM, Dulloo ME, Turok J (eds) Crop wild relative conservation and use. CAB International, Cambridge MA, USA, pp 58–76

Barbaro H (1494) *Castigationes Plinianae*. Pomponius Melam, Venice, Italy

Bartsch D, Brand U (1998) Saline soil condition decreases rhizomania infection of *Beta vulgaris*. J Plant Pathol 80:219–223

Bartsch D, Cuguen J, Biancardi E, Sweet J (2003) Environmental implications of gene flow from sugar beet to wild beet–current status and future research needs. Environ Biosafety Res 2:105–115

Bartsch D, Lehnen M, Clegg J, Pohl-Orf M, Schuphan I, Ellstrand NC (1999) Impact of gene flow from cultivated beet on genetic diversity of wild sea beet populations. Mol Ecol 8:1733–1741

Bateson W (1902) Mendel's principles of heredity. Cambridge Univ Press, Cambridge, UK

Bauhin G (1623) Pinax theatri botanici … etc. Basel, Switzerland

Bauhin H (1731) Kräuter Buch. Jacobus Theodorus, Basel, Switzerland

Baxter W (1837) British phaenerogamous botany. Parker, London, UK

Becker-Dillingen J (1928) Die Wurzelfrüchte (Rüben). In: Parey P (ed) Handbuch des Hackfrucht-baues und Handelapflanzbaues. Berlin, Germany

Berti-Pichat C (1866) Corso teorico e pratico di agricoltura. Unione Tipografico-Editrice, Torino, Italy

Biancardi E, Campbell LG, Skaracis GN, de Biaggi M (eds) (2005) Genetics and breeding of sugar beet. Science Publishers, Enfield NH, USA

Biancardi E, de Biaggi M (1979) *Beta martima* L. in the Po Delta. In: ISCI (ed) Proc Convegno Tecnico Internazionale in Commemorazione di Ottavio Munerati. Rovigo, Italy, pp 183–185

Biancardi E, Lewellen RT, de Biaggi M, Erichsen AW, Stevanato P (2002) The origin of rhizomania resistance in sugar beet. Euphytica 127:383–397

Biancardi E, McGrath JM, Panella LW, Lewellen RT, Stevanato P (2010) Sugar Beet. In: Bradshaw JE (ed) Root and tuber crops. Springer Science+Bussiness Media, LLC, New York, pp 173–219

Blackwell E (1765) Sammlung der Gewachse. de Launoy, Nürnberg, Germany

Bock H (1552) *De stirpium maxime earum quae in Germania ... etc.* Manuscript

Bock H (1560) Kreuter Buch. Gedrucht zu Strassburg, Germany

Bois D (1927) Les plantes alimentaires chez tous les peuples et a travers les ages. France, Paris

Bottema S (2010) Pollen profile of sediment core Agköl Adabag, Turkey and Lake Urmia (Iran). European Pollen Database. https://doi.org/10.1594/pangaea.738949-doi:10.1594/pangaea.739926

Brotero FA (1804) *Flora Lusitanica.* Ex Typographia Regia, Lisbona, Portugal

Bruhnfels O (1534) *In hoc volumine contenitur insignium medicorum etc.* Strasbourg, France

Buschan G (1895) Vorgeschichtliche Botanik der Cultur-und Nutzpflanzen der alten Welt auf Grund prähistorischer Funde. Kern, Breslau, Germany

Campbell GKG, Russell GE (1964) Breeding in sugar beet. Ann Rept Plant Breed Inst Cambridge 1–31

Cavalli-Sforza LL, Edwards AWF (1967) Phylogenetic analysis: models and estimation procedures. Am J Hum Genet 19:233–257

Chabray D (1666) *Stirpium sciatigraphia et icones ex musaeo Dominici Chabraei.* Colonia, Germany

Ciarallo C (2004) Flora pompeiana. Erma di Bretschneider, Rome, Italy

Cheng D, Yoshida Yu, Kitazaki K, Negoro S, Takahashi H, Xu D, Mikkami T, Kubo T (2011) Mitocondrial genome diversity in *Beta vulgaris* L. subsp. *vulgaris* (Leaf and Garden Beet Groups) and its implications concerning the dissemination of the crop. Genet Resour Crop Evol 58:553–560

Coakley L (1787) An account of the culture and use of the mangel wurzel or root scarcity. 4th edn. Charles Dilly, London, UK

Coles W (1657) Adam in Eden or natures paradise. Printed by F Streater, London, UK

Collins M (2000) Medioeval herbals, the illustrative traditon. British Library and Univ Toronto Press, London, UK and Toronto, Canada

Coons GH (1954) The wild species of *Beta*. Ibidem 8:142–147

Dalby A (2003) Food in the ancient world from A to Z. Routledge, UK

Dalechamps J (1587) *Historia generalis plantarum.* Lyon, France

de Bock (1986) The genus *Beta*: domestication, taxonomy, and interspecific hybridization for plant breeding. Acta Hort 182:333–343

de Candolle A (1884) Der Ursprung der Culturpflazen. Brockhaus, Lipsia, Germany

de Candolle A (1885) Origin of cultivated plants. Appleton and Company, New York, USA

de Candolle A (1904) The origin of cultivated plants, 2nd edn. Paul Kegan and Co, London, UK

de Crescenzi P (1605) Trattato dell'agricoltura. Florence, Italy

de Divitiis E, Cappabianca P, de Divitiis O (2004) The *Scola medica salernitana*": the forerunner of the modern university medical schools. Neurosurgery 55:722–745

de Lobel M (1576) *Plantarum seu stirpium historia ... etc.* Ex Officina Christophori Plantinii Anterwep, Belgium de Lobel M (1591) *Icones stirpium seu plantarum tam exoticarum quam indigenarum ... etc.*, Anterwep, Belgium

de Toni E (1925) Il libro dei semplici di Benedetto Rinio. Memorie Pontificia Romana Accademia Nuovi Lincei 8:123–264

de Tournefort JP (1700) *Institutiones rei herbariae*. Thypographia Regia, Paris, France

de Vilmorin JL (1923) L' hérédité de la betterave cultivée. Gauthier-Villars, Paris, France

de Vilmorin L (1850) Note sur un projet d'experence ayant pur but d' augmenter la richesse saccarine de la betterave. Sociètè Imperiale Centrale d' Agricolture 6:169

de Vilmorin L (1856) Note sur la création d'une nouvelle race de betteraves à sucre. - Considérations sur l'hérédité dans les végétaux. Comptes Rendus de l'Académie des Sciences 43:113

Diderot M, D'Alambert M (1751) Encyclopèdie, ou dictionnaire raisonné des sciences, des arts et des metiérs. Briasson, Paris, France

Dodoens R (1553) *Stirpium historia etc*. Ex Officina Iohannis Loci, Anterwep, Belgium

Dorsten T (1540) *Botanicon, continens herbarum aliorumque simlicium*. Frankfurth, Germany

Durante C (1635) Herbario nuovo. Jacomo Bericchi et Jacomo Ternierij, Rome, Italy

East E (1912) The Mendelian notation as a description of physiological facts. Amer Nat 46:633–643

East EM, Jones DF (1919) Inbreeding and outbreeding: their genetic and sociological significance. J. B. Lippincott Company, Philadelphia PA, USA

Fahmy AG (1997) Evaluation of the weed flora of Egypt from Predynastic time to Graeco-Romans time. Veget Hist Archaeobot 6:241–242

Feemster-Jashemsky W, Meyer FG (2002) The natural history of Pompeii. Cambridge Univ Press, Cambridge, UK

Fehr WR (1987) Principles of cultivar development, 1st edn. Macmillan Publishing Company, New York, USA

Fellner S (1879) Compendium der Naturwissenschaften an der Schule zu Fulda ... etc. T. Grieben, Berlin, Germany

Fischer-Benzon R (1894) Alteutsche Gardenflora, Untersuchungen über die Nutzpflanzen des deutschen Mittelater. Verlag Lipsius & Tischer, Lipsia, Germany

Ford-Lloyd BV, Hawkes JG (1986) Weed beets, their origin and classification. Acta Horticult 82:399–404

Ford-Lloyd BV, Williams ALS, Williams JT (1975) A revision of *Beta* section *Vulgares* (Chenopodiaceae), with new light on the origin of cultivated beets. Bot J Linn Soc 71:89–102

Francisco-Ortega J, Santos-Guerra A, Seung-Chul K, Crawford DJ (2000) Plant genetic diversity in the Canarian Islands: a conservation perspective. Am J Bot 87:909–919

Fuchs L (1551) *De historia stirpium commentarii insignes*. Arnolletum, Lyon, France

Funari PP (1998) Le rire populaire a Pompeii. Colloquie Int. "Le rire chez les anciennes". Grenoble, France

Gasparrini-Leporace T, Paolucci G, Maffei SL (1952) Un inedito erbario farmaceutico medioevale. Leo S Olschki Editore, Florence, Italy

Gerard P, Poggi G (1636) The herbal, or general history of plants. Ex Officina Harnoldi Hatfield. London, UK

Geschwind L, Sellier E (1902) La Betterave. Association des Chemistes de Sucrèrie, Paris, France

Gray SF (1821) A natural arrangement of British plants. vol 2. Printed for Baldwin, Kradook, and Joy. London, UK

Greene EL (1909a) Carolus Linnaeus. Chrstopher Sower Company, Philadelphia PA, USA

Greene EL (1909b) Linnaeus as an evolutionist. Proc Washington Academy of Science 9:17–26

Hardwicke HJ (1887) Evolution and creation. A monthly record of science. Published by the author, Chicago MI, USA

Harlan JR (1992) Crops and man. American Society of Agronomy, Madison WI, USA

Harris JA (1917) Biometric studies on the somatic and genetic physiology of the sugar beet. Amer Nat 51:857–865

Hill J (1775) The vegetable system etc. Trueman, London, UK

Hohmann S, Kadereit JW, Kadereit G (2006) Understanding Mediterranean-Californian disjunctions: molecular evidence from Chenopodiaceae-Betoideae. Taxon 55:67–78

Hooker WJ (1835) The British flora. George Walker Arnott, Glasgow, UK

Jackson BD (1881) Guide to the literature of botany; being classified selection of botanical works. Longmans & Green, London, UK

Jaeger W (1952) Diokles von Karystos und Aristoxenos von Tarent über die Prinzipien. Winter, Heidelberg, Germany

Johnson T (1636) The Herbal or general history of plants by J. Gerard (Revised and enlarged by Thomas Johnson). Dover Publications, New York, USA

Kajanus R (1910) Uber die Vererbungsweise gewisser Merkmale der *Beta* und *Brassica* Ruben. Z. fiir Pflanzenzuchtung 1:125–186

Kennedy DJ (1913) Albertus Magnus catholic encyclopedia. Robert Appleton Co, New York, USA

Kirby WF (1906) British flowering plants. Sidney Appleton, London

Kircher A (1643) *Lingua aegiptiaca restituta: opus tripartitum quo linguae coptae sive idiomatis illius primaevi Aegiptiorum Pharaonici plena instauratio continetur* … Rome, Italy

Kleiner M, Hacker M (2010) Grüne Genetechnik. Wiley-VCH Verlag, Weinheim, Germany

Koch DG (1858) *Synopsis florae Germanicae et Helveticae*. Gebhardt & Reisland, Lipsia, Germany

Körber-Grohne U (1987) Nutzpflanzen in Deutschland. Thesis Stuttgart, Germany

Krasochkin VT (1959) Review of the species of the genus *Beta*. Trudy Po Prikladnoi Botanike. Genetik i Selektsii 32:3–35

Krasochkin VT (1960) Beet. Gos. Izdat. S.H. Lit. Moskow-Leningrad, Russia

Kubiak-Martens L (1999) The plant food component of the diet at the late Mesolithic (Ertebolle) settlement at Tybrind Vig, Denmark. Veg Hist Archaeobot 8:117–127

Kubiak-Martens L (2002) New evidence for the use of root foods in pre-agrarian subsistence recovered from the late Mesolithic site at Halsskov, Denmark. Veget Hist Archaeobot 11:23–31

Lamarck K (1810) Encyclopedie methodique botanique etc. France, Paris

Lazzarini E, Di Vito M, Segre RV (2004) *Historia Plantarum transcription*. Panini ed., Modena, Italy

Li J, Schulz B, Stich B (2010) Population structure and genetic diversity in elite sugar beet germplasm investigated with SSR markers. Euphytica 175:35–42

Linnaeus (1735) *Systema Naturae*. Thipis Joh. T. Trattnern, Vienna, Austria

Mabberley DJ (1997) The plant book. Cambridge University Press, Cambridge, UK

Magnol P (1636) *Botanicum Montspelliense*. Ex Officina Danielis Pech, Montpellier, France

Malpighi M (1675) *Anatome plantarum*. Regia Societate ad Scientiam Naturalem Promovendam. London, UK

Malpighi M (1688) *Opera omnia*. Apud Ventrum Vander, Lyon, France

Margraaf AS (1907) Chemischen Versuche einen wahren Zucker aus verschiedenen Pflanzen die in unseren Ländern wachsen zu ziehen. Reprinted in: Ostwald's Klassiker der exacten Wissenschaft (1599) Engelmann, Lipsia, Germany

Mattioli PA (1557) I discorsi di Pietro Andrea Mattioli, medico senese. Venice, Italy

Mattioli PA (1565) *Commentarij in sex libros Pedacij Dioscoridis Anazarbei, De medica materia*. Ex Oficina Valgrisiana, Venice, Italy

Mattioli PA (1586) *De plantis epitome utilissima*. Frankfurt, Germany

McGrath JM, Panella L, Frese L (2011) Beta. In: Cole C (ed) Wild crop relatives: genomic and breeding resources. Industrial crops. Springer, Berlin, Heidelberg, Germany

Meissner B (1926) Könige Babyloniens und Assyriens; Charakterbilder aus der altorientalischen Geschichte. Quelle & Meyer, Lipsia, Germany

Mendel G (1865) Versuche über Plflanzenhybriden. Verhandlungen des naturforschenden Vereines in Brünn, Bd. IV für das Jahr 1865. pp 3–47

Meyer EHF (1849) Die Insel Tenerife. Lipsia, Germany

Michiel PA (1510) I cinque libri di piante. Reprinted by Reale Isituto di Lettere, Scienze ed Arti (1940). Venice, Italy

Miller P (1768) Gardener's Dictionary. Francis Rivington et al, London, UK

Monteiro F, Frese L, Castro S, Duarte MC, S. Paulo O, Loureiro J, Romeiras MM (2018) Genetic and genomic tools to assist sugar beet improvement: The value of wild crop relatives. Frontiers Plant Sci 9 Art. 74

Munerati O, Mezzadroli C, Zapparoli TV (1913) Osservazioni sulla *Beta maritima* L. nel triennio 1910–1912. Staz Sper Ag Ital 46:415–445

Munerati O, Zapparoli TV (1915) di alcune anomalie della *Beta vulgaris* L. Atti Regia Accademia dei Lincei 25:1239

Musa Brassavola A (1537) *Examen omnium simplicium medicamentorum, quorum in officinis usus est.* Jean & François Frellon, Lyon, France

Naccari FL (1826) Flora veneta. Bonvecchiato Editore, Venezia, Italy

Nieuwhof A (2006) Changing landscape and grazing: macroremains from the terp Peins-east, province of Friesland, the Netherlands. Veget Hist Archaeobot 15:125–136

Ober ES, Luterbacher MC (2002) Genotypic variation for drought tolerance in *Beta vulgaris*. Ann Bot 89:917–924

Pabst G (1887) Kohler's medizinal Pflanzen Atlas. Kohler's Verlag, Gera-Huntermhaus, Germany

Pals JP (1984) Plant remains from Aartswoud, a neolithic settlement in the coastal area. In: van Zeist W, Casparie WA (eds) Plants and an cient man: studies in palaeoethnobotany. Proceedings of the sixth symposium of the International Work Group for Palaeoethnobotany, A. Balkema, Rotterdam, pp 313–321

Parkinson J (1629) Paradisi in sole paradisus terrestris, or a garden of all sorts of pleasant flowers. Printed by Humfrey Lownes and Robert Young, London, UK

Parkinson J (1655) *Matthiae de L'Obel Stirpium Illustrationes.* Warren, London, UK

Pena P, de Lobel M (1576) *Stirpium adversaria nova ... etc.* Ex Officina Christophori Plantinii Anterwep, Belgium

Perry D (1999) Vegetative tissues from Mesolithic sites in the northern Netherlands. Curr Anthropol 40:231–232

Pezzella F (2007) Un erbario indedito veneto (sec. XV) svela i segreti delle piante medicinali. Perugia, Italy

Pezzella S (1993) Gli erbari: i primi libri di medicina. Grifo, Perugia, Italy

Piccoli F (2000) Storia ed evoluzione degli erbari. Gli erbari ferraresi. In: Chendi A (ed) Erbe ed erbari a Ferrara dal '400 ai giorni nostri. TLA Editrice, Ferrara, Italy

Pitacco P (2002) Un prestito mai refuso: la vicenda del *Liber de Simplicibus* di Benedetto Rini. In: Borean L, Mason S (eds) Figure di collezionistia Venezia tra Cinquecento e Seicento. Forum, Udine, Italy, pp 11–12

Pitard J, Proust L (1909) Les Iles Canaries, flore de l'archipel. France, Paris

Poehlman JM (1987) Breeding field crops. Van Nostrand Reinhold, New York, USA

Poiret JLM (1827) Historie des plantes de l' Europe. Ladrange et Verdierère, Paris, France

Ray J (1693) *Historia plantarum generalis.* Smith & Walford, London, UK

Ray J (1703) *Methodus plantarum emendata et aucta.* Smith & Walford, London, UK

Ray J (1724) *Synopsis methodica stirpium Britannicarum etc.*, 3rd edn. Innis, London, UK

Reichenbach L, Reichenbach HG (1909) *Icones florae Germanicae et Helveticae.* Sumptibus Federici de Zezschwitz, Lipisia, Germany

Rimpau W (1891) Kreuzungproducte landwirtschaftlicher Kulturpflanzen. Landwirtschaft Jahrbuch, vol 20, Berlin, Germany

Riva E (2010) The XV Century Venetian illuminated herbaria. http://www.cfs-cls.cz/Files/nastenka/page_3024/Version1/The%20XV%20Century%20Venetian%20Illuminated%20Herbaria.pdf

Rivera D, Obón C, Heinrich M, Inocencio C, Verde A, Farajado J (2006) Gathered Mediterranean food plants—In: Ethanobotanical investigators and historical development. In: Heinrich M, Müller WE, Galli C (eds) Local Mediterranean food plants and nutraceuticals. Forum Nutr, Karger, pp 18–74

Robinson DE, Harild JA (2002) The archeobotany of an early Hertebolle (late mesolithic) site at Hallskow, Denmark. In: Mason SUR, Hather JA (eds) Hunter - gatherer archeobotany. Institute of Archeology, London, UK, pp 50–76

Roccabonella N (1457) *Liber simplicibus*. Manuscript SS. Giovanni e Paolo, Venice, Italy

Santoni S, Bervillè A (1992) Two different satellite DNAs in *Beta vulgaris* L: Evolution, quantification and distribution in the genus. Theor Appl Genet 84:1009–1016

Santoni S, Bervillè A (1995) Characterization of the nuclear ribosomal DNA units and phylogeny of Beta L. wild forms and cultivated beets. Theor Appl Genet 83:533–542

Schneider JG (1794) *Scriptorum rei rusticae veterum latinorum*. Fritsch, Lipsia, Germany

Schultes JA (1817) Gründniss einer Geschichte and Literatur der Botanik. Schaumburg, Vienna, Austria

Shaw B, Thomas TH, Cooke DT (2002) Response of sugar beet (*Beta vulgaris* L.) to drought and nutrient deficiency stress. Plant Growth Regul 37:77–83

Shun ZF, Chu SY, Frese L (2000) Study on the relationship between Chinese and East Mediterranean *Beta vulgaris* L. subsp. *vulgaris* (leaf beet group) accessions. In: Maggioni L, Frese L, Germeier CU, Lipman E (eds) Report of a Working Group on *Beta*. First meeting, 9–10 Sept 1999, Broom's Barn, Higham, Bury St. Edmunds, UK. IPGRI, Rome, Italy, pp 65–69

Silvatico M (1523) *Opus pandectarum medicinae*. Venice, Italy

Simmonds NW (1976) Evolution of crop plants. Longman, London, UK

Smith JE (1803) English botany. Taylor Printer, London, UK

Soldano A (2003) l'erbario di Ulisse Aldrovandi. Istituto Veneto di Lettere Scienze ed Arti, Venice, Italy

Sontheimer G (1845) Heilmittel der Araber. Frieburg, Germany

Squalermo L (1561) Liber de simplicibus … etc. Valgrisi, Venice, Italy

Stokes J (1812) *Botanical materia medica*, vol 2. Johnson & Co. London, UK

Sturtevant J (1919) Notes on edible plants. JB Lyon and Co, Albany NY, USA

Tanara V (1674) Economia del cittadino in villa. Curti Stefano, Venice, Italy

Teza E (1898) "*De Simplicibus*" di Benedetto Rinio nel Codice marciano. Atti Regio Istituto Veneto Lettere Scienze Arti 9:18–29

Thornton RJ (1812) Elements of Botany. J. Whiting, London, UK

Throop P (1998) Physica by Hildegard von Bingen. Reprinted and translated by P. Throop. Healing Arts Press, Rochester VE, USA

Tschermak-Seysenegg E (1951) The rediscovery of the Gregor Mendel's work. J Heredity 42:163–174

Turner N (1995) Food plants of coastal first people. Royal British Columbia Museum handbook. UBC Press, Vancouver, Canada

Ulbrich E (1934) Chenopodiaceae. In: Engler A, Harms H (eds) Die Natürlichen Pflanzenfamilien. Wilhelm Engelmann, Leipzig, Germany, pp 375–584

van Raamsdonk LWD (1993) Wild and cultivated plants: the parallelism between evolution and domestication. Evol Trend Plants 7:73–84

van Zeist W, de Roller GJ (1993) Plant remains from Maadi, a predynastic site in Lower Egypt. Veget Hist Archaeobot 2:1–14

Ventura J (1998) Il"*De materia medica*" nel Medioevo: madiazione araba e ricezione occidentale. In: Speer A, Wagener L (eds) Wissen über Grenzen. De Gruyter, Berlin, Germany

Veyssiere de la Croze C (1755) *Lexicon aegiptiaco-latinum*. Typographus Clarendonianus Oxonii, Oxford, UK

Voigt R, Grüger E, Beier J, Meischner D (2008) Seasonal variability of Holocene climate: a paleolimnological study on varved sedimens in Lake Jues (Harz Mountains, Germany). J Paleolimnol 40:1021–1052

von Boguslawski E (1984) Zur Geschichte der Beta-Rübe als Kulturpflanze bis zum Beginn des 19 Jahrhundert. Institut für Zuckerrübenforschung, Göttingen, Berlin, Germany

von Lippmann EO (1925) Geschichte der Rübe (*Beta*) als Kulturpflanze. Verlag Julius Springer, Berlin, Germany

von Lippmann EO (1929) Geschichte des Zuckers, 2nd edit. Verlag Julius Springer, Berlin, Germany

von Proskowetz E (1910) Über das Vorkommen der Wildformen der Zuckerrüben am Quarnero. Österreiche-Ungarische Zeitschrift für Zuckerindustrie und Landwirtschaft 47:631–640

Weinmann JG (1737) *Phytanthoza iconographia, sive conspecus.* Hieronimum Lentium, Ratisbona, Germany

Weitzmann K (1979) Illustrations in rolls and codex. Princeton Univ Press, Princeton NJ, USA

Winner C (1993) History of the crop. In: Cooke DA, Scott RK (eds) The sugar beet crop: science into practice. Chapman & Hall, London, pp 1–35

Woenig F (1866) Die Pflanzen in alten Aegypten. Verlag von Wilhelm Friedrich, Lipsia, Germany

Ximenez F (1615) De la naturaleza, y virutes de las plantas. Diego Lopez, Mexico

Zanichelli G (1735) Storia delle piante che nascono ne'lidi attorno a Venezia. Venice, Italy

Zohary D (2004) Unconscious selection and the evolution of domesticated plants. Econ Bot 58:5

Zohary D, Hopf M (1973) Domestication of pulses in the Old World. Science 182:887–894

Zohary D, Hopf M (2000) Domestication of plants in the old world: The origin and spread of cultivated plants in West Asia. Oxford Univ Press, Oxford, UK, Europe and the Nile Valley

Chapter 2
Range of Distribution

Lothar Frese and Brian Ford-Lloyd

Abstract *Beta maritima* is the most widespread taxon within the genus *Beta*. The plant also named "sea beet" can be found quite easily along the seashores of the Mediterranean Sea, the European Atlantic Ocean and in the western part of the Baltic Sea. Here, countless locations have been reported in the literature beginning in the early 1700s. The frequency of sea beet populations decreases as one goes inland, where the origin of the populations is more likely due to hybridization between sea beet and cultivated beet crops. Infrequently, the presence of sea beet has been reported on the shores of the North Sea, the Middle East, India, China, Japan and California. In North America, wild populations of *Beta maritima*, *Beta macrocarpa* and respective hybrids with cultivated beet likely originated from contaminated seed stocks.

Keywords *Beta maritima* · Habitat · Global distribution · Classification · Inland sites · Historical reports

The recognition of a plant species in wild habitats often is difficult because the traits required for the purpose of classification may not be displayed at the time of observation. The best period for the sea beet (*Beta vulgaris* L. subsp. *maritima* (L.) Arcang.) identification is at early to late flowering when the growth habit, distribution of bracts on the seed stalks, type of perianth segments of the flower and characteristics of the maturating glomerule can be recognized. At flowering time, plants emit a typical fragrance which helps experts identify growing sites. The morphological traits of taxa within genus *Beta* section *Beta* are quite typical and therefore wild *maritima* beets can be easily distinguished from other associated wild species and types during this period.

L. Frese (✉)
Julius Kühn-Institut, Institute for Breeding Research on Agricultural Crops, Quedlinburg, Germany
e-mail: lothar.frese@julius-kuehn.de

B. Ford-Lloyd
University of Birmingham, Birmingham, UK

E. Biancardi et al. (eds.), *Beta maritima*,
https://doi.org/10.1007/978-3-030-28748-1_2

49

Reliable recognition of species is necessary for accurate description of its distribution area. Misclassification of *Beta macrocarpa* or *Beta patula* rarely occurs. As a rule of thumb, *Beta macrocarpa* is distinguished by erect and often spongy perianth segments of the flower as compared to the thin perianth segments of *Beta maritima*. *Beta patula* only occurs on islands close to Madeira. The very limited distribution area, the small leaf rosette diameter and, in particular, the high number of flowers per glomerule clearly differentiate *Beta patula* from all other taxa of section *Beta*. Compared to *Beta macrocarpa* and *Beta patula,* the unambiguous classification of sea beet is more difficult.

Due to the high morphological variation of *Beta maritima*, it is difficult to delineate *Beta vulgaris* subsp. *adanensis* from the eastern Mediterranean sea beet types. Under experimental conditions, the flowering period of *Beta vulgaris* subsp. *adanensis* partly overlaps with the flowering period of the sea beet (Letschert 1993). Gene flow between subspecies may be possible. Progeny resulting from crosses between *Beta maritima* and *Beta vulgaris* subsp. *adanensis* may blur the morphological differences between both taxa. Andrello et al. (2017) applied 9724 SNPs to describe genetic diversity among a collection of *Beta* section *Beta* individuals. The correlation between genetic clustering and the predefined taxonomic groups was investigated by the K-means ex nihilo clustering method. In this genetic dataset, many individuals named *Beta maritima* in germplasm collections were allocated to genetic groups including "*adanensis*" and "Leaf Beet" indicating a more general problem of classification. *Beta vulgaris* sensu lato is not only a genetically diverse species but also a species with a high plasticity. Thus, it is not always possible to unambiguously delineate the subspecies purely by morphological characters. Today, genetic marker technologies (Andrello et al. 2017) can be applied either to confirm classification based on taxonomic keys or to determine causes of misclassification problems.

Hybridization between the sea beet and *Beta vulgaris* crop types presents additional sources of uncertainty. The sea beet has an impressive colonizing capacity as compared to either *Beta macrocarpa* or *Beta patula*. The distribution of the sea beet is not confined to the sea coast habitats. Likely due to human-distributed seeds, the species has established inland populations, which naturalize, and may hybridize with cultivated beet. Such crop–wild–weed complexes (i.e. ruderal beets) have likely been formed many times in various locations throughout the geographic ranges where wild and crop types are endemic.[1] In particular, in rural areas located close to the sea shore (such as on the many Greek islands) where seed of leaf and garden beets are produced by home gardeners, crop–wild–weed complexes are formed.

Accurate classification is always important, but it is particularly so when sea beet germplasm is intended to be used in breeding programs or if there is concern of gene flow from the crop into sea beet populations. Sometimes, the accessions stored in *Beta* germplasm collections bear an incorrect taxonomic name due to misclassification. Seed sample gathered in the wild habitat, unbeknownst to the collector, may include

---

[1]Species complex is a cluster of closely related species, subspecies, cultivated, wild and feral forms, which are able to exchange genetic material in natural conditions (Coyne 1989; Driessen 2003; Fénart et al. 2008; Pernès 1984).

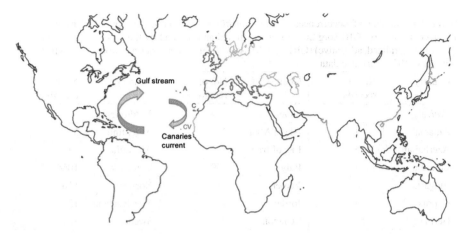

Fig. 2.1 World map showing the distribution of *Beta maritima* along the seashores (red: frequent; blue: sparse; green: rare). For the sea currents see the text. A: Azores, C: Canary Islands, CV: Cape Verde Islands

hybrids generated from a crop–wild–weed complex. Letschert et al. (1994) proposed classifying naturalized populations descended from the introgression of cultivated germplasm into sea beet as *Beta maritima*. They argue that plant breeders will always select against unwanted wild traits so that the taxonomically relevant (morphological) boundaries between subsp. *vulgaris* and subsp. *maritima* will always be maintained. One could add that natural selection will equally act against cultivar traits in non-cultivated hybrid populations. Morphology of such admixed populations may be similar to wild types.

The genus *Beta* likely evolved in the Mediterranean region (Romeiras et al. 2016), which can be called the primary distribution area of the species. Sea beet started colonizing the Atlantic coasts of Europe at the end of the last ice age and populated southern areas of Norway and Sweden as well as the western part of the Baltic Sea coasts. The species was recently observed on a Russian island located in the Eastern Gulf of Finland (Glazkova 2006), in a climate which is not well suited for a frost susceptible species such as sea beet. This striking finding illustrates the strong long-range dissemination and colonizing ability of the species. Today, *Beta maritima* is a characteristic plant of vegetated sea cliffs of the Atlantic and Baltic coasts (e.g. Natura 2000 code 1230) and Atlantic salt meadows (Natura 2000 code 1330; EC, 2013). Beyond the primary distribution area (Fig. 2.1), *Beta maritima* can be found in many countries of the world. The Global Biodiversity Information Facility (GBIF.org 2018) currently reports 10,493 observations of *Beta maritima* in the period from 1800 to 2018.

Table 2.1 compiles all countries where the sea beet has been found. The countries were categorized according to GRIN (2018) into natural, adventive and naturalized. Please note that the number of records per country shown in the table depends on several factors: the number of sites where the species can grow, the interest of experts

Table 2.1 Number of occurrences of *Beta maritima* by country. Data provided by the global biodiversity facility (GBIF.org 2018) were compiled and ordered by the type of distribution area (natural, naturalized, adventive) (GRIN 2018). A "?" refers to countries mentioned by GRIN (2018) without GBIF occurrence data

Country	No. of records	Country	No. of records	Country	No. of records
Native		Israel	2	*Naturalized*	
Albania	1	Isle of Man	2	Argentine	3
Azerbaijan	5	I.R. of Iran	1	Australia	77
Belgium	1127	Italy	259	Denmark	1064
Bulgaria	4	Yemen	11	Norway	116
Cyprus	44	Jordan	1	New Zealand	13
Germany	57	Lebanon	?	Sweden	887
Algeria	29	Libya	3		
Egypt	55	Morocco	101	*Adventive*	
Spain	875	Montenegro	?	Chile	1
France	1658	The Netherlands	177	Finland	5
United Kingdom	3049	Portugal	99	Mexico	1
Guernsey	8	Tunisia	26	Poland	?
Greece	395	Turkey	16	Russia	1
Croatia	3	Syria	?		
Ireland	252	Slovenia	?		

in documenting observations of the species in national information systems and the number of records uploaded by the responsible authorities of the country to GBIF. The data should therefore not be understood as the number of distribution sites or abundance.

The literature, starting with Linnaeus (1753), mentions countless locations of sea beet populations. Early summaries of growing sites were published by Becker-Dillingen (1928) and Ulbrich (1934). Most locations are close to the sea, where the species occurs in a narrow band between high tide and close vegetation (Doney et al. 1990; Doney 1992). *Beta maritima* is common and numerous on seashores and along estuaries (Hohenacker 1838; Frese et al. 1990; Viard et al. 2004; Fievet et al. 2007), however, it avoids sandy beaches and almost completely disappears as one moves inland, which demonstrates the environmental preferences of the species (Biancardi 1999).

According to the currently prevailing schools of thought, it is likely that some inland populations classified as sea beet are, in reality, feral beets, *Beta macrocarpa*, or hybrids and derivatives of wild *Beta* sections and domesticated forms. van Dijk and Boudry (1992) found few wild beet populations growing far from the sea side, but close to the sugar beet production fields in southern France. These populations are genetically close to the sea beet but show some introgression from sugar beet (Desplanque et al. 1999). van Dijk (1998) asserts that only a few true sea beet populations exist in inland regions between southern France (43° N) and The Netherlands (52° N). According to Frese (2010) the altitude of collection sites, corresponding to 798 accessions currently stored in gene banks and entered into the International Data Base for *Beta*, ranges from 280 m below sea level around the Dead Sea (Post 1869) to 1300 m in Sicily (a single plant, collection number IBPGR/SI/81 100, notes taken by Toll and Hendriksen (1981)). The collectors also sampled a population labeled *Beta vulgaris* (IBPGR/SI/81 101) at 1150 m close to Cesaro in Sicily, which was later classified by Letschert and Frese (1993) as *Beta maritima*. In this case, the overall morphological difference between the coastal and inland populations was small, indicating a recent migration from the coastal area to the inland location. Sea beet also displays a wide latitudinal range, which varies from about 15° N (Cape Verde Islands) to about 58° N (southern Norway and southern Sweden).

Information concerning the range of *Beta maritima* is limited and generic up through the end of the Middle Ages. Bauhin (1622) reported the presence of sea beet near Basel (Switzerland). Parkinson (1655), in an edition of the de Lobel's *"Stirpium illustrationes"*, wrote that *Beta maritima syl(vestris) minor* and *Beta maritima syl(vestris) spontana* are spread along the Atlantic coast of France, UK and Scotland. Linnaeus (1753) confirmed that *"Beta maritima habitat Angliae, Belgii littoribus maris"* (sea beet grows in English and Belgian sea shores).

Beginning in the early sixteenth century, the ease of shipping and traveling favored long-range exploration organized by botanical societies, which had become numerous in all European countries by then. The scientific curiosity of botanists was expensive; especially travel to the unexplored territories of the New World, East Asia and Australia. This spirit of research was supported by governments not only out of scientific interest but also for political and commercial purposes. Ray (1738) in collaboration with other local botanists, wrote, "Travels Through the Low-countries, Germany, Italy, and France", and catalogued the plants encountered during long journeys in Spain, Sicily, Germany and so on. Among the botanists cited by Ray, Donati (1826) did not detect the presence of *Beta maritima* in the Lagoon of Venice, and no populations of *Beta cretica* were reported in the Greek islands. Different types of wild beets were described in Portugal and named as *Beta alba maxima*, *Beta radice rubra* and *Beta marina semine aculeato*. The latter was found, together with *Beta marina semine aculeato minor* (sea beet minor with thorny seed), on the island and the promontory of Pachino and Pozzallo (Sicily). Sea beet was described in several parts of the Italian peninsula and included in botanical gardens there (Micheli 1748; Tenore 1851), as well as in other parts of Europe (Desfontaines 1829).

Hooker (1835) wrote *"Beta maritima* Linn. is in sea shores, especially in muddy soil, England; and in the south, principally of Scotland". Bunge, cited by von Proskowetz (1895), listed the localities where the species of genus *Beta* were widespread. He highlighted that 9 out of 14 species belonging to genus *Beta* were identified on the Canary Islands. Boissier (1879) listed the shores of the following locations: *Zacynthus*, *Attica*, *Pontus Exinius* (Black Sea), *Maris Caspii*, *Cyprus*, *Syriae littorals* (Syrian seashores) and *interioribus umidis Babyloniae"* (wet sites around

Babylon). In addition to areas of the North Sea listed by other authors, Reichenbach and Reichenbach (1909) asserted that sea beet also grew at Eppendorf and on the shores of Aerø Island, Denmark. Darwin (1899), cited by von Lippmann (1925), observed that an indigenous wild *Beta* in India developed better than other European varieties. This wild variety, named "*palung*" and "*mitha*", could have been a locally adapted sea beet (Watt 1899). von Proskowetz (1896) summarized several observations and locations made by his contemporaries (Hehn, Willkomm, Bunge, Freyn, Engler and Prantl, de Candolle). The *Beta* species were named according to de Tournefort (1700).

de Vries (1905) wrote "Beets are even now found in large quantity along the shores of Italy. They prefer the vicinity of the sea, as do so many other members of the beet family, and are not limited to Italy, but are found growing elsewhere on the littoral of the Mediterranean, in the Canary Island, and through Persia and Babilonia to India. In most of their native localities they occur in great abundance". Ascherson and Graebner (1919) limited the range of sea beet to Denmark, the British Islands, France, Italy (and its islands), Spain, Albania, Greece (and its islands), Bulgaria, Central and southern Russia, the Near East up to Western India, North-Africa and Canary Islands. Becker-Dillingen (1928) listed the areas into which *Beta maritima* had spread and stated that "the species is clearly halophytic. It is widespread not only along the seacoasts, but also in soils more or less recently submerged by salty water". von Lippmann (1925) summarized the locations of sea beet populations reported in literature at the time, which were divided among the three continents facing the Mediterranean Sea:

Asia   along coasts of the Caspian Sea, Talysh, Caucasus, Dagestan, Transcaucasia, the Black Sea, Armenia, Asia Minor, Syria, Mesopotamia, Red Sea, Persia, India, Turkestan.

Africa in Egypt, Atlantic Isles (Canary, Madeira, Cape Verde).

Europe in Norway, Lapland, Finland, Karelia, Sweden southern coasts of the North Sea, Schleswig (Germany), Holland, England, Ireland, France, Portugal, Spain, Italy, Balkan countries, Malta, Cyprus.

After mentioning the different synonyms for *Beta maritima* used by various authors at the time, Ulbrich (1934) sketched the area of dispersal, confirming much of the range described by Becker-Dillingen (1928), excluding only the most northern parts of Europe cited by von Lippmann (1925). According to Ulbrich (1934), the range extended from the shores of the Mediterranean, the Canary Islands and Iberian Peninsula, to the Middle East and Western India. The species is widespread on the Atlantic coasts of France, England, Holland, Denmark and Germany, and on the southern coasts of Sweden and other Nordic countries bordering the Gulf Stream. As observed by Strobl, cited by Ulbrich (1934), the sea beet grows on the slopes of the volcano Etna (Italy) up to 850 m in altitude.

According to de Candolle (1884), the plant was common in sandy places near the seas of Mediterranean Europe, Africa, Asia Minor and in the Azores and Canary Islands. It was also present in Algeria, Egypt, Persia, India and Eastern Europe. Moquin-Tandon, cited by de Candolle (1884), extended the localization of the species to the Caspian Sea and eastern India. The distribution, according to the USDA-ARS Germplasm Resource Information Network, includes also Azerbaijan, Belgium, Ireland, Morocco and Cape Verde Islands. Castro et al. (2013) confirmed the presence of the species on the Azores and on Madeira. Trotter, quoted by Munerati et al. (1913), detected the presence of *Beta maritima* near Aquiloni, Italy, at 350 m in altitude and about 60 km from the sea.

Zossimovitch (1934) cited by Coons (1954) reported the presence of *Beta maritima* in the Russian steppes, especially in depressions characterized by salty and alkaline soils. Hermann (1937) located sea beet on the East coasts of England, also observing a different flowering behavior and an elevated diversity among the populations. Similar observations (polymorphism in habit, pigmentation, number of flowers/cluster and incidence of male sterility) were recorded by Jassem (1985) along the French and British shores of the English Channel. These populations were widely studied beginning in 1990 (Cuguen et al. 1992). An accurate description of the environments preferred by sea beet was made by Doney and McFarlane (1985) after a survey along the coasts of Southern Italy: "The best collections were near ancient ruins and undisturbed beaches. Near Capo Colonna, Sicily, the wind creates a constant sea water spray on the *Beta maritima* growing in the rocky cliffs along the shore. The intensive farming along with the increased tourism appeared to have driven much of the native flora to fence lines and roadsides".

The first written mention of the presence of sea beet on the Baltic seashores was at the end of the seventeenth century at Marstal, Denmark (Christensen 1996). Hehn and Hück, cited by von Lippmann (1925), reported German locations. Further locations noted were on Samsø Island, Denmark, and in the southern coasts of Sweden and Norway (Engan 1994; Batwik 2000; Often and Svalheim 2001; Pedersen 2009). Since 1967, several new populations have been found on German shores (Driessen 2003), although considered doubtful by Karsten (1880), the presence of sea beet here has been established dating back to prehistoric times (Kubiak-Martens 1999). According to Griesebach, Schübeler, Müller, Kempe, Hehn (cited by von Proskowetz (1895) and von Lippmann (1925)), different species of the genus *Beta*, including sea beet, were present in Lapland, Finland, Karelia, Central Australia, and so on. These locations were not confirmed by later authors. Makino (1901) reported *Beta maritima* at Musashii and Hiranuma on Yokohama Island, Japan, but only as very rare. In China, some populations of sea beet were mentioned by Doney and Whitney (1990). Carsner (1928, 1938) reported of wild beets along the Pacific coasts of California (Santa Clara, Ventura, San Bernardino and Los Angeles) and in the Imperial Valley near the Mexican border. In the first case, he speculated that these beets were either *Beta maritima* or natural crosses between this species and the cultivated varieties. In the latter case, the wild populations were classified as more or less composite crosses with *Beta macrocarpa* and sugar beet varieties (McFarlane 1975; Bartsch and Ellstrand 1999; Bartsch et al. 2003). Calflora (2018) displays more recent observed occurrences on a web-based map. Recently, some sea beet sites have been reported in Australia (near Melbourne) and New Zealand (near Wellington), that is in localities where *Beta maritima* was never previously detected (see GBIF.org 2018).

Confirming the countless locations reported in the literature, *Beta maritima* is widespread on almost all Mediterranean and Black Sea coasts, if the site fits the needs of the species. Those needs are disturbed sites with patchy vegetation, presence of stones, limited periods of drought, full sunlight and a location close to saltwater (Biancardi, unpublished). In Fig. 2.1 note the absence of colonization not only along the east coasts of the Americas, but also in the southern hemisphere. Absence from the African shores below the southern border of Morocco (21° N), or perhaps better demarked as south of 15° N (the latitude of the Cape Verde Islands), could be explained by the prevailing direction of the Canary Ocean current that flows westward toward the Caribbean Islands.

Letschert (1993) confirmed the occurrence of *Beta maritima* at the sites mentioned above, except for China, Japan, Lapland, Karelia, Finland and Australia. Recent surveys did not report the presence of sea beet in China (Shun et al. 2000), Czech Republic (Stehno et al. 2000), Latvia (Rashal and Kazachenko 2000), Belarus (Svirshchevskaya 2000) or Georgia (Aleksidze et al. 2009). Presence was confirmed in Iran (Srivastava et al. 1992), Azerbaijan (Frese et al. 2001) and Armenia (Ghandilyan and Melikyan 2000). *Beta maritima* is relatively widespread on the western coasts of the Caspian Sea, Slovenia, Romania and Crimea (Ukraine), but is very rare in Bulgaria (IPGRI 2004).

The northern limit of *Beta maritima* seems to be the isotherm of 14 °C in July (Villain 2007). Grogan (2009) asserts that Ireland seems to be at the limit of the sea beet habitat, since populations were located only on the southern and central part of the island; that is, near the sea warmed by the Gulf Stream, in well sunny sites. However, other authors have located sea beets along the northern shores of Northern Ireland (Anon. http://habitas.org.uk). Sea beet currently appears to be expanding its range on the German coast of Baltic Sea, perhaps due to global warming (Driessen 2003). On the West-Adriatic coast, a reduction in size of populations has been observed; caused both by the decreasing amount of summer rain and increasing tourist activities along the seashores (Pignone 1989; Stevanato, unpublished). At these locations, the number of plants within undisturbed populations seems to be correlated with an even distribution and amount of rainfall the previous year. In extended drought periods, the number of plants decreases dramatically (Bartsch and Schmidt 1997). If this occurs, the older plants, that is, those with more developed and deeper root systems, survive better because of the very low water holding capacity of the sandy soils along the seashores (Biancardi, unpublished). Doney et al. (1990) wrote "The current (1990) distribution of sea beet (in Ireland) was similar to earlier observation in 1962. However, many small populations were in danger of elimination, or had disappeared. Factors threatening or causing extinction of local populations included livestock grazing (particularly sheep), slippage of mud cliffs, coastal erosion, increase of sea level, industrialization of sea ports, and recreational and touristic activities".

References

Aleksidze G, Akparov Z, Melikyan A, Arjmand MN (2009) Biodiversity of *Beta* species in the caucasus region (Armenia, Azerbaijan, Georgia, Iran). In: Frese L, Maggioni L, Lipman E (eds) Report of a working group on *Beta* and the world *Beta* network. Third Joint Meeting, 8–11 March 2006, Puerto de la Cruz, Tenerife, Spain. Bioversity International, Rome, Italy, pp 38–44

Andrello M, Henry K, Devaux P, Verdelet D, Desprez B, Manel S (2017) Insights into the genetic relationships among plants of *Beta* section *Beta* using SNP markers. Theor Appl Genet. https://doi.org/10.1007/s00122-017-2929-x

Ascherson P, Graebner P (1919) Synopsis der mitteleuropaischen Flora. Verlag von Gebrüder Borntraeger, Lipsia, Germany

Bartsch D, Cuguen J, Biancardi E, Sweet J (2003) Environmental implications of gene flow from sugar beet to wild beet-current status and future research needs. Environ Biosafety Res 2:105–115

Bartsch D, Ellstrand NC (1999) Genetic evidence for the origin of Californian wild beets (genus *Beta*). Theor Appl Genet 99:1120–1130

Bartsch D, Schmidt M (1997) Influence of sugar beet breeding on populations of *Beta vulgaris* ssp. *maritima* in Italy. J Veg Sci 8:81–84

Batwik JI (2000) Strandbete Beta vulgaris L. ssp. maritima (L.) Arc. er trolig borte fra ϕstfold i dag på grunn av barfrost. Natur Φstfold 1–2:38–42

Bauhin G (1622) *Catalogus plantarum circa Basileam sponte nascentium etc.* Basilea, Switzerland

Becker-Dillingen J (1928) Die Wurzelfrüchte (Rüben). In: Parey P (ed) Handbuch des Hackfruchtbaues und Handelspflanzenbaues. Parey, Berlin, Germany

Biancardi E (1999) Miglioramento genetico. In: Casarini B, Biancardi E, Ranalli P (eds) La barbabietola da zucchero in ambiente Mediterraneo. Edagricole, Bologna, Italy, pp 45–57

Boissier E (1879) *Flora orientalis sive enumeratio plantarum in Oriente a Graecia et Aegiptoas Indias fines etc.* Georg, Apud H., Lyon, France & Ginevra, Switzerland

Calflora (2018) Information on California plants for education, research and conservation, with data contributed by public and private institutions and individuals, including the Consortium of California Herbaria. [web application]. Berkeley, California: The Calflora Database. http://www.calflora.org/. Accessed 12 Sept 2018

Carsner E (1928) The wild beet in California. Facts About Sugar 23:1120–1121

Carsner E (1938) Wild beets in California. Proc Am Soc Sugar Beet Technol 1:79

Castro S, Romeiras MM, Castro M, Duarte MC, Loureiro J (2013) Hidden diversity in wild *Beta* taxa from Portugal: insights from genome size and ploidy level estimations using flow cytometry. Plant Sci 207:72–78. https://doi.org/10.1016/j.plantsci.2013.02.012

Christensen E (1996) Neuer Fund der Betarübe an Schleswig-Holsteins Osteeküste. Kieler Notizen zur Pflanzenkunde in Schleswig-Holstein und Hamburg 24:30–38

Coons GH (1954) The wild species of *Beta*. Proc Am Soc Sugar Beet Technol 8:142–147

Coyne JA (1989) Speciation and its consequences. Sinauer Associates, Suderland MA, USA

Cuguen J, Saumitou-Laprade P, Spriet CVP (1992) Male sterility and DNA polymorphism in *Beta maritima*. In: Frese L (ed) International *Beta* genetic resources network. A report on the 2nd international *Beta* genetic resources workship held at the Institute for Crop Science and Plant Breeding, Braunschweig, Germany, 24–28 June 1991. IBPGR, Rome, Italy, pp 49–54

Darwin C (1899) Das Variieren der Tiere und Pflanzen. Carus, Stuttgart, Germany

de Candolle A (1884) Der Ursprung der Culturpflanzen. Brockhaus, Lipsia, Germany

de Tournefort JP (1700) *Institutiones rei herbariae*. Thypographia Regia, Paris, France

de Vries U (1905) Species and varieties. Open Court Publishing CO, Chicago IL, USA

Desfontaines R (1829) *Catalogus plantarum horti regii Parisiensis*. France, Paris

Desplanque B, Boudry P, Broomberg K, Saumitou-Laprade P, Cuguen J, van Dijk H (1999) Genetic diversity and gene flow between wild, cultivated and weedy forms of *Beta vulgaris* L. (Chenopodiaceae), assessed by RFLP and microsatellite markers. Theor Appl Genet 98:1194–1201

Donati A (1826) Flora veneta o descrizione delle piante che nascono nella provincia di Venezula. Presso Leone Bonvecchiato, Venice, Italy

Doney DL (1992) Morphology of North Atlantic *Beta*. In: Frese L (ed) International *Beta* genetic resources network. A report on the 2nd international *Beta* genetic resources workship held at the Institute for Crop Science and Plant Breeding, Braunschweig, Germany, 24–28 June 1991. IBPGR, Rome, pp 17–28

Doney DL, McFarlane JS (1985) Sugar beet exploration Italy and France. Unclassified USDA Report

Doney DL, Whitney E (1990) Genetic enhancement in *Beta* for disease resistance using wild relatives: a strong case for the value of genetic conservation. Econ Bot 44:445–451

Doney DL, Whitney E, Terry J, Frese L, Fitzgerald P (1990) The distribution and dispersal of *Beta maritima* germplasm in England, Wales and Ireland. J Sugar Beet Res 27:29–37

Driessen S (2003) *Beta vulgaris* subsp. *maritima* an Deutschlands Ostseeküste. PhD RWTH Aachen, Germany

Engan NC (1994) Stranbete *Beta vulgaris* ssp. *maritima* funnet sponton i Norge. Blyttia 52:33–42

Fénart S, Arnaud JF, de Cauwer I, Cuguen J (2008) Nuclear and cytoplasmic genetic diversity in weed beet and sugar beet accessions compared to wild relatives: new insights into the genetic relationships within the *Beta vulgaris* complex species. Theor Appl Genet 116:1063–1077

Fievet V, Touzet P, Arnaud JF, Cuguen J (2007) Spatial analysis of nuclear and cytoplasmic DNA diversity in wild sea beet (*Beta vulgaris* ssp. *maritima*) populations: do marine currents shape the genetic structure? Mol Ecol 16:1847–1864

Frese L (2010) Conservation and access to sugar beet germplasm. Sugar Tech 12:207–219

Frese L, de Meijer E, Letschert JPW (1990) New wild beet genetic resources from Portugal and Spain. Zuckerindustrie 115:950–955

Frese L, Akbarov Z, Burenin VI, Arjmand MN, Hajiyev V (2001) Plant exploration in the Talysch Mountains of Azerbaijan and Iran. PGR Newsl 126:21–26

Glazkova EA (2006) *Beta maritima* (Chenopodiaceae), a new species to the flora of Russia. Botaniceskij Zurnal 91:34–45

GBIF.org (16 August 2018) GBIF Occurrence Download. https://www.gbif.org/occurrence/download/0008049-180730143533302

Ghandilyan PA, Melikyan ASh (2000) *Beta* genetic resources in Armenia. In: Maggioni L, Frese L, Germeier CU, Lipman E (eds) Report of a working group on *Beta*. First meeting, 9–10 September 1999, Broom's Barn, Higham, Bury St. Edmunds, United Kingdom. IPGRI, Rome, Italy, pp 15–17

Grogan D (2009) Surwey of *Beta vulgaris* subsp. *maritima* populations in Ireland. In: Frese L, Maggioni L, Lipman E (eds) Report of a working group on *Beta* and the world *Beta* network. Third Joint Meeting, 8–11 March 2006, Puerto de la Cruz, Tenerife, Spain. Bioversity International, Rome, Italy, pp 53–58

GRIN (2018) USDA, Agricultural Research Service, National Plant Germplasm System. 2018. Germplasm Resources Information Network (GRIN-Taxonomy). National Germplasm Resources Laboratory, Beltsville, Maryland. https://npgsweb.ars-grin.gov/gringlobal/taxon/taxonomysearch.aspx. Accessed 16 Aug 2018

Hermann C (1937) Wild beets on the East Coast. Brit Sugar Beet Rev 11:105–108

Hohenacker M (1838) Pflanzen der Provinz Talish. Soc. Imp, Naturalists, Moskow, Russia

Hooker WJ (1835) The British Flora. London, UK

IPGRI (2004) Report of a working group on *Beta* and world *Beta* network. Second joint meeting, 23–26 October 2002, Bologna, Italy. Frese L, Germeier C, Lipman E, Maggioni L, compilers. IPGRI, Rome, Italy

Jassem B (1985) Variation of maritime beet (*Beta maritima* L.). Genetica Polonica 26:463–469

Karsten H (1880) Deutsche Flora, pharmaceutisch-medicinische Botanik. Berlin, Germany

Kubiak-Martens L (1999) The plant food component of the diet at the late Mesolithic (Ertebolle) settlement at Tybrind Vig, Denmark. Veg Hist Archaeobot 8:117–127

Letschert JPW (1993) *Beta* section *Beta*: biogeographical patterns of variation, and taxonomy. PhD Wageningen Agricultural University, Papers 93-1

Letschert JPW, Frese L (1993) Analysis of morphological variation in wild beet (*Beta vulgaris* L.) from Sicily. Genet Res Crop Evol 40:15–24

Letschert JPW, Lange W, Frese L, van den Berg RG (1994) Taxonomy of *Beta* Section *Beta*. J Sugar Beet Res 31:69–85

Linnaeus (1753) *Systema vegetabilium secundum classes, ordines, genera, species, etc.*, 15th edn. Typis et impensis Io., Göttingen, Germany

Makino T (1901) Observations on the flora of Japan. Botanical Magazine Tokio 15:1–494

McFarlane JS (1975) Naturally occurring hybrids between sugar beet and *Beta macrocarpa* in the Imperial Valley of California. J Am Soc Sugar Beet Technol 18:245–251

Micheli PA (1748) *Catalogus plantarum Horti Cesarei Florentini etc.* Florence, Italy

Munerati O, Mezzadroli G, Zapparoli TV (1913) Osservazioni sulla *Beta maritima* L., nel triennio 1910-1912. Stazioni Sperimentali Agricole Italiane 46:415–445

Often A, Svalheim E (2001) Seashore beet *Beta vulgaris* subsp. *maritima*. Blyttia 59:192

Parkinson J (1655) *Matthiae de L'Obel Stirpium Illustrationes*. Warren, London, UK

Pedersen O (2009) Strand plants – new records of strand plants, especially at Lista. Blyttia 67:75–94

Pernès J (1984) Gestion des resources génétique des plantes, vol 2. Agence Cooperation Culturelle Technique, Paris, France

Pignone D (1989) Wild *Beta* germplasm under threat in Italy. FAO/IPGRI Plant Genetic Resources Newsletter 77:40

Post GE (1869) Flora of Syria, Palestine, and Sinai. Syrian Protestant College, Beirut, Syria

Rashal I, Kazachenko R (2000) *Beta* genetic resources in Latvia. In: Maggioni L, Frese L, Germeier CU, Lipman E (eds) Report of a working group on *Beta*. First meeting, 9–10 September 1999, Broom's Barn, Higham, Bury St. Edmunds, United Kingdom. IPGRI, Rome, Italy, pp 35–36

Ray J (1738) Travels through the low-countries, Germany, Italy, and France, 2nd edn. London, UK

Reichenbach L, Reichenbach HG (1909) *Icones florae Germanicae et Helveticae*. Sumptibus Federici de Zezschwitz, Lipsia, Germany

Romeiras MM, Vieira A, Silva DN, Moura M, Santos-Guerra A, Batista D, Duarte MC, Paulo OS (2016) Evolutionary and biogeographic insights on the Macaronesian *Beta-Patellifolia* species (Amaranthaceae) from a Time-Scaled Molecular Phylogeny. PLoS ONE 11(3):e0152456. https://doi.org/10.1371/journal.pone.0152456

Shun ZF, Chu SY, Frese L (2000) Study on the relationship between Chinese and East Mediterranean *Beta vulgaris* L. subsp. *vulgaris* (leaf beet group) accessions. In: Maggioni L, Frese L, Germeier CU, Lipman E (eds) Report of a working group on *Beta*. First meeting, 9–10 September 1999, Broom's Barn, Higham, Bury St. Edmunds, United Kingdom. IPGRI, Rome, Italy, pp 65–69

Srivastava HM, Sun Y-C, Arjmand MN, Masutani T (1992) *Beta* genetic resources activities in Asia. In: Frese L (ed) International *Beta* genetic resources network. A report on the 2nd international *Beta* genetic resources workship held at the Institute for Crop Science and Plant Breeding, Braunschweig, Germany, 24–28 June 1991. IBPGR, Rome, pp xli–lxi

Stehno Z, Chyilova V, Faberova V (2000) Status of the *Beta* collection in the Czech Republic. In: Maggioni L, Frese L, Germeier CU, Lipman E (eds) Report of a working group on *Beta*. First meeting, 9–10 September 1999, Broom's Barn, Higham, Bury St. Edmunds, United Kingdom. IPGRI, Rome, Italy, pp 21–22

Svirshchevskaya A (2000) Germplasm collections in Belarus. In: Maggioni L, Frese L, Germeier CU, Lipman E (eds) Report of a working group on *Beta*. First meeting, 9–10 September 1999, Broom's Barn, Higham, Bury St. Edmunds, United Kingdom, IPGRI, Rome, Italy, p 20

Tenore M (1851) *Sylloge plantarum vascularum florae neapolitanae etc.* Ex Typographia Fibreni, Naples, Italy

Toll J, Hendricksen A (1981) Beet in Sicily. IBPGR Newsl 49:2–4

Ulbrich E (1934) Chenopodiacae. In: Engler A, Harms H (eds) Die Natürlichen Pflanzenfamilien. Wilhelm Engelmann, Leipzig, pp 375–584

van Dijk H (1998) Variation for developmental characters in *Beta vulgaris* subsp. *maritima* in relation to latitude: the importance of *in situ* conservation. In: Frese L, Panella L, Srivastava HM, Lange W (eds) A report on the 4th international *Beta* genetic resources workshop and world *Beta* network conference held at the Aegean Agricultural Research Institute, Izmir, Turkey, 28 February–3 March, 1996. IPGRI, Rome, Italy, pp 30–38 van Dijk H, Boudry P (1992) Genetic variability for life-histories in *Beta maritima*. In: Frese L (ed) International *Beta* genetic resources network. A report on the 2nd international *Beta* genetic resources workship held at the Institute for Crop Science and Plant Breeding, Braunschweig, Germany, 24–28 June 1991. IBPGR, Rome, Italy, pp 9–16

Viard F, Arnaud J-F, Delescluse M, Cuguen J (2004) Tracing back seed and pollen flow within the crop-wild *Beta vulgaris* complex: genetic distinctiveness vs. hot spots of hybridization over a regional scale. Mol Ecol 13:1357–1364

Villain S (2007) Histoire évolutive de la section *Beta*. PhD Université des Sciences et Technologies de Lille, France

von Lippmann EO (1925) Geschichte der Rübe *(Beta)* als Kulturpflanze. Verlag Julius Springer, Berlin, Germany

von Proskowetz E (1895) Über die Culturversuche mit *Beta* im Jahre 1894 und über Beobachtungen an Wildformen auf natürlichen Standorten. Oesterreichische-Ungarische Zeitschrift fur Zuckerindustrie und Landwirtschaft 32:227–275

von Proskowetz E (1896) Über die Culturversuche mit *Beta* im Jahre 1895. Ibidem 33:711–766

Watt JR (1899) Dictionary of the economic products of India. Joret, London, UK

Zossimovitch V (1934) Wild species of beets in Transcaucasia. Naučnye Zapiski VNIS 44:30

Chapter 3
Morphology

Enrico Biancardi and Marco de Biaggi

Abstract Since few references on the morphology of *Beta maritima* are available, most information for this chapter comes from the cultivated forms of *Beta vulgaris*. This is justified by the fact that the two species or subspecies are very similar to each other, so that their safe classification only on the basis of the phenotype is rather difficult. A striking feature of *Beta maritima* gleaned from this review is how variable and adaptive it is. The species is fairly plastic allowing it to live in many different and sometimes in extreme environments. This capacity for adaptation has been correlated with its breeding system, which allows the rapid change of the reproduction systems, flowering time, life span, pollen release, and so on, according to the modified local conditions. This is evident observing the differences between the Mediterranean populations (bolting, short life cycles, form of the root, etc.) and those living on the North-Atlantic coasts of Europe. This chapter provides an overview of the phenotypic features of sea beet.

Keywords *Beta maritima* · Seed · Root · Flower · Pollen · Isolation

The great genetic and environmental variability for morphological traits existing in the genus *Beta* was highlighted first by Owen (1944). The variability, present not only among but also within the wild beet populations, has evolved due to interactions of genetic, climatic, and soil factors (von Proskowetz 1896; Baxter 1837). Genetic variance is considered an essential mean for the survival of *Beta maritima* in hostile environments (de Vilmorin 1923).

Few references are available on the anatomy of sea beet, because the major attention was dedicated on the cultivated types (Baxter 1837). Appreciating the similarity among sea beet and cultivated types, some of the following information was taken from the classical papers on the anatomy of sugar beet (Artschwager 1926, 1927a,

E. Biancardi (✉)
Stazione Sperimentale di Bieticoltura, Rovigo, Italy
e-mail: enrico.biancardi@alice.it

M. de Biaggi
Sugar Beet Breedng Consultant, Massalombarda, Italy

E. Biancardi et al. (eds.), *Beta maritima*,
https://doi.org/10.1007/978-3-030-28748-1_3

b; Esau 1977). Useful references are given by Letschert (1993), Anonymous (1995), and in the reviews edited by Cooke and Scott (1993), Klotz (2005), and Draycott (2006).

3.1 Seed

Beta maritima develops from a hard cluster of fruits fused together and named "seed ball" or more commonly "glomerule" (Figs. 3.1 and 3.2). Each fruit contains a single seed with a single embryo and should be more appropriately called "utricle" (Copeland and McDonald 2001). The mature true seed is composed of the embryo covered by two thin layers (endosperm and the more external perisperm). The embryo is surrounded by a thick pericarp and the operculum which is the upper part of the pericarp. The operculum is considered the major point of entry for water and oxygen needed for germination (Coumans et al. 1976). These parts all together make up what researchers call "fruit". Seed containing a single embryo is called "monogerm", which is a trait relatively frequent in the North Atlantic Sea beet populations (de Vilmorin 1923). The seed ball or multigerm seed is composed of 2–11 (commonly 3–4) fruits fused together, one for each flower that composes the inflorescence.

According to Dale and Ford-Lloyd (1985), the multiple or multigerm seed is a rather rare trait among the angiosperms and is believed to have an important role in the dispersal of the species. The number of flowers per inflorescence is variable even on the same plant. Normally, the largest number of flowers per inflorescence is found at the base of the stalk and decreases toward the apex. The size of the seed ball also is larger on the proximal part of the stalk. Rather rare are plants bearing only single and isolated flowers developing monogerm seed.

Monogermity depends on a pair of alleles designated "*Mm*" and the trait is expressed in the homozygous recessive state (*mm*). Monogermity (*mm*) was selected and introduced into commercial varieties to eliminate the need for hand singling (removing with finger tips and a short-handled hoe, all but one seedling germinating from multigerm seed), which was compulsory for optimal growth of the roots (Savitsky 1952) (Fig. 3.3). Seed of commercial hybrids is usually genotypically multigerm (*Mm*) but phenotypically monogerm because the seed is harvested on monogerm (*mm*) female parent pollinated by a multigerm (*MM*) male parent. Monogerm (*mm*) trait conditions plants produce either a single flower or lateral (bud) branch in the leaf axil, but never both, as is found in multigerm (*MM or Mm*) plants, where the inflorescences are placed in the axils of the leaf bracts. Some maturing *mm* plants produce neither a flower nor a lateral bud in the leaf axil, giving rise to very poor set or poorly formed rosettes in the commercial sugar crop. Fortunately, the pollinators of the current hybrid varieties are always multigerm (*MM, Mm*) endowed with high degree of heterosis (Figs. 3.4, 3.5 and 3.6).

Fully monogerm plants were isolated in the U.S. from parental component of a synthetic *Cercospora* resistant variety, possibly originating from crosses with sea beet (Biancardi et al. 2010). de Candolle (1884) reported that *Beta maritima* grown

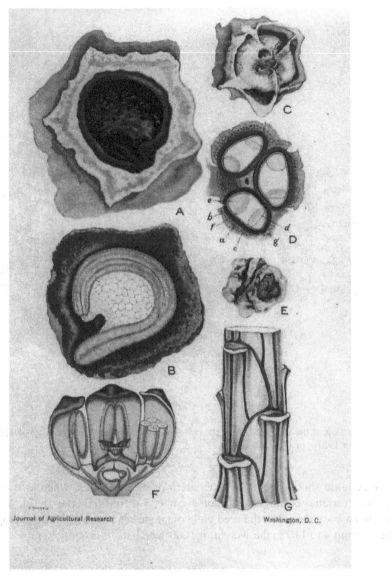

Fig. 3.1 Anatomy of flower and multigerm seed of sugar beet (Artschwager 1927a). A = Pericarp and operculum (dark); B = seed section showing the embryo (bright); C = raw multigerm seed; D = section of the seed: E = polished seed; F = section of flower; G = axis of the flowers

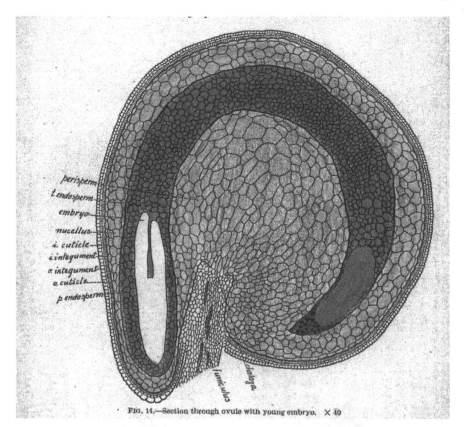

perisperm
t. endosperm
embryo
nucellus
i. cuticle
i integument
o. integument
o. cuticle
p. endosperm

funiculus

chalaza

FIG. 14.—Section through ovule with young embryo. × 40

Fig. 3.2 Section of the true seed, with the embryo (white); perisperm (green); endosperm (red)
(Artschwager 1927b, modified)

in the Atlantic shores produces monogerm or bigerm seeds, whereas the seed of
cultivated varieties is composed of, on average, 3–4 fruits (Baxter 1837). According
to Frese (www.genres.de), the seed yield of single sea beet plant is on average 40 g,
ranging from 4 to 110 g; the weight of 1000 seed balls is around 36 g.

3.2 Leaves

Leaves develop in a close spiral, arranged in 5/13 position, on the crown of the
taproot (Artschwager 1926). The leaves' size is variable and the surface increases
progressively up to the 12–13th pair, at least in sugar beet; then they decrease. The
leaves are opposite (Fig. 3.7) and almost ovate or deltoid; sometimes lanceolate or
rhombic in shape (Fig. 3.8). The form of leaves and petioles changes on the same
plant and in agreement with the order in which they develop. Form and area change

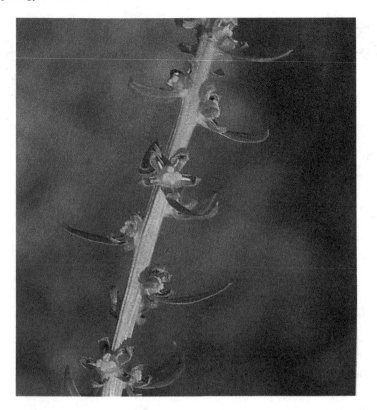

Fig. 3.3 Monogerm and male sterile flowers, which develop into "monogerm" seeds

greatly if the plant is bolting (Letschert 1993). The leaf blade, often asymmetrical, is normally shiny, glabrous, and frequently waxy and waved, with a more or less undulate or curly edge (Fig. 3.9).

Pubescence (presence of hairs) of the leaves is quite rare being currently recognizable only on East-Mediterranean populations (Letschert 1993). The trait (Fig. 3.10) is almost unknown in the Adriatic and Atlantic sea beet (Biancardi, unpublished). Pubescence may occur on both sides of the lamina and on the petiole as well (Letschert 1993). Sometimes the character occurs only in the first emitted leaves, disappearing later with the aging of the plant and varies across environments and years. As in other plants, pubescence protects the plant against biotic and abiotic stresses, including excessive water transpiration and intense solar radiation. Any negative correlation with other traits was detected in the genus *Beta* (Frese, personal communication).

In sea beet, the stomata are more numerous on the lower face of the leaf. Leaves, veins, and petioles may be colored differently. Letschert and Frese (1993) give the following average measures taken on 35 Sicilian sea beet populations: lamina length 6.9 cm, width 5.4 cm, thickness 0.06 mm, petiole length 6.4 cm, width 3.3 mm, stem diameter 1 cm, plant height 74.7 cm, and biomass 1.4 kg.

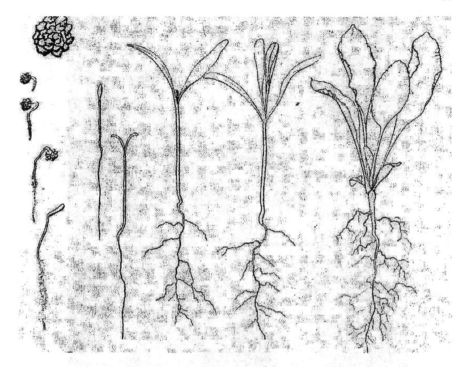

Fig. 3.4 Germination and first development of *Beta maritima* (Schindler 1891)

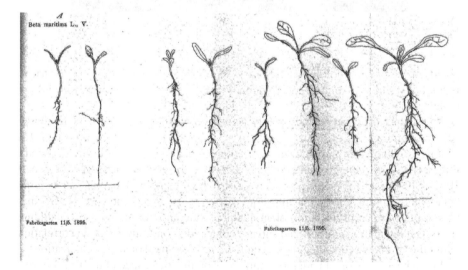

Fig. 3.5 Initial development of *Beta maritima* (von Proskowetz 1896)

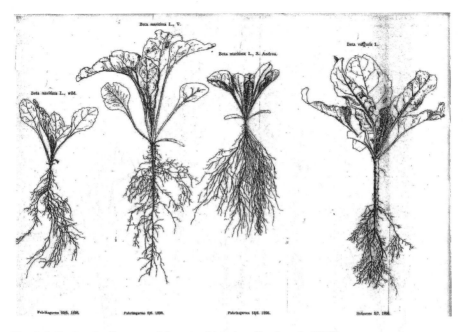

Fig. 3.6 Further development of *Beta maritima* (von Proskowetz 1896)

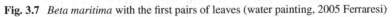

Fig. 3.7 *Beta maritima* with the first pairs of leaves (water painting, 2005 Ferraresi)

Fig. 3.8 Leaves of *Beta maritima*

3.3 Root

The sea beet taproot is single or variously fanged or sprangled (Fig. 3.11), partially depending on soil structure and age (de Vilmorin 1923). "Some are as thick as the arm and edible, others are not thicker than a finger and of a woody composition" (de Vries 1905). The root is composed of three parts: crown, neck, and true root (Fig. 3.11). The crown (also called epicotyl) is the apex of the taproot that bears single or multiple rosettes. The neck (also called hypocotyl) is below the crown or, more precisely, below the insertion of the first two true leaves (Hayward 1938). The true root can be slightly flattened. It comprises the major portion of the taproot and extends downward from the neck to the tail. If there is a single taproot, the secondary roots often emerge at about 90° (Figs. 3.12 and 3.13). The transect of the root shows a variable number of concentric rings of phloem and xylem tissue separated from each other by a zone of interstitial parenchyma (Hayward 1938). The consistency of the root depends on its age. It becomes progressively harder, more fibrous and woody toward the end of the first year: "The fleshiness of the root is very variable" (de Vries 1905). von Proskowetz (1894) reported the following mean measures for sea beet roots: weight 147 g, length 35 cm, diameter 3.7 cm, sugar content 4.5%.

Fig. 3.9 Leaves of *Beta maritima* (Munerati 1920)

Fig. 3.10 Young leaves of sugar beet showing pubescence (Frese, unpublished)

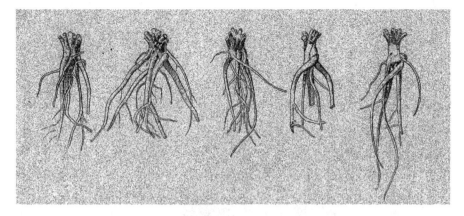

Fig. 3.11 Roots of *Beta maritima* (Schindler 1891)

Fig. 3.12 *Beta maritima*
with secondary roots
emerging perpendicularly to
the tap root (de Vilmorin
1923)

Fig. 3.13 Painting of *Beta maritima* showing the flowers and the root with secondary roots emerging at 90° from the tap root

Srivastava et al. (2000) observed morphological traits from 34 populations of sea beet accessed from different countries and grown in northern India (Table 3.1). The root system of sea beet is very expanded and deep especially under low water availability (Fig. 3.14), or where the plants grow near stones or cliffs. In the void layer between soil and rock, the roots find lower resistance to penetration, allowing them to reach considerable length and depth.

Table 3.1 Mean and range for 8 traits of *Beta maritima* (Srivastava et al. 2000)

Characters	Unit	Mean	Maximum	Minimum
Root weight/plant	g	38.89	115	10
Top weight/plant	g		410	10
Root length	cm	18.09	13	12
Crown size	cm	2.33	6	1
Leaf length	cm		24	2
Leaf width	cm	4.43	17	1
Length/width ratio		1.88	4	1
Petiole length	cm	49.66	101	11

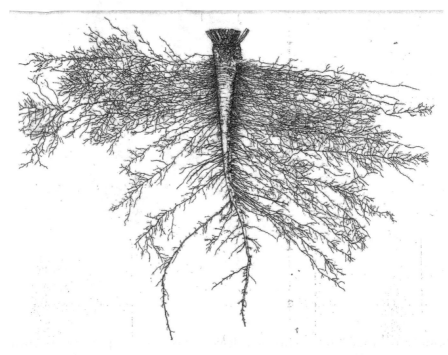

Fig. 3.14 *Beta maritima* root system (von Proskowetz 1896)

3.4 Color

Several authors, including de Vries (1905), emphasized the variability of colors in sea beets: "Some wild population have red leafstalk and veins, others a uniform red or green foliage, some have red or white or yellow roots, or show alternating rings of a red or white tinge on cut surfaces". de Vilmorin (1923) noted that the root of Atlantic sea beet can be colored differently: (i) primary hypocotyl (white, pink,

Fig. 3.15 Drawing of *Beta maritima* with reddish veined stalk

yellow, green, red, etc.); (ii) taproot skin (white ivory, yellow, orange, red, dark red, brown, black, etc.); and (iii) root flesh (white, yellow, orange, red, purple, etc.). The colors are due to the concentration and proportion of some red (betalain, betanin) and yellow (betaxantin) pigments. Munerati et al. (1913) stated that the color of the roots, at least on Adriatic coasts, is normally white-ivory and rarely light pink or yellow (Fig. 3.15). More intense colors seemed due to crosses with cultivated types (Fruhwirth cited by von Lippmann 1925). In these cases, the flesh is also colored and normally shows distinct light and dark rings. The only color observed on the shoot of the Adriatic populations is red or reddish on stalks, veins and hypocotyls.

3.5 Seed Stalk

According to von Proskowetz (1894), the development of the stalk can start very early and rapidly. For the annual types, stalks may appear after the first pair of true leaves. On average, 40–260 days are necessary for emergence of the seed stalk under controlled conditions (http://www.ecpgr.cgiar.org/networks/sugar_starch_fibre_crops/beta.html).

In the transition to the reproduction phase, the new leaves become smaller. The stem or seed stalk starts to elongate from the center of the rosette (Artschwager 1927a; Lexander 1980). On individual plants, one or more stalks can emerge (Figs. 3.16 and 3.17). The number of stalks is likely dependent on the strength of apical dominance. There is normally one stalk per crown or rosette. In the multiple

Fig. 3.16 Picture of *Beta maritima* showing the erect stalks and the fangy root (von Proskowetz 1894)

Fig. 3.17 *Beta maritima*
around Torcello, Venice,
showing erect stalks

crowned beets the stalks are likely to be prostrate (Fig. 3.18). In the Mediterranean
Sea beets studied by von Proskowetz (1896), the mean dimensions of the stalks of
10 plants were 1.25 m long and 0.92 cm thick. Each plant developed from 1 to 4
stalks. Schindler (1891), in different populations counted a mean of seven stalks per
plant, with lengths ranging from 60 to 130 cm. de Cauwer et al. (2010) stated that
each sea beet plant "bears one to several hundred of floral stems".

The first recorded difference between the species *Beta vulgaris* and *Beta maritima*
was "*caule erecto*" (upright stalk) instead of "*caule decumbente*" (procumbent or
prostrate stalk), respectively (Linnaeus 1753). Also the color and the pigmentation
are very variable. The primary stem and all branches terminate in inflorescences
composed of branched spikes bearing the flowers (Artschwager 1927a).

Fig. 3.18 *Beta maritima* around Pescara, Italy, with procumbent stalks

3.6 Flower

Flowering in *Beta maritima* is indeterminate. The stalk grows continuously as does the production of new flowers. On the same plant, it is possible to see newly formed flower buds, ripened and shattered seeds, and all the stages in between (Fig. 3.19). In Mediterranean Sea beets, flowering lasts until the early autumn if the conditions of moisture, light, and temperature remain favorable (Smith 1987; Biancardi, unpublished). For harder bolting and biennial NE-Atlantic accessions, stalk elongation and flowering usually end with the onset of shorter days toward the end of summer. Relatively high temperatures may also cause a reversion to vegetative growth.

The flowers located in the bract axils are sessile, single, or more frequently assembled in clusters (glomerules) of 2–11, which develop multigerm seeds (Artschwager 1927a; Smith 1987). Monogerm flowers similarly develop in the leaf axils. The flower consists of five narrow green sepals surrounded by five stamens (Fig. 3.20) (Hayward 1938). The bases of the sepals and stamens are positioned above the ovary. The pistil is normally tricarpellate with a short style that terminates in a stigma with three or more lobes (Artschwager and Starrett 1933), but "very frequently are three in number" (Smith 1803). The ovary encloses the ovule, which contains the embryo sac and the egg. The position of ovary changes during the flower's development; at the beginning it is superior, becoming inferior in the subsequent phases (Flores-Olvera et al. 2008). Five stamens extend above the pistil, bearing the anthers consisting

Fig. 3.19 *Beta maritima* showing seeds at different stages of ripening (Numana, Italy)

of two loggias, each made up of four pollen-filled sacks (Fig. 3.21) (Artschwager 1927a; McGrath et al. 2007).

Each flower produces up to 85,000 pollen granules. Since a single cultivated beet can develop as many as 20,000 flowers (Knapp 1958; Schneider 1942), a total of several billion of pollen granules per plant seems possible. Each sea beet develops on average 4000 flower clusters, that is 2000 flowers per plant. According to Dufaÿ et al. (2007), in Mediterranean populations, the number of pollen granules pro flower can be 40,000. Therefore, every single flower produces up to 200,000 pollen granules, yielding and releasing billions from each plant.

Flowers begin opening from the base of the stem and from the middle flower of the cluster (Artschwager 1927a). Among three groups of *Beta maritima* collected in oriental sites (Pakistan, Iran, and India), eastern Mediterranean (Turkey and Greece), and central Mediterranean (France, Italy, and Corsica), initial flowering progressed from East to West over two weeks (Letschert 1993).

3.7 Pollen

Based on the observations of Oksijuk (1927) and confirmed by Artschwager and Starrett (1933), the flowers open when the sun rises. The sepals gradually move into

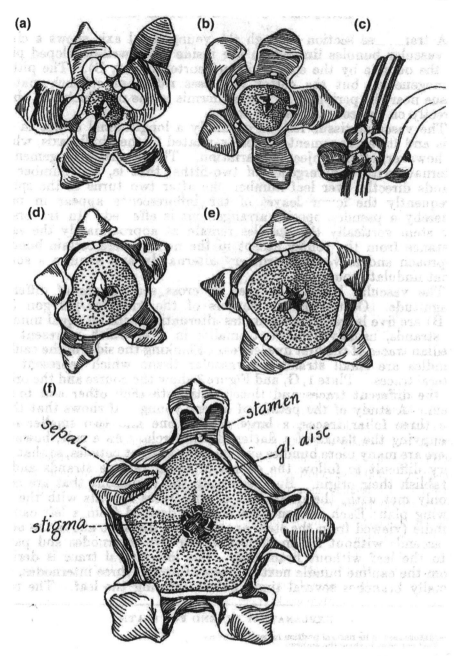

Fig. 3.20 Evolution of *Beta maritima* flower. **a** single mature flower; **b** flower with stamen already dropped; **c** luster of flowers attached to branch axil; **d** flower after fertilization; **e** flower with young embryo; **f** flower with mature embryo (Artschwager Artschwager 1927a, b)

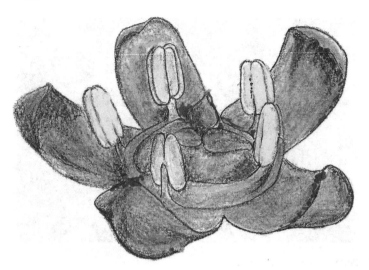

Fig. 3.21 Flower with fertile anthers (pastel, 2011 Ferraresi)

a position perpendicular to the flower's axis, while, simultaneously, the anthers begin to open lengthways. At first, the pollen granules stick together, but as the humidity decreases, they separate and are pushed out of the anthers (Scott and Longden 1970). While the flowers are opening, the stigma's lobes are closed to form a kind of tube. The lobes gradually open on the second or third day after anthesis, becoming receptive to pollen. On the outside of the stigma there are fine papillas, which increase the surface area and the probability of capturing the pollen granules (Artschwager and Starrett 1933).

Beet pollen is almost sphere-shaped and its outside has the characteristic relief and small circular marks for the exit of the pollen tube (Fig. 3.22). Granule diameter varies around an average value of 16–20 μm (Schindler 1891). The mean diameter of sea beet pollen is 2–3 μm less than that of the cultivated varieties (Schindler 1891). Munerati et al. (1913) did not confirm such a difference. Dufaÿ et al. (2007), working on Atlantic sea beets, found a larger variation in pollen size within two populations; the most frequent size classes were between 10 and 20 μm. They also observed a high positive correlation between diameter and viability. The same authors found large individual variation, both for pollen production and viability caused by gynodioecy inside the population. A diameter of about 20 μm was most frequently recorded in a survey carried out on 586 anemophilous species throughout Europe (Stanley and Linskens 1974). Therefore, the authors assumed this dimension was the most suitable for wind pollination and long-distance diffusion.

Pollen release in sugar beet varies during the day; around midday it reaches the maximum, almost in correspondence to minimum humidity and highest air temperature (Scott and Longden 1970). These authors found that pollen concentration in the air is lowered by rain and high relative humidity. In many cases, however, rain at night increased the day-after pollen release. The highest release recorded in UK over

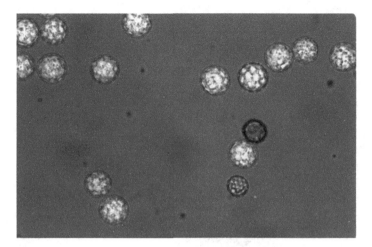

Fig. 3.22 Pollen granules of sugar beet

a three-year period was observed between 1 and 15 July, with high variation among the years. This variability likely depended on the different flower development caused by weather conditions (Scott and Longden 1970).

Wind pollination is the most important method of pollination in beets. This is evidenced by the height and branching of stalks, the enormous production of pollen, and pollen release over long periods, only under favorable weather (van Roggen et al. 1998). Nevertheless, sea beet preserves some ancestral traces of traits for pollination through insects, including joined flowers, nectar secretion, and emission of its characteristic scent (Archimowitsch 1949; van Roggen 1997). The spread of beet pollen by insects happens at a significantly lower rate and also has lower range than wind pollination (Bateman 1947; Free 1975).

Pollen viability is the time span over which it can germinate (Dufaÿ et al. 2007). The same authors found large individual variation, both for pollen production and viability caused largely by gynodioecy. Viability also depends on climatic conditions. With low temperatures and air humidity, pollen germination can take place up to 50 days after release. Under field conditions, viability does not last for more than 24 h (Villain 2007). Intense sunlight further lowers the time viable to no more than 3 h after release, but in this time pollen may to travel up to 100 km or more (Artschwager and Starrett 1933; Knapp 1958). It should be remembered that each sea beet plant can release around one billion pollen grains (Dufaÿ et al. 2007; Schneider 1942).

Several methods have been used to evaluate the spread of pollen. Petri dishes coated by glycerine (Archimowitsch 1949) or automated devices described by Hirst (1952) and Dufaÿ et al. (2007) have been used to capture and count pollen grains. In place of such methods, flowering male sterile (CMS) beets can be employed to check the percentage of cross-pollination. They develop no viable pollen and, therefore, any seed produced must have been with foreign pollen. For these experiments, monogenic and dominant traits such as red skin or red hypocotyls are used as markers in the pollinator beet populations. Archimowitsch (1949) placed groups of sugar beets at increasing distances from the central pollen source of about 500 red beets along eight lines corresponding to the four cardinal points and the intermediate directions. The results varied greatly over the 3 years, due to climate, and particularly, to the wind direction. In 1938, German carried out a key experiment in Russia using airborne Petri boxes coated by glycerine. He discovered the presence of beet pollen up to a height of 2500 m. Meier and Artschwager (1938), using similar methods, flew over an area covering about 400 hectares of beet seed crops located near El Paso, Texas. Large quantities of pollen were found up to an altitude of 1500 m. With strong rising air currents, the pollen was carried up to heights of 10,000 m and beyond. Therefore, considerable distances can be covered: 5 and 8 km according to Smith (1987) and to Harding and Harris (1994), respectively.

Pollen flow was studied from a new point of view in the late 1960s (Chamberlain 1967; Tyldesley 1978). They used updated knowledge gleaned from other fields of research, such as the atmospheric spread and fallout of volcanic or radioactive dusts, with which pollen can be compared, at least in terms of aerodynamic properties. Similar attention has been given to spread of certain types of pollen, due to their allergenic properties on an increasing number of sensitive people. Further stimulus for this type of research was the risk of contamination of wild plant populations by transgenic crops (Bartsch and Pohl-Orf 1996; Bartsch et al. 2001; Boudry et al. 1993; Brand 1997; Ellstrand 2003; Gepts and Papa 2003), given the future need to keep separate and avoid pollen exchanges among conventional, organic, and genetically modified crops (Kapteijns 1993). An indirect and different method of evaluation of the gene flow by means of pollen consisted in evaluating the resulting genetic differentiation in neighboring populations (Tufto et al. 1998).

Because of the ease of long-range cross-pollination and the inability to positively identify the morphological characters that differentiate sea beet, feral beet, weed beet, and cultivated beet, it is not easy to distinguish the pure populations of *Beta maritima* from those partly derived from cross-pollination with other types of beet. Therefore, for populations to be classified as "pure", they should be sufficiently distanced from other sources of pollen. Fénart et al. (2007) referenced the case of about 5% intercrossing between weed beet populations 9.6 km apart. In a commercial hybrid seed crop, de Biaggi (personal communication) observed around 0.5% intercrossing with red beets located 12 km apart. Pollination with foreign pollen is more difficult in large, aggregated populations, because the local (home) pollen creates a sort of barrier on the flower stigma, and, as a consequence, these populations are more protected against polluting pollen coming from outside (Arnaud 2008). For this reason, the safe distance for pollen isolation is hard to quantify, especially due to the variability of wind speed and direction, air humidity, and so on, that greatly influence the dynamic of spread. Moreover, it must be noted that very low percentages (0.1%) of cross-pollination are able to produce severe damage in commercial varieties. This is true in the case of dominant traits, such as annuality and red color, which became evident soon after the emergence in the following F1 generation, as are the commercial varieties.

For seed production of similar varieties, the safe distance between fields varies from 1 to 3.2 km (Campbell and Mast 1971; OECD 2001; Smith 1987). Based on experience made in this research, 10 km would be a reliable isolation distance from sources of contaminating pollen (i.e., annual beet, red beet, leaf beet, etc.), also valid among populations of *Beta maritima*. This distance is considerably less between populations located along the Atlantic coastline, where the direction of dominant wind is quite always perpendicular to the coast line (Middelburg, personal communication).

The following isolation distances are usually applied in Europe for commercial seed production (Treu and Emberlin 2000):

Between pollinators with the same ploidy—300 m;

Between pollinators with different ploidy—600 m;

Between sugar beet seed production and other types of beets—1000 m.

The National Pollen Research Unit (NPRU) proposed the following distances in case of GM sugar beet cultivation (from Treu and Emberlin 2000, modified)

GM sugar beet / Organic sugar beet—3000 m;

Conventional sugar beet / Weed and sea beets—1000 m;

Organic sugar beet seed production / Weed and sea beets contaminated by GM traits—3000 m;

Organic sugar beet / Weed and sea beets contaminated by GM traits—1000 m;

Pollen isolation on sea beets is seen as an important condition for both speciation and fitness to environment of the populations (Wright 1943; 1946). The author points out the importance, as mentioned above, of the number of individuals inside the population in ensuring better condition for pollen isolation.

3.8   Chemical Composition

Only few references regarding the chemical composition of sea beet are available. According to Schindler (1891), the sugar content [sucrose concentration expressed as a percent of the fresh weight of the root, also called polarimetric degree (°S)] of roots harvested in their natural environment is highly variable, ranging from 0.3 to 8.2 °S. Under cultivated field conditions the mean was 11.2%. von Proskowetz (1896) analyzed roots of *Beta maritima* collected in the Sant'Andrea Island, Adriatic Sea. Water was 91%, proteins 1.6%, fat 0.2%, nitrogen-free compounds 3.1%, fiber 1.0%, ash 1.3%, and silica 1.4%.

Munerati et al. (1913) analyzed sea beet roots obtained with seed harvested in the Po Delta sown in different conditions. Of course, higher sugar content occurred under field conditions. The differences may be due to the better soil tilth and availability of water and nutrients (Table 3.2). The same authors observed a rapid increase of sugar content and root weight in the subsequent generations obtained after mass selection. Sugar content in sea beets collected in Atlantic localities was given by von Lippmann (1925).

Table 3.2 Ranges of sugar content (%) of *Beta maritima* in three European localities (von Lippmann 1925)

Location (France)	Minimum	Maximum
Saillard, Bretagne	13.8	19.6
Baudry and Collins, Garonne	9.0	14.0
Vilmorin	13.0	14.6

Moldenhawer (1935) measured the sugar content of the sea beet roots on the Polish coasts of the North Sea. Values ranged from 7.8 to 13.8% with an average of 10.4%. Krasochkin (1936) found sea beets from the north Atlantic shores to have 15% sugar as the maximum value. Stehlik (1937) compared the biomass production of a French sea beet with a standard sugar beet variety. The plots yielded as follows: tops—128 and 365 kg; roots—38 and 385 kg; sugar content—10.8 and 18.2%; sugar yield—4.1 and 70 kg, respectively. The great gain of sugar yield was due to selection. More recently Baydara (2008) analyzed sea beet roots using mass spectrometry and identified 288 proteins having a central role in salt tolerance.

References

Anonymous (1995) Descriptors for *Beta*. CPRO-DLO, Wageningen, The Netherlands and IPGRI, Rome, Italy

Archimowitsch A (1949) Control of pollination in sugar beet. Bot Rev 15:613–628

Arnaud JF (2008) Importance de la dispersion dans la structuration génétique et l'évolution du système de reproduction chez une espèce gynodioique. Université des Sciences et Technologies de Lille, France

Artschwager E (1926) Anatomy of the vegetative organs of sugar beet. J Agr Res 33:143–176

Artschwager E (1927a) Development of flowers and seed in the sugar beet. J Agr Res 34:1–25

Artschwager E (1927b) Micro and macrosporogenesis in sugar beet with special reference to the problem of incompatibility. Memoirs Hort Soc NY, USA 3:295–297

Artschwager E, Starrett R (1933) The time factor in fertilization and embryo development in the sugar beet. J Agr Res 47:823–843

Bartsch D, Pohl-Orf M (1996) Ecological aspects of transgenic sugar beet: transfer and expression of herbicide resistance in hybrids with wild beets. Euphytica 91:55–58

Bartsch D, Brand U, Morak C, Pohl-Orf M, Schuphan I, Ellstrand NC (2001) Biosafety of hybrids between transgenic virus-resistant sugar beet and Swiss chard. Ecol Appl 11:142–147

Bateman A (1947) Contamination in seed crops. Heredity 1:235–246

Baxter W (1837) British phaenerogamous botany. Parker, London, UK

Baydara EP (2008) Salt stress responsive proteins identification in wild sugar beet (*Beta maritima*) by mass spectrometry. M.S. İzmir Institute of Technology

Biancardi E, McGrath JM, Panella LW, Lewellen RT, Stevanato P (2010) Sugar beet. In: Bradshaw JE (ed) Root and tuber crops. Springer, New York, NY USA, pp. 173–219

Boudry P, Mörchen M, Saumitou-Laprade P, Vernet P, Dijk H (1993) The origin and evolution of weed beets: consequences for the breeding and release of herbicide-resistant transgenic sugar beets. Theor Appl Genet 87:471–478

Brand U (1997) Untersuchungen zur Diversität in italienischen Wildpopulationen von *Beta vulgaris* L. subsp. *maritima* – Ein Beitrag zur ökologischen Risikoabschätzung von transgenen Kulturpflanzen. Diplomarbeit RWTH-Aachen, Germany

Campbell SC, Mast AA (1971) Seed Production. In: Johnson RT, Alexander JT, Rush GE, Hawkes GR (eds) Advances in sugarbeet production: principles and practices. The Iowa State University Press, Ames IA, USA, pp 438–450

Chamberlain AC (1967) Cross-pollination between fields of sugar beet. Q J Roy Meteorol Soc 93:509–515

Cooke DA, Scott RK (1993) The sugar beet crop: Science into practice, 1st edn. Chapman & Hall, London, UK

Copeland LO, McDonald MB (2001) Principles of seed science and technology, 4th edn. Kluwer Academic Publishers, Boston MA, USA

Coumans M, Come D, Gaspar T (1976) Stabilized dormancy in sugar beet fruits I. Seed coats as physiochemical barrier to oxygen. Bot Gaz 137:274–278

Dale MFB, Ford-Lloyd BV (1985) The significance of multigerm seedballs in the genus *Beta*. Watsonia 15:265–267

Dale MFB, Ford-Lloyd BV, Arnold MH (1985) Variation in some agronomically important characters in a germplasm collection of beet (*Beta vulgaris* L.). Euphytica 34:449–455

de Candolle A (1884) Der Ursprung der Culturpflazen. Brockhaus, Lipsia, Germany

de Cauwer I, Dufaÿ M, Cuguen J, Arnaud J-F (2010) Effects of fine-scale genetic structure on male mating success in gynodioecious *Beta vulgaris* subsp. *maritima*. Mol Ecol 19:1540–1558

de Vilmorin JL (1923) L' hérédité de la betterave cultivée. Gauthier-Villars, Paris, France

de Vries U (1905) Species and varieties. Open Court Publishing, Chcago, USA

Draycott AP (2006) Sugar Beet, 1st edn. Blackwell Publishing Ltd, Oxford, UK

Dufaÿ M, Touzet P, Maurice S, Cuguen J (2007) Modeling the maintenance of male-fertile cytoplasm in a gyodioecious population. Heredity 99:349–356

Ellstrand NC (2003) Current knowledge of gene flow in plants: implications for transgene flow. Philos Trans R Soc Lond B Biol Sci 358:1163–1170

Esau K (1977) Anatomy of seed plants, 2nd edn. Wiley, New York, USA

Fénart S, Austerlitz F, Cuguen J, Arnaud J-F (2007) Long distance pollen-mediated gene flow at a landscape level: the weed beet as a case study. Mol Ecol 16:3801–3813

Flores-Olvera H, Smets E, Vrijdaghs A (2008) Floral and inflorescence morphology and ontogeny in *Beta vulgaris*, with special emphasis on the ovary position. Ann Botany 102:643–651

Free JW (1975) Insect pollination of sugar beet. Ann Appl Biol 81:127–134

Gepts P, Papa R (2003) Possible effects of (trans)gene flow from crops on the genetic diversity from landraces and wild relatives. Environ Biosafety Res 2:89–103

Harding K, Harris PS (1994) Risk assessment of the release of genetically modified plants: a review. Ministry of Agriculture, Fisheries and Food, London, UK

Hayward DH (1938) The structure of economic plants. MacMillan and Co., New York, USA

Hirst JM (1952) An automatic volumetric spore trap. Ann Appl Biol 39:255–257

Kapteijns AJAM (1993) Risk assessment of genetically modified crops. Potential of four arable crops to hybridize with wild flora. Euphytica 66:145–149

Klotz KL (2005) Anatomy and Physiology. In: Biancardi E, Campbell LG, Skaracis GN, de Biaggi M (eds) Genetics and breeding of sugar beet. Science Publishers Inc., Enfield (NH), USA, pp 9–19

Knapp E (1958) *Beta* rüben. In: Roemer R, Rudorf W (eds) Handbuch der Pflanzenzüchtung. Parey, Paul, Berlin, Germany, pp. 196–284

Krasochkin VT (1936) New facts in beet-root breeding. Bull Appl Bot (Leningrad) 19:27

Letschert JPW (1993) *Beta* section *Beta*: biogeographical patterns of variation, and taxonomy. PhD Wageningen Agricultural University Papers 93-1

Letschert JPW, Frese L (1993) Analysis of morphological variation in wild beet (*Beta vulgaris* L.) from Sicily. Genet Res Crop Evol 40:15–24

Lexander K (1980) Present knowledge on sugar beet bolting mechanisms. In: Proceedings of the international institute sugar beet research 43rd winter congress. IIRB, Brussels, Belgium, pp 245–258

Linnaeus (1753) *Species plantarium exhibentes plantas rite cognitas etc.*, 1st edn. Stockholm, Sweden

McGrath JM, Saccomani M, Stevanato P, Biancardi E (2007) Beet. In: Kole C (ed) Vegetables. Springer, Berlin, pp 191–207

Meier FC, Artschwager E (1938) Airplane collections of sugar beet pollen. Science 88:507–508

Moldenhawer K (1935) Studies on wild beet (*Beta maritima*) of the North Sea Region. Br Sugar Beet Rev 9:47–49

Munerati O (1920) Sulla salita in seme il primo anno delle bietole coltivate. Bollettino Associazione italiana delle Industrie dello Zucchero e dell'Alcool 90–95

Munerati O, Mezzadroli G, Zapparoli TV (1913) Osservazioni sulla *Beta maritima* L., nel triennio 1910–1912. Stazioni Sperimentali Agricole Italiane 46:415–445

OECD. Environment, Health and Safety Publications Series on Harmonization of Regulatory Oversight in Biotechnology Nr. 18. Consensus Document on the Biology of *Beta vulgaris* L. (Sugar beet). 2001. Paris, France, Environment Directorate Organisation for Economic Co-operation and Development. http://www.oecd.org/dataoecd/16/58/46815688.pdf

Oksijuk P (1927) Entwicklungsgeschichte der Zuckerrübe. Bulletin Bot Garden, Kiev, Russia 12:416–450

Owen FV (1944) Variability in the species *Beta vulgaris* L. in relation to breeding possibilities with sugar beets. J Am Soc Agron 36:566–569

Savitsky VF (1952) Monogerm sugar beets in the United States. Proc Am Soc Sugar Beet Technol 7:156–159

Schindler F (1891) Über die Stammpflanze der Runkel- und Zuckerrüben. Botanisches Centralblatt 15:6–16

Schneider F (1942) Züchtung der *Beta* Rüben. In: Parey P (ed) Handbuch der Pflanzenzüchtung. Berlin, Germany

Scott RK, Longden PC (1970) Pollen release by diploid and tetraploid sugar beet planta. Ann Appl Biol 66:129–136

Smith GA (1987) Sugar beet. In: Fehr WR (ed) Principles of cultivar development. Macmillan Publishing Company, New York, pp 577–625

Smith JE (1803) English botany. Taylor Printer, London, UK

Srivastava HM, Shahi HN, Kumar R, Bhatnagar S (2000) Genetic diversity in *Beta vulgaris* ssp.

Stanley RG, Linskens HF (1974) pollen: biology biochemistry management. Springer, Heidelberg, New York

Stehlik V (1937) Die wilde Rübe (*Beta maritima*) verglichen in unserem Klima mit der heutigen veredelten Zuckerrube. Zeitschrift Zuckerindustrie Czek Republ 61:236–240

Treu R, Emberlin J (2000) Pollen dispersal in the crops maize, oilseed rape, potatoes, sugar beet and wheat. Soil Association from the National Pollen Research Unit

Tufto J, Raybould AF, Hinfaar K, Engen S (1998) Analysis of genetic structure and dispersal patterns in a populations of sea beet. Genetics 149:1975–1985

Tyldesley JB (1978) Out-crossing in sugar-beet due to airborne pollen. Agric Meteorol 19:463–469

van Roggen PM (1997) The sex life of sugar beet. Br Sugar Beet Rev 65:28–30

van Roggen PM, Debenham B, Hedden P, Phillips AL, Thomas SG (1998) A model for control of bolting and flowering in sugar beet and the involvement of gibberellins. Flower Newsl 25:45–49

Villain S (2007) Histoire evolutive de la section *Beta*. PhD Universite des Sciences et Technologies de Lille

von Lippmann EO (1925) Geschichte der Rübe (*Beta*) als Kulturpflanze. Verlag Julius Springer, Berlin, Germany

von Proskowetz E (1894) Über die Culturversuche mit *Beta maritima* L. (und *Beta vulgaris* L.) im Jahre 1893. Österreiche-Ungarische Zeitschrift für Zuckerindustrie und Landwirtschaft 31:201–223

von Proskowetz E (1896) Über die Culturversuche mit *Beta* im Jahre 1895. Österreiche-Ungarische
    Zeitschrift für Zuckerindustrie und Landwirtschaft 33:711–766
Wright S (1943) Isolation by distance. Genetics 28:114–138
Wright S (1946) Isolation by distance under diverse systems of mating. Ibidem 31:39–59

Chapter 4
Physiology

Nina Hautekeete, Henk van Dijk, Pascal Touzet and Enrico Biancardi

Abstract Sea beet is an ideal plant species for evolutionary ecology studies. The range of variation is very large among its populations for life cycle and several life-history traits, such as the proportion of dormant seeds, mean life span, age at maturity, flowering time, day length requirement and vernalization requirement for flowering. These traits follow latitudinal patterns in France, of which many correlate with ecological factors. The range of variation of these traits, their geographical patterns, heritability and the ecology have been studied. A large potential for a genetic change in day length sensitivity has been observed experimentally, and a substantial genetic change in the flowering date within two decades has been reported, probably in line with the recent climate change. Sea beet also displays a geographical variation for some reproductive traits, that is, self-incompatibility and male-sterility. Sea beet exhibits two specific characteristics in its mating system when compared with the other species of the section: it is self-incompatible and gynodioecious. Gynodioecy is under cytonuclear control, sterilizing factors being encoded by the mitochondrial genome. Recent advances on the molecular characterization of the different male-sterilizing mitochondrial genomes found in beet are reported.

Keywords *Beta maritima* · Cytoplasmic male sterility · Gynodioecy · Iteroparity · Semelparity · Life span · Life-history traits · Ecology · Delayed maturity · Vernalization requirement · Climate changes · Weed beets

4.1 Seed Germination

Under conditions of favourable temperature, germination of non-dormant seeds starts when the imbibition of water to rehydrate the embryo tissues has reached a sufficient

N. Hautekeete (✉) · H. van Dijk · P. Touzet
Lab. Evolution Ecologie Paleontologie UMR CNRS 8198, Univ. Lille, CNRS, UMR 8198 - Evo-Eco-Paleo, 59000 Lille, France
e-mail: nina.hautekeete@univ-lille.fr

E. Biancardi
Formerly Stazione Sperimentale di Bieticoltura, Rovigo, Italy

level. The resulting expansion of the tissues causes the opening of the operculum after breaking its connections to the pericarp. This is a critical phase, because the germination of the true seed (embryo and perisperm without the pericarp) is much more rapid (Hermann et al. 2007; Richard et al. 1989). This step is followed by the exit of the radicle through the opening allowed by the lifting of operculum. Usually the germination of a single seed is considered completed when the radicle begins to elongate through the operculum (Hermann et al. 2007). Because the embryo is relatively small, the seedling is sensitive, not only to weed competition, pests, and diseases, but above all to unfavourable soil and weather conditions during early growth (Hojland and Pedersen 1994) (Fig. 4.1).

Germination speed depends on temperature, water availability and concentration of endogenous chemical inhibitors, which are much higher in sea beet than in commercial seed (Morris et al. 1984). In favorable conditions, seeds of Mediterranean Beta maritima take 7 to 11 days for germination (von Proskowetz 1894), and about 10 days of imbibition at various temperature regimes before germination could start in Belgian and Dutch sea beet populations (Letschert 1993). Germination of seeds collected in the Balearic Islands began after 11 days at 20 °C and reached a final rate of germination of 31% (Letschert 1993; Galmés et al. 2006). Fourteen days later, germination rate reached 50%. Germination speed in Adriatic populations was proportional to the percentage of annual individuals (Munerati et al. 1913). Analyzing Irish populations, Grogan (2009) found an average germination rate of 7%. Experience made by the editors of this book showed that: (i) freshly harvested seeds of *Beta vulgaris* subsp. *maritima* germinate less than the 2–3 years old seeds;

Fig. 4.1 Monogerm hermaphrodite (left) and male sterile (right) flowers

(ii) seeds produced in rainy season or under overhead irrigation have higher percentage of germination; (iii) decortication (Santos and Pereira 1989) and soaking improved germination rate. All of these factors probably relate to the level of inhibitors in the seed coat.

Seeds of sea beets display a behavior typical of wild plants, germinating at different times even when brought into suitable conditions of humidity and temperature. Germination continues slowly over months. It is thus possible to observe various levels of development among seedlings scattered around the same plant, that is, germinating seeds, well-developed plantlets, and already flowering individuals (Munerati et al. 1913), though it is noteworthy that this pattern is also partly explained by seed dormancy (Sect. 4.2). Over this prolonged time of germination, at least a small percentage of plants can find favorable conditions for their development and successful seed production. von Proskowetz, Dupont, Riffard, Townsend and Beguinot, cited by von Lippmann (1925), confirmed that the prolonged germination during the growing season also resulted in a high variability of morphological traits and flowering behavior, which is of vital importance for survival.

4.2 Seed Dormancy

Seed dormancy is defined as the reversible failure of a viable seed to germinate under conditions that are normally suitable to germination process (Vleeshouwers et al. 1995). Alleviation of seed dormancy requires some precise environmental stimulus, after which germination becomes simply dependent on the suitability of the environmental conditions, just as in non-dormant seeds. In sea beet, climatic factors, in particular cold or drought, alleviate seed dormancy (Letschert 1993; Wagmann et al. 2010, 2012). Seed dormancy is thought to allow seeds to germinate at the right time of the year in terms of seedling fitness.

Seed dormancy in *Beta vulgaris* is thought to be imposed by the maternal parts of the seed. The operculum can differentially limit imbibition of oxygen and water to the embryo, slowing germination based on its thickness (Angevine and Chabot 1979; Bewley and Black 1994). Phenolic substances and inorganic salts present in the pericarp might also inhibit germination. These inhibitors require some time to be deactivated to permit the beginning of germination, which happens if the temperature is above 3 °C (Campbell and Entz 1991). Hermann et al. (2007) reported the influence of abscisic acid (ABA) and similar compounds present in the pericarp to delay germination in sugar beet. These factors were especially active in the northern sea beet populations. Moreover, a mucilage surrounding ovary in the presence of water might limit access to oxygen (Peto 1964).

Most of these mechanisms explain that drought treatment, which might for example break the cement of the operculum, is the best way to alleviate dormancy in sea beets.

In 85 natural populations encompassing different climates over the whole French distribution area, Wagmann et al. (2010, 2012) observed that only about 32% of the germination occurred within four weeks after sowing in optimal conditions, which can be attributed to the non-dormant part of the total germination. About 68% of the total germination happened after dormancy had been alleviated by a stimulus or succession of stimuli, that is, cold or drought. The proportion of dormant seeds covered almost the entire range from 0 to 100%, depending on the accession, following a geographical pattern from lower dormancy at high latitudes to higher dormancy in southern France. 63% of the germination occurred after one single or repeated drought treatments.

Under around 10 °C, germination is very slow, the optimum being around 20–25 °C: low temperatures are not suitable to the germination process of non-dormant seeds. However, low temperatures can play another role and alleviate dormancy of a part of the seeds in some sea beet populations: without any cold stratification, these seeds could not germinate even under optimal germination conditions (Letschert 1993; Wagmann et al. 2010, 2012). Reduced germination before winter helps avoiding establishment problems for the seedlings (Letschert 1993). With untimely germination (e.g. in late summer or autumn), plants could be unable to ripen seed before the beginning of the cold season, or might enter the winter in too early a growth stage for survival. The emergence of the seedlings in beet crops normally happens around a week after planting, depending also on the sowing depth. For sea beet, this phase is slower, even though the seed normally develops at the soil surface. After the separation from the empty pericarp, the hypocotyl elongates and the cotyledons unfold. Cotyledons quickly become green and photosynthetically active, giving rise to shoot and leaf development (Figs. 3.4, 3.5 and 3.6).

Letschert (1993) noted that seeds collected in one accession of the Mediterranean area do not display the patterns of dormancy related with low temperatures. On the contrary, germination could be enhanced by cold stratification in two accessions collected in the Netherlands (Letschert 1993). Wagmann et al. (2010, 2012) observed that 0–34% of the total germination (depending on the accession, average = 5%) required cold stratification, without any obvious geographical pattern.

Seed dormancy in sea beet has obvious effects on germination timing, resulting in effects on fitness (Wagmann et al. 2010, 2012). In their experiments, Wagmann et al. (2012) showed that germination mainly occurred in first autumn, spring and following autumn (summer germination being low whatever the accession). The geographical distribution of dormancy was positively correlated with yearly mean temperatures, especially summer temperatures. Minimum temperatures in winter did not significantly explain the trait variation. Moreover, they found positive associations between: i) immediate germination in the greenhouse and germination during first autumn on the one hand and ii) drought treatments and delayed germination in the field (spring, summer and second autumn) on the other hand. Sea beet dormancy can thus be interpreted as a way to spread germination over autumn and spring, while limiting summer germination, that is, when drought might be unfavorable to the survival of the young seedlings.

Dormancy rate in sea beet varies among individuals (ensuring population persistence) and also within-progeny (a strategy known as bet-hedging, i.e. spreading the mother plant's risk among its progeny). Dormancy spreads risk over two seasons (autumn and spring), which is favorable to the survival of at least a part of the seedlings under latitudes where meteorological conditions vary from year to year. It furthermore spreads germination among first autumn and the following ones, spreading risk among years. Seed dormancy rate is largely heritable; however, the genetic component is probably completed by some level of adjustment to local conditions brought about by maternal adaptive phenotypic plasticity (Wagmann et al. 2010, 2012). Similarly, in weed beets (*Beta vulgaris* resulting from crosses between cultivated and wild beets; see Sect. 9.7) seed dormancy plays a role in the formation of the seed bank (Sester et al. 2006).

4.3 Chromosome Number

Unlike other species of the section *Beta* and several cultivated types, the chromosome number (ploidy) of sea beet cells is always $2n = 2x = 18$. The presence of different degrees of ploidy in sea beet populations is an indicator of hybridization with cultivated tetraploid ($4n = 4x = 36$) pollinators, or with tetraploid *Beta macrocarpa* (Artschwager 1927; Lange et al. 1993; McFarlane 1975). These occurrences are quite rare due to: (i) the currently reduced use of tetraploid pollinators in sugar beet seed crops; (ii) the weaker competitive ability of the pollen released by tetraploid plants when compared to pollen produced by diploid beet (Scott and Longden 1970); (iii) the limited geographic range of tetraploid *Beta macrocarpa* (Bartsch et al. 2003); (iv) the differences in flowering time of sea beet and *Beta macrocarpa* (Villain 2007). The Imperial Valley of California is the only area in which there is slight evidence of composite intercrosses between the species *vulgaris* (including subsp. *maritima*) and *Beta macrocarpa* (Bartsch et al. 1999, 2003; McFarlane 1975). In 17 Danish sea beet populations, only a few 3n individuals were detected, demonstrating low levels of gene flow from tetraploid pollinators (Andersen et al. 2005).

4.4 Self-incompatibility

Sea beet is an allogamous (naturally outcrossing) species due to the combined actions of proterandry and a complex gametophyic self-incompatibility system hampering self-pollination (Panella and Lewellen 2007). Sea beet is normally characterized by a high degree of self-incompatibility, with self-pollination rare. Self-incompatibility, more widespread in the northern European sites (Villain 2007), is a good means for ensuring the allogamy and for maintaining high levels of heterozygosity within the populations. *Beta macrocarpa* and *Beta vulgaris* subsp. *adanensis*, on the other hand, are autogamous (usually self-pollinating), although they belong to section *Beta* (as does sea beet) (Bruun et al. 1995; Touzet et al. 2018).

Self-sterility generally does not prevent the germination of incompatible pollen grains on the surface of the stigma; rather the growth of the pollen tubes is stopped inside the pistils (Savitsky 1950). The author recognizes two other physiological mechanisms capable of explaining the failure of self-fertilization: i) the eventual death of zygotes formed; ii) the abnormal growth of embryos resulting in degeneration. Owen (1942a) studied the heritability of self-sterility in the curly top resistant sugar beet variety "US1", containing a preponderance of self-sterile plants. The self-sterility mechanism can be explained by identical or duplicate multiple alleles $S^1–S^n$ and $Z^1–Z^n$ acting gametophytically and carried on different chromosomes. The hypothesis assumed that a single S or Z factor transported by the pollen results in fertility if not present in the tissue of the stigma. For example, a $S^1S^2Z^1Z^2$ female plant producing S^1Z^1 and S^2Z^2 gametes would be successfully fertilized with S^3Z^3 or S^2Z^2 pollen, because at least one of S- or Z-alleles did not encounter their corresponding allele in the style.

Kroes (1973) postulated another type of self-incompatibility caused by the abnormal growth of the pollen tube due to the lack of nutrients available in the pistil. According to Larsen (1977), the self-sterility system in sugar beet is conditioned by four gametophytic S-loci with complementary interaction. In other words, four S genes in the pollen have to match the corresponding genes in the pistil to cause sterility. The four loci were designated S^a, S^b, S^c and S^d. In some cases, self-sterility is partial or incomplete and the plant produces some seed after selfing. This behavior, named pseudo-compatibility, seems to happen under particular climatic conditions or during the late flowering (Bosemark 1993; Owen 1942a).

4.5 Male Sterility

Male sterility is caused by pollen abortion in the anthers or the inability of the anthers to release viable pollen (Fig. 4.1), thus producing functionally female plants inside a normally hermaphroditic species (Bliss and Gabelman 1965; Halldén et al. 1988). The occurrence of sea beet producing small or non-germinating pollen grains after cross of sea beet with sugar beet was detected by Zajkovskaja (1960). The same was observed in Japan (Kinoshita 1965). In wild populations, the interaction of nucleo-cytoplasmic factors for male sterility (CMS) is much more common than the purely nuclear determination (Boutin-Stadler et al. 1989). The mechanism working in sugar beet was discovered by Owen (Owen 1942b, 1945) and it is due to the interaction of mitochondrial and nuclear genes. He explained it through the existence of a sterile cytoplasm "S", as opposed to the normal or "N" type, which results in male sterility, but only when two nuclear "restorer" loci, designated Xx and Zz, are in homozygous recessive state. In other words, male sterile plants bear the S $xxzz$ genotype, whereas all other combinations produce normal hermaphrodites with fertile or partially fertile anthers. Therefore, nuclear genes counteracting the mitochondrial (cytoplasmic) sterility are called "restorers" in sugar beet. When a line contains them in the homozygous recessive state, it is called "maintainer" (Halldén et al. 1988; Owen 1945). The restorer loci are located on the chromosomes III and IV (Schondelmaier and Jung 1997). In summary, plants with sterilizing mitochondria genes (sterile cytoplasm) produce pollen-fertile progeny only if they are crossed to a line with dominant nuclear restoring genes to counteract the sterilizing cytoplasm (mitochondrial genes) (Ducos et al. 2001a).

The N and S cytoplasms have different mitochondrial DNA profiles named S_{vulg} and N_{vulg} (Saumitou-Laprade et al. 1993). These differences are so numerous that it was impossible to ascribe the CMS trait to any one (Ducos et al. 2001b). Recently, substantial advances have been achieved in the understanding of the molecular mechanism of male sterility of the Owen CMS.

A sequence preceding the mitochondrial atp6 gene called preSatp6 encodes a new polypeptide of 35 kDa (preSATP6) specific to Owen CMS which is integrated in the mitochondrial membrane (Yamamoto et al. 2005). The difficulty to prove that it was the sterilizing factor was, that contrarily to most CMSs, the expression of the preSatp6, which was not reduced in restorer plants. Recently, the same team disentangled the nature and the action of one of the restorer locus (Rf1/X locus) (Matsuhira et al. 2012). At the Rf1 locus, one open reading frame called bvorf20 encodes a protein that is similar to a protease and interacts physically with the pre-SATP6 polypeptide. This protein–protein interaction would inhibit the action of the preSATP6 polypeptide (Kitazaki et al. 2015).

The reproduction of CMS lines requires the use of maintainer lines containing N cytoplasm and the nuclear factors x and z in homozygous and recessive state. In sugar beet, the maintainer lines are called "O-types" (Oldemeyer 1957). Owen's CMS monogerm seed bearers crossed with 2n or 4n pollinators allow produce 2n or 3n hybrids, respectively. Coç (2005) found a positive and strong correlation between the frequencies of CMS and O-type in sugar beet open pollinated populations. Eight commercial CMS analyzed using RFLP gave similar results, which demonstrated the existence of a large degree of variation among the male sterile seed bearers currently used for producing hybrid seed (Weihe et al. 1991).

Nuclear or Mendelian male sterility (NMS) in sugar beet is controlled by a single nuclear gene, Aa. NMS is conditioned by the homozygous recessive alleles aa (Bosemark 1971; Owen 1952). This type of male sterility was not identified in sea beet by Arnaud et al. (2009). Bosemark (1998) remarked: "Nuclear male sterility (NMS) has been found in both sugar beet and wild *Beta* beets. However, in spite of a likely high rate of mutations to ms genes, there are few reports of NMS in beets. This is probably due to i) the mostly recessive nature of such genes in combination with predominantly out-pollinated beet populations, and ii) the limited incentive to search for alternatives to the readily available *a1* gene".

Beta maritima populations contain a variable percentage of the individuals, among the normally hermaphrodite plants, that are more or less male sterile, but male sterility can be associated to various mitochondrial haplotypes (Dale et al. 1985; Boutin et al. 1987, 1988; Boutin-Stadler 1987; Halldén et al. 1988; Boutin-Stadler et al. 1989; Cuguen et al. 1994; Dalke and Szota 1993; Dufaÿ et al. 2009). Among the 20 mitochondrial haplotypes described in beets (Desplanque et al. 2000), the following three mitochondrial types were found to be positively associated with sea beet male sterility: CMS E, CMS G and CMS H (Cuguen et al. 1994). CMS E is similar to I-12CMS(3) (Darracq et al. 2011) for which the sterilizing gene orf129, a chimeric open reading frame with a part of cox2, encodes a novel 12 kDa polypeptide (Yamamoto et al. 2008). CMS G is a variant of the mitochondrial complex IV (COX) and has also a modified subunit 9 of the Complex I (NAD) (Ducos et al. 2001a, b; Meyer et al. 2018). It is not known whether any of these modifications are responsible for male sterility.

The occurrence of Owen's CMS, also called S_{vulg}, is rare (Dufaÿ et al. 2009), and consequently this analysis can be a useful means to distinguish true sea beet from the offspring of accidental hybridizations with the cultivated *Beta* complex (Arnaud et al. 2003; Driessen 2003). Conversely, Coe and Stewart (1977) found some individuals functioning as O-type in a Danish population crossed with commercial sugar beet, thus indicating the presence of similar CMS factors in that sea beet population.

After analyses of progeny in two sea beet populations, Boutin-Stadler et al. (1989) observed plants segregating and non-segregating for male sterility. The first population produced female, intermediate and hermaphrodite individuals, and the second yielded only hermaphrodite offspring. Thus, three different sexual types were present in the sea beet populations analyzed: (i) females carrying the CMS genes; (ii) restored hermaphrodites carrying the CMS genes and the nuclear genetic alleles, mixed in various proportions; and (iii) hermaphrodites with normal cytoplasm (de Cauwer et al. 2010). Hiroshi and Tomohiko (2003) analyzed French sea beets and found that they differed from both Owen and normal cytoplasm and more resembled the CMS G (Ducos et al. 2001a). The ancestral and more widespread cytoplasm seems to be the N_{vulg} (Fénart et al. 2006).

Several attempts have been made to broaden the genetic base of the commercial male sterile lines used as females for beet hybrid production (Coe and Stewart 1977; Coons 1975; Mann et al. 1989; Xie et al. 1996), but apparently without significant progress. In fact, molecular analyses carried out by Duchenne et al. (1989) indicated that all the current CMS females used for hybrid seed multiplication still seem derived from Owen's lines (Ducos et al. 2001b). Notwithstanding the common origin, the mitochondrial DNA displays a considerable variation among the currently employed CMS lines (Weihe et al. 1991).

4.6   Self-fertility

The heritability of self-fertility of sugar beet was studied by Owen (1942a), who demonstrated that self-fertility was controlled by a single Mendelian factor S^F. The genetic composition of homozygous self-fertile plants is $S^F S^F$, and $S^F S^a$ (or $S^F S^b$ or $S^F S^x$) for heterozygous self-fertile plants, while $S^a S^b$ represents the self-sterile condition. The designations S^a, S^b and S^x are necessary because diverse self-sterile sugar beet sources may carry many different S allelomorphs. The progeny of crosses between self-sterile $S^a S^b$ seed parents and heterozygous self-fertile $S^F S^x$ pollen parents segregated in 50% self-fertile 50% self-sterile ratio confirming the hypothesis of a single determining factor.

Lewellen (1989) describes an intense self-fertility expression in a O-type line that made hybridization with other lines very difficult. In some sea beet populations, it is possible to find completely self-sterile plants, completely self-fertile plants, and everything in between. In some populations collected at Kulundborg Fjord (Denmark) a very low level of self-fertility was found (Coe and Stewart 1977). Dale and Ford-Lloyd (1983) ascertained some floral differences associated with the reproductive patterns of *Beta maritima*, that is, self-compatible (autogamous) and self-incompatible (xenogamous) populations. In the autogamous individuals where fertilization was assured through self-pollination, the capturing surface of the stigma is reduced as is the length of the anther (ensuring a larger degree of protandry) and the amount of pollen produced. There were also fewer flowers on each plant than in the self-incompatible populations (Orndruff 1969).

4.7 Flowering

Flowering in *Beta maritima* is indeterminate. The stalk grows continuously as does the production of new flowers. On the same plant, it is possible to see newly formed flower buds, ripened and shattered seeds, and all the stages in between. In Mediterranean Sea beets, flowering lasts until the early autumn if the conditions of moisture, light and temperature remain favorable (Smith 1987; Biancardi, unpublished). For harder bolting and biennial NE-Atlantic accessions, stalk elongation and flowering usually end with the onset of shorter days toward the end of summer. Relatively high temperatures may also cause a reversion to vegetative growth (Smit 1983).

Flowering date of sea beet varies among accessions, as suggested by a common garden experiment comparing ca. 100 accessions in France and adjacent regions (van Dijk et al. 1997). Flowering began gradually earlier going from accessions from the Mediterranean basin to those of the western coast of Brittany, whereas going from Brittany northwards flowering happens later (van Dijk 1998; van Dijk et al. 1997). This seems to be due to the different influences of the photoperiod and high temperature in the south on the one hand, and of cold requirement for flowering induction in the northern part of France on the other hand. The heritability of the trait "flowering time" within accessions was about 0.33, which is quite low. Considerable differences were observed among and within Mediterranean and Atlantic sea beet populations. Some plants flowered as early as 33 days after germination, while others needed over 100 days. Some entries flowered in the first year, others in their second year (van Dijk and Boudry 1992).

Phenology, especially flowering time—in the sense of the day of flowering in the year—is a crucial life-history trait. It will indeed largely determine the plant fitness, due to trade-offs (negative correlations) between resource investment in reproduction vs. survival, reproduction vs. growth and/or current vs. future reproduction. Flowering too early might depress the amount of resources that can be stored for future survival and reproduction. Moreover, plant growth might be decreased, implying the production of less seeds in indeterminate plants. Flowering too late might imply decreased seed production and resource loss, if bad weather prevents seed ripening.

Stimuli for flowering induction are directly related with climate and latitude (see Höft et al. (2018) for the present knowledge of its molecular basis). Climate change might thus affect plant phenology and fitness. Sufficient (and sufficiently rapid) phenotypic plasticity and/or genetic change would then be critical for the species persistence in its original distribution area. In a selection experiment based on genetic material from southwestern France, van Dijk and Hautekèete (2007) showed a large potential for evolutionary change in day length sensitivity in the sea beet. Ten generations were enough to reduce the necessary day length for flowering induction from 13 to 11 h, corresponding to the necessary day length for flowering observed in northern Africa populations.

Sea beets grown from seeds collected in 1989 and 2009 in 73 populations in France and adjacent regions were then compared in a common garden experiment. Confirming the prediction that requirements for flowering might evolve under the current climate change in order to keep flowering at the best moment in the year, a genetic change in the flowering date of sea beet within two decades has been reported (van Dijk and Hautekèete 2014). Populations from the southern part of the latitudinal gradient flowered significantly later (mean value 1.78 days), while populations from the northern part of the gradient flowered significantly earlier (mean value −4.04 days). This was thought to be related to a genetic change for higher temperature and longer photoperiod requirements for flowering to the south, and for lower vernalization requirements to the north. This change would be equivalent to a mean northward shift in phenotypes of about 39 km over two decades. It should be noticed that this result obtained in controlled conditions does not imply that flowering date had changed in situ, nor that its pace had been sufficiently steady to compensate for the environmental change.

Sea beets are perennials: under controlled conditions, they almost always succeed in reproducing repeatedly (van Dijk 2009). First flowering may happen without or with passing a cold period. To keep things simple, those sea beets are often called annual or biennial, respectively, although these qualifications do not apply to their life span length. According to Munerati, de Vilmorin (1923) confirmed the dominance of the Mendelian trait of "annuality" conditioned by the "B" locus, which cancels any vernalization requirement of temperature and photoperiod (Abegg 1936; van Dijk et al. 1997). In other words, annual sea beet does not require vernalization (exposure of the plant to winter conditions) for flowering, because the trait depends only on the genetic factor Beta. Consequently, it was assumed that annuality depends on the presence of the allelic combinations of "BB" or "Bb", and the bienniality on "bb" (Driessen 2003). In reality, the traits are more complex, depending also on the environment, and the fact that fixing the biennial trait seems almost impossible (Munerati 1920; van Dijk 1998; Biancardi 2005). The annual cycle is described as common in Mediterranean populations and less frequent in the North-Atlantic ones (Dahlberg 1938; Dale et al. 1985). According to van Dijk and Hautekèete (2007), the northernmost coastal annual sea beet populations occur in the southern French

Atlantic coasts (Boudry et al. 1994, 2002). Weed beets (see Sect. 9.7), which form a plague in sugar beet field all over Europe, are also "annual" (Boudry et al. 1993).

Early season flowering is useful for avoiding the frequent drought conditions during the Mediterranean summer. More frequent rains on the Atlantic coasts and lower winter temperatures induce a different selection pressure favoring the biennial cycle. Much of this behavior is related to the latitude (van Dijk et al. 1997), altitude (Hautekèete et al. 2002a), mortality risk (either climatic or due to man-made disturbance), climatic factors controlling length of the growing season and resources (Hautekèete et al. 2009). Biennial beet overwinters by storing resources and sucrose in the root, useful for winter survival and for rapid development of the reproductive organs early in the subsequent season (Hautekèete et al. 2009).

Driessen (2003) stated that the "annual" cycle is relatively widespread within the Baltic Sea beet populations as well, but could be "escaped" weed beets. In disagreement with other observations, she affirmed that biennial beets can be heterozygous for the B gene (Driessen 2003). In cold environments with a shorter growing season, the biennial habit is necessary to delay the reproduction phase until the second year, that is, in more favorable conditions for producing seeds (Hautekèete et al. 2002b, 2009).

Flowering habit depends on the time of germination, not only on the interaction of the genotype and the environment. In other words, if annual seed germinates in autumn and the rosette is not sufficiently developed, flowering is delayed after winter, thus the plant behaves as a biennial. Moreover, low spring temperatures may induce young plants to flower in their first year (van Dijk and Desplanque 1999). The effect of the photoperiod is more difficult to quantify exactly, due to the variability of light intensity during the day, variable day length, interactions with temperature, and so on.

The stigma remains receptive for more than two weeks (Crane and Walker 1984). On the outside there are very fine papillas, which increase the probability of capturing pollen grains (Artschwager and Starrett 1933). The pollen on the stigma germinates about 2 h after contact and the tube elongates through the ovary toward the embryo (Artschwager and Starrett 1933). Fertilization involves the fusion of the two sperm cells of the pollen with the egg and the central cell of the embryo sac (Klotz 2005). The male cells are released from the pollen by rupture of the pollen tube after its entry into the embryo (Esau 1977). One sperm cell fuses with the egg to produce the zygote. The second sperm cell fuses with the binuclear embryo sac's central cell, producing the primary endosperm nucleus that develops into the endosperm (Artschwager and Starrett 1933). The true seed develops from the fertilized egg and its ovule. "When the seed is ripe the germen becomes purple and granulated" (Hooker 1835). The true seed is in turn surrounded by a thick woody pericarp, which makes it a fruit. (Hooker 1835). During the ripening, the seed ball can be green (Munerati et al. 1913), but is more frequently veined or entirely reddish-violet (de Vilmorin 1923). The ripen seed shatters around the plant to be sometime transported away by wind, water or animals in search of more favorable conditions for germination.

Within the section *Beta*, species and subspecies theoretically can cross easily, but under natural conditions interspecific hybridization is rather rare (see Sect. 6.4).

Crosses between *Beta vulgaris* subspecies and *Beta macrocarpa* are hindered because of different flowering and seed setting periods (May and July, respectively) (McFarlane 1975; Villain 2007).

4.8 Growth Habit

Cultivated beets are sown in early spring and are harvested after a variable length of time—3 or more months for leaf and garden beets, 5 to 9 months for fodder and sugar beets. There is the possibility of sowing in the autumn (winter beet) if the climate is suitable. Because only storage roots and/or rosette leaves are desired, cultivated beets are harvested before the reproductive phase (only useful for seed production), which normally occurs at the beginning of the second year after a period of cold temperature (vernalization) and increasing photoperiod (Smit 1983).

On the contrary, sea beet displays a large variety of life cycles, from short-lived individuals reproducing only a few times before dying (seemingly semelparous in the wild), to long-lived iteroparous ones, reproducing repeatedly in a life time. Life cycle in sea beet varies among populations. In the wild, where environmental conditions might be harsh, sea beet populations are thus frequently a mixture of apparently annual and perennial plants, flowering in first or second year. The presence of different life cycles in the same population ensures survival under severe conditions (Sect. 3.21) because the behavior exhibited by each population (or plant) has been expressed in response to the environment of the site (Letschert 1993). Short-lived sea beets are favored in difficult environments, such as in southwestern France, whereas long-lived perennial beets thrive in the more favorable conditions of the Atlantic coasts (Hautekèete et al. 2002a, 2009; Villain 2007).

Some sea beet populations are described as annual but some authors prefer to describe them as virtually annual (Hautekèete et al. 2002a) or prevalently annual, since under protected conditions, for example, in a greenhouse they almost always succeed in reproducing repeatedly (van Dijk 2009). Some amount of confusion in classifying the growth habit is due to a lack of precision in the terms annual, biennial, perennial, and so on (Smit 1983). A more coherent attempt at a definition is given by Letschert (1993): "Assuming that seed germination takes place in spring, an annual beet can be defined as a plant which flowers and sets seed in the year of germination and does not survive the first year. Biennial beets are vegetative in the first year, flower in the second and die after flowering once. Perennial beets are able to flower repeatedly during several years". These definitions are invalid when germination occurs after spring. Munerati (1920) asserted that the right definition should be "prevalently annual, prevalently biennial, etc.", not only due to the different types co-existing in the same population but also because "it is absolutely impossible to fix definitely the annuality or the bienniality trait".

4.9 Life Span

Because *Beta maritima* normally is an iteroparous species (i.e. it reproduces repeatedly once a year), another important fitness trait is the life span. The trait has obvious effects on the number of reproductive episodes of a single plant (Hautekèete et al. 2002a). Longevity is therefore subjected to natural selection. Given the co-existence in the same population of individuals with different behaviors, observations are frequently subjected to error. Around the Mediterranean Sea, the following behaviors can be observed: (i) plants which normally do not die after flowering and setting seeds, but live up to 8–11 years producing seeds every year (Munerati et al. 1913); (ii) plants which flower some months after emergence and die soon after seed ripening; (iii) plants which behave as (ii), but survive the first flowering and produce seeds for one or two years (Biancardi, unpublished). According to Letschert (1993), in cases (i) and (ii) the plants can be defined as perennial and annual, respectively. However, in the third case, what would be the correct definition?

Life span is a genetic trait but in the wild it depends largely on environmental conditions (drought, frost, salty soils, etc.) and biotic factors (parasites, grazing, cutting for seed cleaning, or for food and fodder harvest, disturbance by human activity) (Low 2007). Studies in controlled conditions are therefore necessary. After studying 104 accessions (collected from different locations from the Netherlands and Great Britain to southern France) in controlled and uniform conditions, Hautekèete et al. (2002a) suggested that inland populations of sea beet had the shortest life span (around 2 years). In this study, mean life span increased with latitude from 2 years in south-western France to at least 11 years in north Brittany, before decreasing to about 5 years in northern France (Hautekèete et al. 2002a). Life span in French sea beet was found to be positively associated with (i) habitat stability (from highly disturbed habitats like field margins, roads, and buildings to mostly undisturbed sites like seawalls and cliffs), as well as with (ii) climatic stability, while it is (iii) negatively associated with the length of the growing season (Hautekèete et al. 2002a, 2009). In highly disturbed habitats, that is, due to man-made disturbance or climatic risks and variability, natural selection may favor individuals reproducing early and with a very short life span. On the contrary, more stable conditions give the opportunity to invest resources into growth and survival, thus allowing a higher reproduction in the future. Bartsch et al. (2003) found remarkable changes in population size due to the mortality following prolonged drought periods in nine years of continued observations along the northwest coasts of Adriatic Sea, supporting the hypothesis that climatic risks of mortality might constitute a major selective force.

Life span in French sea beet is also negatively associated with the length of the growing season, which might influence directly the amount of acquired resources. This result thus supports a model suggesting that the amount of available resources is, beside mortality probability, one possible important factor selecting for shorter life span in sea beet (Hautekèete et al. 2009).

In a composite population based on 93 sea beet populations sampled on the shores of Brittany, van Dijk (2009) found aging effects—in the sense of a decline in reproduction and root storage—combined with a gradual change in flowering date along plant lifetime that could be related to a normal physiological basis (e.g. increasing plant size). Each year, plants flowered, on average, 1.3 days later than in the year before, and seed production decreased continuously with age. This trend increased until the year before the plant died, when plants flowered, on average, 3.3 days later than the year before, and seed and root production was even more rapidly reduced. The negative relationship (trade-off) between reproduction and root invest-ment (Hautekèete et al. 2001, 2002b) became positive near plant death. Effects of aging combined with later flowering were thus particularly pronounced in the year before death.

4.10   Age at Maturity

Age at maturity and life span are intimately related traits, due to trade-offs between survival and reproduction (Hautekèete et al. 2001, 2002b, 2009): in iteroparous plants, postponing first reproduction is supposed to allow the reallocation of resources to survival and thus increase life span and future reproduction. This expected correla-tion between age at first reproduction and life span has not been observed in northern Atlantic or in Mediterranean locations (Munerati 1910). As previously mentioned, the normal growth of sea beet is delayed in the cold and short growing season of the northern areas. Therefore, first flowering happens only when the plant is developed enough to produce the highest amount of ripe seeds, that is, in the second year (after the first overwintering). Plants that flower under the very first favorable conditions, that is, early autumn, would severely reduce the resources needed for overwintering survival (Hautekèete et al. 2009). On the other hand, in southern environments, sea beet must flower rapidly to avoid the summer drought or the related increasing salt concentration in the soil. Therefore, the time between emergence and seed set can be very short.

This correlation, however, appears among regions. Along a latitudinal gradient in France, a positive correlation was found in *Beta maritima* between life span and the age of first reproduction (also called "age at maturity"), mainly driven by a strong association of short life span with first year flowering in southwestern France. This suggested common selection pressures for shorter life cycles in this area (Hautekèete et al. 2002a). This result supports the theoretical expectation obtained by modeling that longer life span and later maturity should be both optimal in poorer and safer environments, and thus might be selected at the same time in such environments (Hautekèete et al. 2009). In sea beet, long life span and later maturity are associated with more stable environments and climatic conditions as well as with a longer growing season (Hautekèete et al. 2002a, 2009).

In the case of global climate change, Jones et al. (2003) and Jaggard et al. (2010) hypothesized that sugar beet yield may benefit from longer growing season and from the opportunity of sowing earlier without the risk that beet seedlings be vernalized. In sea beet however, vernalization requirement varies with latitude. Climate change impact on age at first flowering will thus depend on the individual vernalization requirement. By comparing plants grown from seeds collected in 1989 and 2009 in controlled conditions, van Dijk and Hautekèete (2014) observed the change in vernalization requirement among sea beet populations along a latitudinal gradient along French coasts. In the northern part of the gradient, all individuals required vernalization for flowering in 1989 and this rate had apparently not changed in 2009. In the southern part, the rate of vernalization requirement was low in 1989 and had not changed in 2009 either. Along Atlantic coasts, there apparently had been a genetic change toward a decreased requirement for exposure to cold temperatures. However, this does not imply that this genetic change lead to any change in the plant life cycle in situ.

4.11 Reproductive Systems

Three sexual types co-exist in sea beet: (i) females with non-restored CMS cytoplasm; (ii) hermaphrodites with restored CMS; and (iii) normal or non-restored hermaphrodites (Cuguen et al. 1994; de Cauwer et al. 2010). In 33 populations collected in Brittany, the rate of male sterile or female plants varied from 0 to 0.43, male sterility being mainly due to CMS E and CMS G. The level of restoration, that is, the ratio of the number of restored plants on a given CMS on the total of individuals carrying this CMS, differed between CMS E and CMS G with a level of 50 and 15%, respectively. A study carried out on the two types of hermaphrodites (restored on CMS E and non-restored) revealed that even though restored hermaphrodites were poorer pollen producers that non-restored ones, they sired more seedlings than the non-restored plants, due to their closer proximity to females (de Cauwer et al. 2010).

4.11.1 Gynodioecy

Gynodioecy is defined as the occurrence in the same population of female (or male sterile) and hermaphrodite plants (Darwin 1877; Dufaÿ et al. 2007), and is thought to be initiated when a female mutant enters a hermaphroditic population. The relatively low frequency (7%) of gynodioecious species among the flowering plants indicates that the reproduction system is advantageous for only a few species in some environments (Delph and Bailey 2010). Gynodioecy in the section *Beta* is common only in *Beta maritima* (Villain 2007). The co-existence of female and hermaphroditic individuals is usually the result of interactions between CMS factors causing the pollen abortion and nuclear genes, which are able to restore male functions (Dufaÿ et al. 2009). Gynodioecy is mentioned as a means for understanding the evolution of wild populations; it is considered an intermediate or transitory phase from hermaphroditism (male and female functions on the same plant) to dioecy (functions in separate individuals) (Charlesworth and Charlesworth 1978; Barrett 2002). This evolution in some sea beet populations must depend on the female function (seed production, seed size, germination quality, etc.), which is expected to be better in CMS females than in hermaphrodite plants, consequently providing a fitness selective advantage (Dufaÿ et al. 2007). Another advantage for female (or male-sterile) plants that cannot be applied for a self-incompatible species as beet is the avoidance of inbreeding (Dornier and Dufay 2013; Thompson and Tarayre 2000). So far, such gain of fitness in females on the main CMS in beet populations (CMS E) has not been found (Boutin et al. 1988; de Cauwer et al. 2011).

4.11.2 Sex Ratio

The co-existence of sexually polymorphic plants is frequent in wild plant populations. Because the males transmit their genes to the offspring only through pollen, the female only through seed and the hermaphrodites through a combination of these two means, each sexual phenotype utilizes different strategies for survival inside the population (Barrett 2002). In 30 sea beet populations sampled on the coasts of Normandy, the frequency of females ranged from 0.02 to 0.46 (Dufaÿ et al. 2007). This large variation was explained through different frequency dependent selection pressures working on the genes determining CMS and fertility restoration. The distribution of the sexual polymorphic plants inside the population is not by chance: where the hermaphroditic individuals are locally rare and the pollen release is scarce, female plants are, consequently, rare as well. Under these conditions, the female plants produce fewer seeds, becoming disadvantaged when compared to hermaphroditic plants in natural selection for survival (de Cauwer et al. 2010)

References

Abegg FA (1936) A genetic factor for the annual habit in beets and linkage relationship. J Agr Res 53:493–511
Andersen NS, Siegismund HR, Jørgensen B (2005) Low level of gene flow from cultivated beets (*Beta vulgaris* L. ssp. *vulgaris*) into Danish populations of sea beet (*Beta vulgaris* L. ssp. *maritima* (L.) Arcangeli). Mol Ecol 14:1391–1405
Angevine MW, Chabot BF (1979) Seed germinaion syndromes in higher plants. In: Solbrig OT, Jain S, Johnson GB, Raven PH (eds) Topics in plant population biology. Columbia University Press, New York, U.S
Arnaud JF, Fénart S, Godé C, Deledicque S, Touzet P, Cuguen J (2009) Fine-scale geographical structure of genetic diversity in inland wild beet populations. Mol Ecol 18:3201–3215
Arnaud JF, Viard F, Delescluse M, Cuguen J (2003) Evidence for gene flow via seed dispersal from crop to wild relatives in *Beta vulgaris* (Chenopodiaceae): consequences for the release of genetically modified crop species with weedy lineages. Proc R Soc B Biol Sci 270:1565–1571
Artschwager E (1927) Development of flowers and seed in the sugar beet. J Agr Res 34:1–25
Artschwager E, Starrett R (1933) The time factor in fertilization and embryo development in the sugar beet. J Agr Res 47:823–843

Barrett SCH (2002) The evolution of plant sexual diversity. Nat Rev Genet 3:274–284

Bartsch D, Clegg J, Ellstrand N (1999) Origin of wild beets and gene flow between *Beta vulgaris* and *B. macrocarpa* in California. Proc Br Crop Prot Counc Symp 72:269–274

Bartsch D, Cuguen J, Biancardi E, Sweet J (2003) Environmental implications of gene flow from sugar beet to wild beet–current status and future research needs. Environ Biosaf Res 2:105–115

Bewley JD, Black M (1994) Seeds: physiology of development and germination Plenum Press

Biancardi E (2005) Objectives of sugar beet breeding—morphological and physiological iraits. In: Biancardi E, Campbell LG, Skaracis GN, de Biaggi M (eds) Genetics and breeding of sugar beet. Science Publishers Inc, Enfield, NH, USA, pp 72–73

Bliss FA, Gabelman WH (1965) Inheritance of male sterility in beets, *Beta vulgaris* L. Crop Sci 5:403–406

Bosemark NO (1971) Use of Mendelian male sterility in recurrent selection and hybrid breeding in beets. In: Report of meeting of the Eucarpia fodder crops section at Lusignan, France, pp 127–136

Bosemark NO (1993) Genetics and breeding. In: Cooke DA, Scott RK (eds) The sugar beet crop: science into practice. Chapman & Hall, London, pp 66–119

Bosemark NO (1998) Genetic diversity for male sterility in wild and cultivated beets. In: Frese L, Panella L, Srivastava HM, Lange W (eds) International beta genetic resources network. A report on the 4th International. Beta genetic resources workshop and world beta network conference held at the Aegean Agricultural Research Institute, Izmir, Turkey, 28 Feb–3 Mar 1996. International Plant Genetic Resources Institute, Rome, pp 44–56

Boudry P, Mccombie H, van Dijk H (2002) Vernalization requirement of wild beet *Beta vulgaris* subsp. *maritima*: Among population variation and its adaptive significance. J Ecol 90:693–703

Boudry P, Mörchen M, Saumitou-Laprade P, Vernet P, van Dijk H (1993) The origin and evolution of weed beets: consequences for the breeding and release of herbicide-resistant transgenic sugar beets. Theor Appl Genet 87:471–478

Boudry P, Wieber R, Saumitou-Laprade P, Pillen K (1994) Identification of RFLP markers closely linked to the bolting gene *B* and their significance for the study of the annual habit in beets (*Beta vulgaris* L.). Theor Appl Genet 88:852–858

Boutin V, Jean R, Valero M, Vernet P (1988) Gynodioecy in *Beta maritima*. Oecol Plant 9:61–66

Boutin V, Pannenbecker G, Ecke W, Schewe G, Saumitou-Laprade P, Jean R, Vernet P, Michaelis G (1987) Cytoplasmic male sterility and nuclear restorer genes in a natural population of *Beta maritima*: genetical and molecular aspects. Theor Appl Genet 73:625–629

Boutin-Stadler V (1987) Sélection sexuelle et dynamique de la stérilité male dans les populations naturelle de *Beta maritima*. Université des Sciences et Technologies de Lillel

Boutin-Stadler V, Saumitou-Laprade P, Valero M, Jean R, Vernet P (1989) Spatio-temporal variation of male sterile frequencies in two natural populations of *Beta maritima*. Heredity 63:395–400

Bruun L, Haldrup A, Petersen SG, Frese L, de Bock TSM, Lange W (1995) Self-incompatibility reactions in wild species of the genus Beta and their relation to taxonomical classification and geographical origin. Genet Resour Crop Evol 42:293–301

Campbell LG, Entz GW (1991) Temperature effects on sugar beet seedling emergence. J Sugar Beet Res 28:129–140

Charlesworth B, Charlesworth D (1978) A model for the evolution of dioecy and gynodioecy. Am Nat 112:975–997

Coç H (2005) Investigation of male sterility in sugar beet populations. Akadeniz Universitesi Ziraat Fakultesi Dergisi 18:195–300

Coe GE, Stewart D (1977) Cytoplasmic male sterility, self fertility, and monogermness in *Beta maritima* L. J ASSBT 19:257–261

Coons GH (1975) Interspecific hybrids between *Beta vulgaris* L. and the wild species of *Beta*. J ASSBT 18:281–306

Crane E, Walker P (1984) *Beta Vulgaris* L. Pollination directory for world crop. IBRA, Hill House, UK

Cuguen J, Wattier R, Saumitou-Laprade P, Forciolo D, Mörchen M, van Dijk H (1994) Gynodioecy and mitochondrial DNA polymorphism in natural populations of *Beta vulgaris* ssp *maritima*. Genet Sel Evol 26(Suppl 1):87s–101s

Dahlberg HW (1938) Some observations on the wild beet (Beta maritima). Proc Am Soc Sugar Beet Technol 1:76–79

Dale MFB, Ford-Lloyd BV (1983) Reproductive characters associated with breeding behaviour in *Beta* sect. *Beta* (*Chenopodiaceae*). Plant Syst Evol 143:277–283

Dale MFB, Ford-Lloyd BV, Arnold MH (1985) Variation in some agronomically important characters in a germplasm collection of beet (*Beta Vulgaris* L.). Euphytica 34:449–455

Dalke L, Szota M (1993) Utilizing male sterility from *Beta maritima* in sugarbeet breeding. J Sugar Beet Res 30:253–260

Darracq A, Varré JS, Maréchal-Drouard L, Courseaux A, Castric V, Saumitou-Laprade P, Oztas S, Lenoble P, Vacherie B, Barbe V, Touzet P (2011) Structural and content diversity of mitochondrial genome in beet: a comparative genomic analysis. Genome Biol Evol 3:723–736

Darwin C (1877) The different form of flowers on plants of the same species. Murray, London, UK

de Cauwer I, Dufaÿ M, Cuguen J, Arnaud J-F (2010) Effects of fine-scale genetic structure on male mating success in gynodioecious *Beta vulgaris* ssp. *maritima*. Mol Ecol 19:1540–1558

de Cauwer I, Arnaud J-F, Courseaux A, Dufaÿ M (2011) Sex-specific fitness variation in gynodioecious Beta vulgaris ssp. maritima: do empirical observations fit theoretical predictions? J Evol Biol 24:2456–2472

de Vilmorin JL (1923) L' hérédité de la betterave cultivée. Gauthier-Villars, Paris, France

Delph LF, Bailey MF (2010) The nearness of you: the effect of population structure on siring success in a gynodioecious species. Mol Ecol 19:1520–1522

Desplanque B, Viard F, Bernard J, Forcioli D, Saumitou-Laprade P, Cuguen J, van Dijk H (2000) The linkage disequilibrium between chloroplast DNA and mitochondrial DNA haplotypes in Beta vulgaris ssp. maritima (L.): the usefulness of both genomes for population genetic studies. Mol Ecol 9:141–154

Dornier A, Dufay M (2013) How selfing, inbreeding depression, and pollen limitation impact nuclear-cytoplasmic gynodioecy: a model. Evolution 67:2674–2687

Driessen S (2003) *Beta vulgaris* ssp. *maritima* an Deutschlands Ostseeküste. PhD, RWTH Aachen, Germany

Duchenne M, Lejeune B, Fouillard P, Quetler F (1989) Comparison of the organization and expression of mtDNA of fertile and male-sterile sugar beet varities (*Beta vulgaris* L.). Theor Appl Genet 78:633–640

Ducos E, Touzet P, Boutry M (2001a) The male sterile G cytoplasm of wild beet displays modified mitochondrial respiratory complexes. Plant J 26:171–180

Ducos E, Touzet P, Saumitou-Laprade P, Vernet P, Cuguen J (2001b) Nuclear effect on mitochondrial protein expression of the CMS Owen cytoplasm in sugar beet. Theor Appl Genet 102:1299–1304

Dufaÿ M, Cuguen J, Arnauld JF, Touzet P, Shykoff J (2009) Sex ratio variation among gynodioecious populations of sea beet: can it be explained by negative frequency-dependent selection? Evolution 63:1483–1497

Dufaÿ M, Touzet P, Maurice S, Cuguen J (2007) Modeling the maintenance of male-fertile cytoplasm in a gyodioecious population. Heredity 99:349–356

Esau K (1977) Anatomy of seed plants, 2nd edn. Wiley, New York, NY, USA

Fénart S, Touzet P, Arnaud J-F, Cuguen J (2006) Emergence of gynodioecy in wild beet (*Beta vulgaris* ssp. *maritima* L.): a genealogical approach using chloroplastic nucleotide sequences. Proc R Soc B Biol Sci 273:1391–1398

Galmés J, Medrano H, Flexas J (2006) Germination capacity and temperature dependance in Mediterranean species of Balearic Islands. Invest Agrar: Sist Recur For 15:88–95

Grogan D (2009) Survey of Beta vulgaris subsp. maritima populations in Ireland. In: Frese L, Maggioni L, Lipman E (eds) Report of a Working Group on Beta and the World Beta Network. Third Joint Meeting, 8–11 March 2006, Puerto de la Cruz, Tenerife, Spain. Bioversity International, Rome, Italy, pp 38–44

Halldén C, Bryngelsson T, Bosemark NO (1988) Two new types of cytoplasmic male sterility found in wild *Beta* beets. Theor Appl Genet 75:561–568

Hautekèete NC, Piquot Y, van Dijk H (2001) Investment on survival and reproduction along a semelparity-iteroparity gradient in the *Beta* species complex. J Evol Biol 14:795–804

Hautekèete NC, Piquot Y, van Dijk H (2002a) Life span in *Beta vulgaris* ssp. *maritima*: the effects of age at first reproduction and disturbance. J Ecol 90:508–516

Hautekèete NC, Piquot Y, van Dijk H (2002b) Variation in ageing and meristemic activity in relation to flower-bud and fruit excision in the *Beta* species complex. New Physiol 179:575–579

Hautekèete NC, van Dijk H, Piquot Y, Teriokhin A (2009) Evolutionary optimization of life-history traits in the sea beet *Beta vulgaris* subsp. *maritima*: comparing model to data. Acta Oecol 35:104–116

Hermann K, Meinhard J, Dobrev P, Linkies A, Pesek B, Hess B, Machackova I, Fischer U, Leubner-Metzger G (2007) 1-Aminocyclopropane-1-carboxylic acid and abscisic acid during the germination of sugar beet (*Beta vulgaris* L.): a comparative study of fruits and seeds. J Exp Bot 58:3047–3060

Hiroshi S, Tomohiko K (2003) Genetical and molecular analyses of male sterility found in *Beta maritima* accession FR4-31. Proc Jpn Soc Sugar Beet Technol 44:21–25

Höft N, Dally N, Hasler M, Jung C (2018) Haplotype variation of flowering time genes of sugar beet and its wild relatives and the impact on life cycle regimes. Front Plant Sci 2211

Hojland JG, Pedersen L (1994) Sugar beet, beetroots, and fodder beet (*Beta vulgaris* L. ssp. *vulgaris*): dispersal, establishment and interactions with the enviornment. The National Forest and Nature Agency, Copenhagen, Denmark

Hooker WJ (1835) The british flora. London, UK

Jaggard KW, Qi A, Ober ES (2010) Possible change to arable crop yields by 2050. Phil Trans R Soc B 365:2835–2851

Jones PD, Lister DH, Jaggard KW, Pidgeon JD (2003) Future climate impact on the productivity of sugar beet (*Beta vulgaris* L.) in Europe. Clim Change 58:93–108

Kinoshita T (1965) Male sterility in *Beta maritima* L. Tensai Kenkyu Hokoku/Bull Sugar Beet Res Suppl. 5:60–63

Kitazaki K, Arakawa T, Matsunaga M1, Yui-Kurino R, Matsuhira H1, Mikami T, Kubo T (2015) Post-translational mechanisms are associated with fertility restoration of cytoplasmic male sterility in sugar beet (Beta vulgaris). Plant J. 83(2):290–299

Klotz KL (2005) Anatomy and physiology. In: Biancardi E, Campbell LG, Skaracis GN, de Biaggi M (eds) Genetics and breeding of sugar beet. Science Publishers Inc, Enfield, NH, USA, pp 9–19

Kroes HW (1973) An enzyme theory of self-incompatibility. Incomp Newsl Assoc 5–14

Lange W, Oleo M, de Bock TSM, D'Haeseleer M, Jacobs M (1993) Chromosomal assignment of three enzyme coding loci (*Icd1*, *Nad-Mdh1* and *Aco1*) using primary trisomics in beet (*Beta vulgaris* L.). Plant Breed 111:177–184

Larsen K (1977) Self-incompatibility in *Beta vulgaris* L. I. Four gametophytic, complementary S-loci in sugar beet. Hereditas 85:227–248

Letschert JPW (1993) *Beta* section *Beta*: biogeographical patterns of variation, and taxonomy. PhD Wageningen Agricultural University Papers 93-1

Lewellen RT (1989) Registration of cytoplasmic male sterile sugarbeet germplasm C600CMS. Crop Sci 29:246

Low EG (2007) Shingle biodiversity and habitat disturbance. University of Sussex, Brighton, UK

Mann V, McIntosh L, Theuer C, Hirschberg J (1989) A new cytoplasmic male sterile type in the sugar beet *Beta vulgaris* L.: a molecular analysis. Theor Appl Genet 78:293–297

Matsuhira H, Kagami H, Kurata M, Kitazaki K, Matsunaga M, Hamaguchi Y, Hagihara E, Ueda M, Harada M, Muramatsu A, Yui-Kurino R, Taguchi K, Tamagake H, Mikami T, Kubo T (2012) Unusual and typical features of a novel restorer-of-fertility gene of sugar beet (Beta vulgaris L.). Genetics 192(4):1347–1358

McFarlane JS (1975) Naturally occurring hybrids between sugarbeet and *Beta macrocarpa* in the Imperial Valley of California. J ASSBT 18:245–251

Meyer EH, Lehmann C, Boivin S, Brings L, de Cauwer I, Bock R, Kühn K, Touzet P (2018) CMS-G from *Beta vulgaris* ssp. *maritima* is maintained in natural populations despite containing an atypical cytochrome c oxidase. Biochem J 475:759–773

Morris PC, Grierson D, Whttington WJ (1984) Endogenous inhibitors and germination of *Beta vulgaris*. J Exp Bot 35:994–1002

Munerati O (1910) Osservazioni sulla bietola selvaggia (*Beta maritima* L.). Staz Sper Ag Ital

Munerati O (1920) Sulla salita in seme il primo anno delle bietole coltivate. Bollettino Associazione italiana delle Industrie dello Zucchero e dell'Alcool 90–95

Munerati O, Mezzadroli G, Zapparoli TV (1913) Osservazioni sulla *Beta maritima* L., nel triennio 1910–1912. Stazioni Sperimentali Agricole Italiane 46:415–445

Oldemeyer RK (1957) Sugar beet male sterility. J ASSBT 9:381–386

Orndruff R (1969) The importance or reproductive biology for taxonomy. Taxon 18:133–212

Owen FV (1942a) Inheritance of cross- and self-sterility in *Beta vulgaris* L. J Agr Res 64:679–698

Owen FV (1942b) Male sterility in sugar beet produced by complementary effects of cytoplasmic and Mendelian inheritance. Am J Bot 29:692

Owen FV (1945) Cytoplasmically inherited male-sterility in sugar beets. J Agr Res 71:423–440

Owen FV (1952) Mendelian male sterility in sugar beets. Proc ASSBF7:371–376

Panella l, Lewellen RT (2007) Broadening the genetic base of sugar beet: introgression from wild relatives. Euphytica 154:382–400

Peto FH (1964) Methods of loosening tight seed caps in monogerm seed to improve germination. J Am Soc Sugar Beet Technol 13:281–286

Richard P, Raymond P, Curbineau F, Pradet A (1989) Effect of the pericarp on sugar beet (*Beta vulgaris* L.) seed germination: study of the energy metabolism. Seed Sci Technol 17:485–498

Santos DSB, Pereira MFA (1989) Restrictions of the tegument to the germination of *Beta vulgaris* L. seeds. Seed Sci Technol 17:601–612

Saumitou-Laprade P, Rouwendal GJA, Cuguen J, Krens FA, Michaelis G (1993) Different CMS sources found in *Beta vulgaris* ssp *maritima*: mitochondrial variability in wild populations revealed by a rapid screening procedure. Theor Appl Genet 85:529–535

Savitsky H (1950) A method of determining self-fertility of self sterility in sugar beet based upon the stage of ovule development shortly after flowering. Proc ASSBT 6:198–201

Schondelmaier J, Jung C (1997) Chromosomal assignment of the nine linkage groups of sugar beet (*Beta vulgaris* L.) using primary trisomics. Theor Appl Genet 95:590–596

Scott RK, Longden PC (1970) Pollen release by diploid and tetraploid sugar beet planta. Ann Appl Biol 66:129–136

Sester M, Dürr C, Darmency H, Colbach N (2006) Evolution of weed beet (Beta vulgaris L.) seed bank: quantification of seed survival, dormancy, germination and pre-emergence growth. Eur J Agron 24:19–25

Smit AL (1983) Influence of external factors on growth and development of sugar beet (*Beta vulgaris)* (in English with Dutch summaries). Pudoc, Wageningen, The Netherlands

Smith GA (1987) Sugar Beet. In: Fehr WR (ed) Principles of Cultivar Development. Macmillan Publishing Company, New York, pp 577-625

Thompson JD, Tarayre M (2000) Exploring the genetic basis and proximate causes of female fertility advantage in gynodioecious *Thymus vulgaris*. Evolution 54:1510–1520

Touzet P, Villain S, Buret L, Martin H, Holl A-C, Poux C, Cuguen J (2018) Chloroplastic and nuclear diversity of wild beets at a large geographical scale: toward an evolutionary history of the *Beta* section. Ecol Evol 8:2890–2900. https://doi.org/10.1002/ece3.3774

van Dijk H (1998) Variation for developmental characters in *Beta vulgaris* subsp. *maritima* in relation to latitude: The importance of *in situ* conservation. In: Frese L, Panella L, Srivastava HM, Lange W (eds) A report on the 4th international *Beta* genetic resources workshop and world *Beta* network Conference held at the Aegean Agricultural Research Institute, Izmir, Turkey, 28 Feb–3 Mar 1996. International Plant Genetic Resources Institute, Rome, Italy, pp 30–38

van Dijk H (2009) Ageing effects in an iteroparous plant species with a variable life span. Ann Bot 104:115–124 van Dijk H, Boudry P (1992) Genetic varitaion for life histories in *Beta maritima*. International *Beta* Genetic Resoures Network. A report on the 2nd international WBN workshop, Institute for Crop Science and Plant Breeding, Braunschweig, Germany, 24–28 June 1991. IPGRI, Rome, pp 9–16

van Dijk H, Boudry P, McCombie H, Vernet P (1997) Flowering time in wild beet *(Beta vulgaris* ssp. *maritima)* along a latitudinal cline. Acta Oecol 18:47–60

van Dijk H, Desplanque B (1999) European *Beta:* crops and their wild and weedy relatives. In: van Raamsdonk LWD, Den Nijs JCM (eds) Plant evolution in man-made habitats. Hugo de Vries Laboratory, Amsterdam, The Netherlands, pp 257–270

van Dijk H, Hautekèete NC (2007) Long day plants and the response to global warming: Rapid evolutionary change in day length sensitivity is possible in wild beet. Evol Biol 20:349–357

van Dijk H, Hautekèete NC (2014) Evidence of genetic change in the flowering phenology of sea beets along a latitudinal cline within two decades. J Evol Biol 27:1420–9101

Villain S (2007) Histoire evolutive de la section *Beta*. PhD Universite des Sciences et Technologies de Lille, France

Vleeshouwers LM, Bouwmeester HJ, Karssen CM (1995) Redefining seed dormancy: an attempt to integrate physiology and ecology. J Ecol 83:1031–1037

von Lippmann EO (1925) Geschichte der Rübe *(Beta)* als Kulturpflanze. Verlag Julius Springer, Berlin, Germany

von Proskowetz E (1894) Über die Culturversuche mit *Beta maritima* L. (und *Beta vulgaris* L.) im Jahre 1893. Österreiche-Ungarische Zeitschrift für Zuckerindustrie und Landwirtschaft 31:201–223

Wagmann K, Hautekèete NC, Piquot Y, van Dijk H (2010) Potential for evolutionary change in the seasonal timing of germination in sea beet *(Beta vulgaris* ssp. *maritima)* mediated by seed dormancy. Genetica 138:763–773

Wagmann K, Hautekèete NC, Piquot Y, Meunier C, Schmitt E, van Dijk H (2012) Seed dormancy distribution: explanatory ecological factors. Ann Bot 110:1205–1219

Weihe A, Dudareva NA, Veprev SG, Maletsky SI, Melzer R, Salganik RI, Borner T (1991) Molecular characterization of mitochondrial DNA of different subtypes of male-sterile cytoplasms of the sugar beet *Beta vulgaris* L. Theor Appl Genet 82:11–16

Xie W, Kang C, Wang J, Wang B, Guo D (1996) Molecular characterization of a new type of cytoplasmic male sterile sugar beet. Sci China Ser C Life Sci 39:53–62

Yamamoto MP, Kubo T, Mikami T (2005) The 5'-leader sequence of sugar beet mitochondrial atp6 encodes a novel polypeptide that is characteristic of Owen cytoplasmic male sterility. Molecular genetics and genomics. Mol Genet Genomics 273:342–349

Yamamoto MP, Shinada H, Onodera Y, Komaki C, Mikami T, Kubo T (2008) A male sterility-associated mitochondrial protein in wild beets causes pollen disruption in transgenic plants. Plant J 54:1027–1036

Zajkovskaja NE (1960) Pollen sterility in sugar beet (in Russian). Agrobiologia (Agrobiology) 778–780

Chapter 5
Ecology

Detlef Bartsch and Enrico Biancardi

Abstract The traits of *Beta maritima* have been reviewed and summarized from a number of recent and classical publications dealing with the ecology, morphology, and whole plant physiology of the species. Because few papers have been written only on *Beta maritima*, most information comes from cultivated forms of *Beta vulgaris*. A striking feature of *Beta maritima* gleaned from this review is how variable and adaptive it is. The species is fairly plastic allowing it to live in many different environments. This capacity for adaptation to the local environmental conditions has been correlated with breeding system and with the rapid change in reproduction systems. This is evident in the differences between the Mediterranean populations (easy bolting, short life cycle, sprangled taproot) and those growing the sea coasts of northwest Europe or other parts of the world. This chapter provides the reader with a comprehensive overview of the plant and populations to answer the question: What is *Beta maritima*?

Keywords *Beta maritima* · Habitat · Survival · Seed dispersal · Floating seed · Gene flow

5.1 Survival Strategies

The current climate changes require adequate fitness in the surviving individuals, species, and genotypes (Wagmann et al. 2010). In other words, to improve its chance of survival, every wild population can optimize the fitness to environment by modifying its timing of germination, reproduction time, life span, etc. (Hautekèete et al. 2009; van Dijk 2009b). These strategies in reproduction are crucial when rapid adaptations are required (van Dijk 2009a), particularly in the current situation of climate

D. Bartsch (✉)
Federal Office for Consumer Protection and Food Safety, Mauerstrasse 39-41, 10117 Berlin, Germany
e-mail: detlef.bartsch@bvl.bund.de

E. Biancardi
Formerly Stazione Sperimentale di Bieticoltura, Rovigo, Italy

change toward higher temperature and reduced rainfall, at least in Europe (Jones et al. 2003; Jaggard et al. 2010). If temperatures rise, it may require variation in the day length needed for flowering induction in biennial sea beet (van Dijk and Hautekèete 2007). From this perspective, sea beet could reduce rapidly such as its day length requirement for entering in advance the reproductive phase. This involvement of population genetics (Crow and Kimura 1970; Hartl and Clark 1997) will be briefly summarized.

Seed dormancy plays a significant role in the survival of individuals within wild populations. Germination in nondormant seed depends only on current conditions. On the other hand, dormant seed undergoes a long-lasting exchange of information with the environment to remove the inhibition factors which hinder germination. Seeds subjected to drought and cold periods delay time and rate of germination, demonstrating the existence of inhibiting mechanisms (Wagmann et al. 2010). In field and greenhouse experiments, about 40% of the total sea beets seedlings germinated and developed from dormant seeds. The dormancy trait seems maternally inherited, is highly variable and have a narrow-sense heritability of $h^2 = 0.40$, which may indicate a sufficient ability of sea beet populations to react in the presence of rapid environmental changes (Wagmann et al. 2010).

Some of these traits play an important role in survival of sea beet populations. For example, the relatively large shape of the seed ball and embryos observed in Afghanistan and Iran could improve the seedling's chances of survival during the critical first stages of pre-germination and germination in difficult environments (Krasochkin 1959). According to Hautekèete et al. (2009), the factors influencing the life history strategies are (i) mortality; (ii) availability of resources; (iii) age at maturity; and (iv) climate.

(i) Mortality due to abiotic stresses and diseases plays a central role in population fitness. The dynamics in 21 Adriatic Sea beet populations were studied by Bartsch and Schmidt (1997). They demonstrated that, under favorable conditions, some populations doubled the number of plants present the year before. In this case, it means that only one out of about 10,000 seeds produced by each plant developed an average of one plant surviving the first year. Under such extremely severe selection pressure and in the presence of long-lasting diseases, it is believed that individuals endowed with some degree of genetic resistance or tolerance should be favored in reproduction and survival in presence of that specific disease. In other words, sea beet undergoes, year after year, a sort of natural selection in situ against adverse agents. The fittest plant reproduces faster than the rest of population and rapidly replace the susceptible individuals. This seems not always to be true. It is well known that the Danish sea beet accessions WB41 and WB42 displayed good rhizomania resistance even though they were sampled in fully BNYVV free locations (de Biaggi et al. 2003; Gidner et al. 2005). In soils of Adriatic shores, where sea beet population developed the first source of monogenic rhizomania resistance (Biancardi et al. 2005), Bartsch and Brand (1998) ascertained the absence of BNYVV in the soil. Notwithstanding, some populations have proven very resistant. The foreign origin of wild populations could explain this disagreement.

(ii) Concerning the availability of resources, Hautekèete et al. (2009) stated that the availability of water, nutrients, light, as well as the length of the growing season can influence the photosynthate accumulation and life tactics of *Beta maritima* populations. Increasing resources should hasten the reproduction cycles, whereas the reduced resources could require more time from the plant for flowering and setting seed.

(iii) Age at maturity (age at first reproduction) is also influenced by the available resources. Inadequate resources delay the time until first reproduction and reduce the vegetative growth as well. The seed bearer plant needs adequate time to store enough energy for successful seed yield (Hautekèete et al. 2009).

(iv) Of course, the climate factors—latitude, altitude, distance to the sea, and so on—play a key role in both ages at maturity and survival strategies.

For survival, wild plants such as sea beet must allocate their photosynthate either for reproduction, or for survival, or both. The annual individuals "do not store a large quantity of food in their roots" (De Vries 1905), which remain thin even at the time of flowering. Reproductive effort is higher and invariable for annual or semelparous plants (i.e., they die soon after the very first flowering and setting seed). Normally sea beet is iteroparous, living two or more years, but the behavior can be strictly semelparous in annual plants. The possibility of producing seed once in some period of the year is a successful strategy of reproduction in unpredictable and difficult environments, like the Mediterranean seashores (Hautekèete et al. 2001). On the other hand, the need to survive is more important in an iteroparous plant (living several years and producing seed annually), which is much more influenced by the environment and, above all, by nutrient availability (Hautekèete et al. 2001). Allocation for reproduction and for survival are inversely correlated in iteroparous beets, the opposite happens for the annual and semelparous sea beet. Reproductive effort is inversely correlated also with the life span (Hautekèete et al. 2001).

The genes can be used to increase the local genetic variation (Viard et al. 2004). Transmitting only the male traits, pollen is the prevalent means of dispersal, but seed, which carries both male and female factors, should not be discounted, especially because of the easy movement of sea beet seed by seawater and other means (Ennos 1994). An analysis of the gene dispersal patters in *Beta maritima* was attempted by Tufto et al. (1998). The dispersal into new localities happens in different ways: (i) unintentional or natural introduction of seeds; (ii) naturalization of cultivated genotypes; and (iii) combinations of the former processes with composite intercrosses via pollen among the *Beta vulgaris* complex (Driessen 2003). The dispersal of sea beet along the marine sites happens mainly through the corky multi-seeded glomerule, obviously adapted to drift dispersal by means of seawater (Dale and Ford-Lloyd 1985; Sauer 1993; Wagmann 2008). The seed, also fitted to spreading by wind (Hautekèete et al. 2002; Smartt 1992), is washed away from the beaches during storms and can float and be transported by the sea currents covering up to 50 km per day (Fievet et al. 2007). The wind also can move the seeds carried into new environments by the seawater out of the splash zone to where they can germinate and grow. Tjebbes (1933) confirmed that "the seed can float for days without losing germination capacity". According to Driessen et al. (2001), after 20–25 weeks in salty water, the seed retained 2% of its germination ability. The sea beet populations located on the southern coast of Norway originated probably from the English Islands (Engan 1994). The same was hypothesized by Rasmussen (1933) for few populations located on the Swedish shores. Andersen et al. (2005) evaluated the genetic distance and found that the Danish and Swedish populations are closely related. Both are more similar to the Irish than the French and Italian sea beet populations. The presence of very small and isolated populations in remote, in other ways inaccessible shores of the North Sea, Baltic Sea, and British Islands, is evidence of the dispersal of sea beet via seawater (Dale and Ford-Lloyd 1983; Letschert et al. 1994). This is true also for the Mediterranean and Adriatic populations (Biancardi, unpublished).

5.2 Dispersal

The multigermity of sea beet seed is believed to be essential for the species dispersal in new and remote sites (Dale and Ford-Lloyd 1985). In fact, the trait is necessary to overcome the normally high degrees of self-sterility, which could hinder the reproduction of isolated plants in new localities. It is well known that the beets developed from the same seed ball are genetically different because each embryo originated from different pollen grains and most likely from different male parents, thus allowing the cross reproduction in the new site by the first plants, termed founder population, originating from a single seed ball (Dale and Ford-Lloyd 1985). These authors demonstrated the interfertility of beets developed from the same glomerule. The normal level of genetic variability necessary to better fitting the new environment can be guaranteed by pollen coming from the same source of the seed. Obviously, the chances of stable colonization in this way are extremely low since it reduced according to the square of the distance. As written above, of the several thousand seeds produced by a plant, only few plantlets survive around the source. But in nature, the time is almost never a limiting factor (Biancardi, unpublished).

There is also the possibility of seed dispersal by means of animals (Driessen 2003). Indeed, beet seed is attractive to birds, especially if monogerm or bigerm seed. The seed ball easily can be opened with the beak to separate the edible embryo from the woody pericarp. Some seed may be swallowed entire and pass unharmed through the digestive system. In this way, it may be transported for considerable distances. This possibility of dispersal could explain the presence of sea beets in continental areas otherwise inaccessible, such as Mount Etna (Letschert and Frese 1993), or up to 1,800 m altitude in Caucasian Mountains (Aleksidze et al. 2009), or Mount Olympus (Greece) for *Beta nana* (Frese et al. 2009). On the Adriatic coasts, sea beet is spread only in sites always located near the sea, confirming that the seed dispersal happens mainly through the saltwater. In fact, usually, the sea beet can be found only in the last 150–250 meters in the banks of the river estuaries (Biancardi unpublished).

Dahlberg and Brewbaker (1948) hypothesized that the wild beets growing in Santa Clara County, California, USA, were introduced by the Franciscan Fathers between 1779 and 1780, mixed together with beet or other kinds of seed (Fig. 5.1). Another mean of long-distance dispersal of sea beet might have been the sand or soil ballast used some centuries ago in the sailing vessels (Bartsch and Ellstrand 1999). The sand was collected near the harbors, possibly containing sea beets, and put on board for improving the stability of the empty ships. The ballast was discharged once the ship had arrived before loading merchandise. In agreement with this hypothesis, some pure sea beet populations that were identified around the harbor of Santa Barbara, California USA, and analyzed with allozymes (UPGMA), showed a close relationship to Spanish accessions. In fact, ships came frequently at that time from Cartagena, Spain, after sailing the Pacific Ocean and both *Beta maritima* and *Beta macrocarpa* are fairly widespread on the Spanish Atlantic and Mediterranean coasts (Christensen 1996). Driessen et al. (2001) and Poulsen and Dafgård (2005) explained in a similar way the dispersal of sea beet from the British Islands to the Baltic Sea, and from the Danish to the German coasts. The same could have happened for sea beet, currently very widespread in the lagoon of Venice, through long-established ship trade with the eastern Mediterranean harbors. Carsner (1928) speculated that the wild beets present in several Californian localities were either *Beta maritima* or crosses between sea beet and cultivated varieties. Commercial seed containing unwanted F_1 crosses with sea beet pollen is another mean for long-distance dispersal of *Beta maritima* germplasm.

Fénart et al. (2008) and Villain et al. (2009) explained the spread of sea beet into the current locations and into remote sites as a consequence of the last Quaternary glaciations and the subsequent plant recolonization. The introduction of sea beet at Østvold, Norway, a location quite far from the sea seems due to glaciations as well

Fig. 5.1 Flowering sea beet in salt marsh environment of California (USA)

(Batwik 2000). Villain et al. (2009), based on molecular analysis, speculated that the *Beta maritima* had two different evolutionary lineages: (i) European, carrying the mutation "LF 118", and (ii) Balkanic–Adriatic, with the mutation "LF 124". After the last quaternary glaciations, the North Atlantic coasts were colonized by the plants that survived in the North African and Spanish refuges (Villain 2007). Those that survived in the eastern refuges expanded into the Mediterranean basin. In other words, the species coming from their southern refuges, spread toward the European areas, which became free of ice in the late upper Neolithic (Rivera et al. 2006). Villain et al. (2009) hypothesized also that the sea beet colonization of the western Mediterranean basin should have happened more recently than the Eastern region.

Krasochkin (1960) considered the Mediterranean sea beet as the primary form of the populations adapted to grow far from the sea. In agreement with this hypothesis, the distribution patterns of the specific allozyme *Acpl*-2 (Letschert 1993) suggested the existence of two distinct gene pools (Atlantic and Mediterranean), with different morphological traits as well. The first form flowers preferably later (if not in the second year), the leaves are more succulent and thick, the seed stalks are more prostrate, and the morphology is much more uniform than the Mediterranean (Letschert and Frese 1993). In the last one, the monogerm seeds are rather rare. The genetic diversity evaluated with the same allozyme is quite similar among the plants of the same population and between neighboring populations (Letschert 1993). This polymorphism seems caused by the variable habitat. Shen et al. (1996) confirmed that "sea beet can broadly be subdivided into northern and southern European forms, the first being biennial and the many of the second being annual".

5.3   Gene Flow

Cases of pollen flow from crop to wild beet have been noted in France (Lavigne et al. 2002; Viard et al. 2002; Arnaud et al. 2003). Pollen produced by the large seed crop area (around 3,000 ha each containing around 10,000 flowering male-fertile beets) located in Emilia-Romagna, Italy, did not seem to have contaminated the sea beet populations along the Adriatic coast ranging from 2 to 90 km (Bartsch and Schmidt 1997; Bartsch et al. 2003). According to Schneider (1942), one hectare of beet seed crop with around 25,000 flowering beets produces approximately 25 trillion pollen grains.

The gene flow in the opposite direction (wild to crop) also seems low (Bartsch and Brand 1998). Andersen et al. (2005) analyzed 18 sea beet populations collected in different localities and confirmed that the introgression of cultivated genotypes into the wild ones was not extensive. In the USA, wild beets have been reported along the California coast from San Francisco to San Diego (Carsner 1928, 1938). Carsner speculated that these were either *Beta maritima* or natural crosses between this species and the cultivated types. Wild beets have also been reported in the Imperial Valley of California; these have been classified as *Beta macrocarpa* and, perhaps,

Fig. 5.2 Seed stalk with fully developed seeds of *Beta macrocarpa* (Martinez, California, USA)

crosses between *Beta macrocarpa* (Fig. 5.2) and cultivated beet (McFarlane 1975; Bartsch and Ellstrand 1999; Bartsch et al. 2003).

According to de Cauwer et al. (2010), around 40% of successful pollinations happen inside 15 meters from the pollen source. However, 2.5% of pollinations were detected some kilometers away. Although the general study of the pollen flow is very frequent in other anemophilous species, given the specificity of the single species, the best thing to do is to avoid generalizations and comparisons (de Cauwer et al. 2010).

The extensive genetic and genotypic variability among sea beet populations has been associated with the adaptability of the species under various conditions of environmental stress (Hanson and Wyse 1982). This enables sea beet to flower in inhospitable environments, often characterized by high salinity, limited water availability, and low soil fertility (Stevanato et al. 2001). In these environments, the wild populations are subjected to selection pressures very different from those present in beet cultivation. Faced with gene flow and the pressure of human activities in the areas colonized by sea beet, the genetic conservation of wild germplasm can be seen as securing a source of genetic resistance to biotic and abiotic stresses, to be used in future genetic improvement programs (Doney and Whitney 1990; Luterbacher et al.1998; Frese et al. 2001). The ability of sea beet to hybridization with cultivated beet easily and without genetic abnormalities has facilitated a number of substantial improvements to commercial varieties. The phenomena of spontaneous intercrossing or gene flow from cultivated to wild poses a serious threat to the future conservation of the wild genetic resources (Bartsch et al. 2002), especially in the case of introduction of transgenic varieties (Bartsch and Schuphan 2002; Lelley et al. 2002).

Surveys carried out by Bartsch et al. (1999) helped to identify two alleles (Mdh2-1 and Aco1-2), normally present both in cultivated sugar beet and wild populations in the vicinity of areas devoted to commercial reproduction of seed. This evidence indicated interaction among the wild populations and commercial varieties. Crop-to-wild gene flow could reduce the native allelic diversity and introgress domesticated traits that lower fitness to environment into the wild populations (Arnaud et al. 2009). Such hybridization could lead to extinction of some sea beet populations, especially those located in environmentally challenging sites. Similar unfavorable gene exchange might happen through wild or feral beets, which grow between the cultivated crop and sea beets in some areas (Viard et al. 2004; Ellstrand et al. 2013).

As mentioned, beet crops have been selected for a biennial life cycle. Under certain conditions, plants (normally not exceeding 0.1% of the crop) can return to their ancestral state and flower in the first year. The seed produced by the bolted plants can give rise to weed beets. When this happens, the population gradually diverges from the original morphology, but even after many generations, does not approach the morphology of sea beet (Greene 1909; Ford-Lloyd and Hawkes 1986; Hanf 1990). Sometimes weed beets can originate from hybridization with sea beet or, rarely, with *Beta macrocarpa* (Lange et al. 1993; Bartsch et al. 2003). The effects of gene flow between wild and cultivated beets tend to homogenize the genetic variability in the populations, if not sufficiently isolated. This gene flow may be responsible for highly heterogeneous genotypes called "feral", because they colonize sites affected by human activities (dams, ditches, street borders, etc.) outside of cultivated fields (Mücher et al. 2000). In many European countries, weed beets, mainly derived from bolted beets, can create difficulties for the beet crop because of their high competitiveness (Desplanque et al. 1999). Control of weed beets inside sugar beet fields using the usual herbicides is impossible because they are as sensitive as the beet crop. Only the use of transgenic resistant varieties is effective against weed beets (Coyette et al. 2005).

Gene flow via seed and pollen is an important process in plant evolution. Bartsch et al. (2003) and Viard et al. (2004) observed evidence of gene flow among sea beet, wild beet, and sugar beet, the sea beet located along the Northern France coasts, the sugar beet inland, the weed beet in between. In some sea beet populations and in weed beets in their vicinity, the presence of Owen CMS was detected, indicating that reciprocal crosses had occurred. Therefore, weed beet may be considered a bridge plant for gene flow between cultivated and sea beet. To avoid gene transfer between sea beet and crops and vice versa, it would be necessary to keep the isolation distance on the order of several kilometers (Viard et al. 2004). Evans and Weir (1981) observed an increased salt tolerance in annual weed beets, which could have resulted from pollen flow from the coastal *Beta*. Gene flow also can happen through seed dispersal, as was observed by Arnaud et al. (2003) (see chap. 3). To significantly minimize gene transfer between sea beet and crops and vice versa, it would be necessary to keep the isolation distance on the order of several hundred meters up to kilometers (Viard et al. 2004) or to establish management measures like bolter control.

References

Aleksidze G, Akparov Z, Melikyan A, Arjmand MN (2009) Biodiversity of *Beta* species in the Caucasus Region (Armenia, Azerbaijan, Georgia, Iran). In: Frese L, Maggioni L, Lipman E (eds) Report of a Working Group on *Beta* and the World *Beta* Network. Third Joint Meeting, 8–11 Mar 2006, Puerto de la Cruz, Tenerife, Spain. Bioversity International

Andersen NS, Siegismund HR, Jørgensen B (2005) Low level of gene flow from cultivated beets (*Beta vulgaris* L. ssp. *vulgaris*) into Danish populations of sea beet (*Beta vulgaris* L. ssp. *maritima* (L.) Acangeli). Mol Ecol 14:1391–1405

Arnaud JF, Fénart S, Godé C, Deledicque S, Touzet P, Cuguen J (2009) Fine-scale geographical structure of genetic diversity in inland wild beet populations. Mol Ecol 18:3201–3215

Arnaud JF, Viard F, Delescluse M, Cuguen J (2003) Evidence for gene flow via seed dispersal from crop to wild relatives in *Beta vulgaris* (Chenopodiaceae): consequences for the release of genetically modified crop species with weedy lineages. Proc R Soc B Biol Sci 270:1565–1571

Bartsch D, Brand U (1998) Saline soil condition decreases rhizomania infection of *Beta vulgaris*. J Plant Pathol 80:219–223

Bartsch D, Clegg J, Ellstrand N (1999) Origin of wild beets and gene flow between *Beta vulgaris* and *B. macrocarpa* in California. Proc Br Crop Prot Counc Symp 72:269–274

Bartsch D, Cuguen J, Biancardi E, Sweet J (2003) Environmental implications of gene flow from sugar beet to wild beet–current status and future research needs. Environ Biosaf Res 2:105–115

Bartsch D, Ellstrand NC (1999) Genetic evidence for the origin of Californian wild beets (genus *Beta*). Theor Appl Genet 99:1120–1130

Bartsch D, Schmidt M (1997) Influence of sugar beet breeding on populations of *Beta vulgaris* ssp. *maritima* in Italy. J Veg Sci 8:81–84

Bartsch D, Schuphan I (2002) Lessons we can learn from ecological biosafety research. J Biotechnol 98:71–77

Bartsch D, Stevanato P, Lehnen M, Mainolf A, Mücher T, Moschella A, Driessen S, Mandolino G, Hoffmann A, de Biaggi M, Wehres U, Saeglitz C, Biancardi E (2002) Biodiversity of sea beet in Northern Italy. In: Proceedings of the 65th IIRBB Congress, Feb 2002, Brussels, Belgium, pp 171–180

Batwik JI (2000) Strandbete *Beta vulgaris* L. subsp. *maritima* (L.) Arc. er trolig borte fra Østfold i dag på grunn av barfrost. Natur Østfold 1–2:38–42

Biancardi E, Campbell LG, Skaracis GN, de Biaggi M (2005) Genetics and breeding of sugar beet. Science Publishers, Enfield HN, USA

Carsner E (1928) The wild beet in California. Facts About Sugar 23:1120–1121

Carsner E (1938) Wild beets in California. Proc Am Soc Sugar Beet Technol 1:79

Christensen E (1996) Neuer Fund der Betarübe an Schleswig-Holsteins Osteeküste. Kieler Notizien zur Pflanzenkunde in Schleswig-Holstein und Hamburg 24:30–38

Coyette B, Tencalla F, Brants I, Fichet Y, Rouchouze D, Pidgeon J, Molard MR, Wevers JDA, Beckers R (2005) Effect of introducing glyphosate-tolerant sugarbeet on pesticide usage in Europe. In: Pidgeon J, Molard MR, Wevers JDA, Beckers R (eds) Genetic modification in sugar beet. International Institute for Beet Research, Brussels, Belgium, pp 73–81

Crow JF, Kimura M (1970) An introduction to population genetics theory. Harper and Row, New York, USA

Dahlberg HW, Brewbaker HE (1948) A promising sugar beet hybrid of the Milpitas wild type x commercial. Proc Am Soc Sugar Beet Technol 5:175–178

Dale MFB, Ford-Lloyd BV (1983) Reproductive characters associated with breeding behaviour in *Beta* sect. *Beta* (*Chenopodiaceae*). Plant Syst Evol 143:277–283

Dale MFB, Ford-Lloyd BV (1985) The significance of multigerm seedballs in the genus *Beta*. Watsonia 15:265–267

Dale MFB, Ford-Lloyd BV, Arnold MH (1985) Variation in some agronomically important characters in a germplasm collection of beet (*Beta Vulgaris* L.). Euphytica 34:449–455 de Biaggi M, Erichsen AW, Lewellen RT, Biancardi E (2003) The discovery of rhizomania resistance traits in sugar beet. In: 1st joint IIRB-ASSBT Congress, pp 131–147

de Cauwer I, Dufaÿ M, Cuguen J, Arnaud J-F (2010) Effects of fine-scale genetic structure on male mating success in gynodioecious *Beta vulgaris* ssp. *maritima*. Mol Ecol 19:1540–1558

de Vries U (1905) Species and varieties. Open Court Publishing CO, Chcago, USA

Desplanque B, Boudry P, Broomberg K, Saumitou-Laprade P, Cuguen J, van Dijk H (1999) Genetic diversity and gene flow between wild, cultivated and weedy forms of *Beta vulgaris* L. (Chenopodiaceae), assessed by RFLP and microsatellite markers. Theor Appl Genet 98:1194–1201

Doney D, Whitney E (1990) Genetic enhancement in *Beta* for disease resistance using wild relatives: a strong case for the value of genetic conservation. Econ Bot 44:445–451

Driessen S (2003) *Beta vulgaris* ssp. *maritima* an Deutschlands Ostseeküste. PhD RWTH Aachen, Germany

Driessen S, Pohl M, Bartsch D (2001) RAPD-PCR analysis of the genetic origin of sea beet (*Beta vulgaris* ssp. *maritima*) at Germany's Baltic Sea coast. Basic Appl Ecol 2:341–349

Ellstrand NC, Meirmans P, Rong J, Bartsch D, Ghosh A, de Jong TJ, Haccou P, Lu BR, Snow AA, Stewart CN, Strasburg JL, Van Tienderen PH, Vrieling K, Hooftman D (2013) Introgression of crop alleles into wild or weedy populations. Ann Rev Ecol Evol System 44:325–345

Engan NC (1994) Stranbete *Beta vulgaris* ssp. *maritima* funnet sponton i Norge. Blyttia 52:33–42

Ennos RA (1994) Estimating the relative rates of pollen and seed migration among plant populations. Heredity 72:250–259

Evans A, Weir J (1981) The evolution of weed beet in sugar beet crops. Genet Res Crop Evol 29:301–310

Fénart S, Arnaud J-F, de Cauwer I, Cuguen J (2008) Nuclear and cytoplasmic genetic diversity in weed beet and sugar beet accessions compared to wild relatives: new insights into the genetic relationships within the *Beta vulgaris* complex species. Theor Appl Genet 116:1063–1077

Fievet V, Touzet P, Arnaud JF, Cuguen J (2007) Spatial analysis of nuclear and cytoplasmic DNA diversity in wild sea beet (Beta vulgaris ssp. maritima) populations: Do marine currents shape the genetic structure? Mol Ecol 16:1847–1864

Ford-Lloyd BV, Hawkes JG (1986) Weed beets, their origin and classification. Acta Hort 82:399–404

Frese L, Desprez B, Ziegler D (2001) Potential of genetic resources and breeding strategies for base-broadening in *Beta*. In: Cooper HD, Spillane C, Hodgkin T (eds) Broadening the genetic base of crop production. FAO, IBPRGI jointly with CABI Publishing, Rome, Italy, pp 295–309

Frese L, Hannan R, Hellier B, Samaras S, Panella L (2009) Survey of *Beta nana* in Greece. In: Frese L, Maggioni L, Lipman E (eds) In: Report of a working group on beta and the world beta network. Third joint meeting, 8–11 Mar 2006, Puerto de la Cruz, Tenerife, Spain. Bioversity International, Rome, Italy, pp. 45–52

Gidner S, Lennefors BL, Nilsson NO, Bensefelt J, Johansson E, Gyllenspetz U, Kraft T (2005) QTL mapping of BNYVV resistance from the WB41 source in sugar beet. Genome 48:279–285

Greene EL (1909) Linnaeus as an evolutionist. Proc Wash Acad Sci 9:17–26

Hanf M (1990) Ackerunkräuter Europas. BLV Verlaggesellschaft, Munich, Germany

Hanson AD, Wyse R (1982) Biosynthesis, translocation, and accumulation of betaine in sugar beet and its progenitors in relation to salinity. Plant Physiol 70:1191–1198

Hartl DL, Clark AG (1997) Principles of population genetics. Sinauer Associates Inc., Sunderland, MA, USA

Hautekèete NC, Piquot Y, van Dijk H (2001) Investment on survival and reproduction along a semelparity-iteroparity gradient in the *Beta* species complex. J Evol Biol 14:795–804

Hautekèete NC, Piquot Y, van Dijk H (2002) Life span in *Beta vulgaris* ssp. *maritima*: the effects of age at first reproduction and disturbance. J Ecol 90:508–516

Hautekèete NC, van Dijk H, Piquot Y, Teriokhin A (2009) Evolutionary optimization of life-history traits in the sea beet *Beta* vulgaris subsp. *maritima*: Comparing model to data. Acta Oecol 35:104–116

Jaggard KW, Qi A, Ober ES (2010) Possible change to arable crop yields by 2050. Phil Trans R Soc B 365:2835–2851

Jones PD, Lister DH, Jaggard KW, Pidgeon JD (2003) Future climate impact on the productivity of sugar beet (*Beta vulgaris* L.) in Europe. Clim Change 58:93–108

Krasochkin VT (1959) Review of the species of the genus *Beta*. Trudy Po Prikladnoi Botanike. Genetik i Selektsii 32:3–35

Krasochkin VT (1960) Beet. Gos. Izdat. S.H. Lit. Moskva-Leningrad

Lange W, de Bock TSM, Speckmann GJ, de Jong JH (1993) Disomic and ditelosomic alien chromosome additions in beet (*Beta vulgaris*), carrying an extra chromosome of *B. procumbens* or telosome of *B. patellaris*. Genome 36:261–267

Lavigne C, Klein EK, Couvet D (2002) Using seed purity data to estimate an average pollen mediated gene flow from crops to wild relatives. Theor Appl Genet 104:139–145

Lelley T, Balàzs E, Tepfer M (2002) Ecological Impact of GMO dissemination in agro-ecosystems. Facultas Verlag, Vienna, Austria

Letschert JPW (1993) *Beta* section *Beta*: biogeographical patterns of variation, and taxonomy. Ph.D. Wageningen Agricultural University Papers 93-1

Letschert JPW, Frese L (1993) Analysis of morphological variation in wild beet (*Beta vulgaris* L.) from Sicily. Genet Res Crop Evol 40:15–24

Letschert JPW, Lange W, Frese L, van Der Berg RG (1994) Taxonomy of *Beta* selection *Beta*. J Sugar Beet Res 31:69–85

Luterbacher MC, Smith JM, Asher MJC (1998) Sources of disease resistance in wild *Beta* germplasm. Aspects App Biol 52:423–430

McFarlane JS (1975) Naturally occurring hybrids between sugarbeet and *Beta macrocarpa* in the Imperial Valley of California. J Am Soc Sugar Beet Technol 18:245–251

Mücher T, Hesse P, Pohl-Orf M, Ellstrand NC, Bartsch D (2000) Characterization of weed beets in Germany and Italy. J Sugar Beet Res 37:19–38

Poulsen G, Dafgård SN (2005) Microsatellites as a model for decision making in in-situ management of sea beets. In: First International Conference of Wild Relatives, 14–17 Sept, Agrigento, Italy

Rasmussen J (1933) [Some observations on *Beta maritima*]. Bot Notiser, pp 316–324

Rivera D, Obón C, Heinrich M, Inocencio C, Verde A, Farajado J (2006) Gathered mediterranean food plants—ethanobotanical investigators and historical development. In: Heinrich M, Müller WE, Galli C (eds) Local mediterranean food plants and nutraceuticals. Forum Nutr. Karger, Basel, pp 18–74

Sauer JD (1993) Historical geography of crop plants: a select roster. CRC Press, Boca Raton, Florida

Schneider F (1942) Züchtung der *Beta* Rüben. In: Parey P (ed) Handbuch der Pflanzenzüchtung. Berlin, Germany

Shen Y, Newbury HJ, Ford-Lloyd BV (1996) The taxonomic characterisatoin of annual *Beta* germplasm in a genetic resources collection using RAPD markers. Euphytica 91:205–212

Smartt J (1992) Ecogeographical differentiation and ecotype formation. In: Frese L (ed) International *Beta* genetic resources network. A report on the 2nd international *Beta* genetic resources workshop held at the institute for crop science and plant breeding, Braunschweig, Germany, 24–28 June 1991. IBPGR, Rome

Stevanato P, Biancardi E, de Biaggi M, Colombo M (2001) Leaf dynamic in sugar beet under cercospora leaf spot attacks. In: Proceedings of the 31st meeting of the American society of sugar beet technologists. ASSBT, Denver, USA

Tjebbes K (1933) The wild beets of the North Sea region. Bot Notiser 14:305–315

Tufto J, Raybould AF, Hinfaar K, Engen S (1998) Analysis of genetic structure and dispersal patterns in a populations of sea beet. Genetics 149:1975–1985

van Dijk H (2009a) Ageing effects in an iteroparous plant species with a variable life span. Ann Bot 104:115–124

van Dijk H (2009b) Evolutionary change in flowering phenology in the iteroparous herb Beta vulgaris ssp. maritima: a search for the underlying mechanisms. J Exp Bot 60:3143–3155 van Dijk H, Hautekèete NC (2007) Long day plants and the response to global warming: rapid evolutionary change in day length sensitivity is possible in wild beet. J Evol Biol 20:349–357

Viard F, Bernard J, Desplanque B (2002) Crop-weed interactions in the *Beta vulgaris* complex at a local scale: allelic diversity and gene flow within sugar beet fields. Theor Appl Genet 104:688–697

Viard F, Arnaud J-F, Delescluse M, Cuguen J (2004) Tracing back seed and pollen flow within the crop-wild *Beta vulgaris* complex: genetic distinctiveness vs. hot spots of hybridization over a regional scale. Mol Ecol 13:1357–1364

Villain S (2007) Histoire evolutive de la section *Beta*. PhD Universite des Sciences et Technologies de Lille

Villain S, Touzet P, Cuguen J (2009) Reconstructing the evolutionary history of *Beta* section *Beta* with molecular data. A focus on the Canary Islands. In: Frese L, Germeier CU, Lipman E, Maggioni L (eds) Report of the 3rd joint meeting of the ECP/GR *Beta* working group and world *Beta* network, 8–10 Mar 2006. Tenerife, Spain. Bioversity International, Rome, Italy. Nn K (2008) La dispersion des graines dans le temp (dormance) et dans l'espace chez la betterave maritime (*Beta vulgaris* ssp. *maritima*) quel potentiel èvolutif pour répondre au changement climatique globale. Université des Sciences et Technologies de Lille

Wagmann K (2008) La dispersion des graines dans le temp (dormance) et dans l'espace chez la betterave maritime (*Beta vulgaris* ssp. *maritima*) quel potentiel èvolutif pour répondre au changement climatique globale. Université des Sciences et Technologies de Lille

Wagmann K, Hautekèete NC, Piquot Y, Van Dijk H (2010) Potential for evolutionary change in the seasonal timing of germination in sea beet (*Beta vulgaris* ssp. *maritima*) mediated by seed dormancy. Genetica 138:763–773

Chapter 6
Taxonomy, Phylogeny, and the Genepool

Lothar Frese and Brian Ford-Lloyd

Abstract The presence of conspecific and interfertile wild and cultivated forms impeded the development of a stable taxonomic system of wild beet species (genus *Beta* and *Patellifolia*) in the past. Further difficulties have been caused by the proliferation of synonyms and the confusion this has caused in the nomenclature, which was only slightly reduced after Linnaeus. The frequent errors in classification solely based on plant morphology are being reduced by studying the structures of genetic diversity with molecular genetic markers. Genetic markers as well as results from interspecific crossing experiments have helped revealing the past and ongoing evolutionary processes, as well as phylogenic relationships between taxa. Despite some unresolved taxonomic questions, wild beet plants can be determined and classified with sufficient reliability today which allowed categorizing the wild taxa into the primary, secondary, and tertiary genepool of the crop.

Keywords *Beta maritima* · Taxonomy · Genus · Section · Species · Fingerprinting · Phylogenetics · Crop genepool

6.1 Pre-Linnaean Systems

The first list of types of beet is attributed to Hippocrates and Theophrastus, and it distinguished the cultivated beets (white and black) from the wild (*Limonium*, *Blitum*). The list was confirmed by Pliny (who called the wild type *Beta sylvestris*), by Dioscorides, and by other authors (see Chap. 1). This nomenclature remained almost unchanged for nearly 1500 years, as has the division, introduced by Theophrastus, of the vegetable kingdom into trees, shrubs, bushes, and herbs, within which the cultivated and wild plants are classified.

In his treatise *"De Plantis"*, Cesalpino (1583) mentions the following types of beet: (i) *vulgaris* (with short and green leaves); (ii) *cum caudicantibus foliis* (with prostrate leaves); (iii) *rubra* (with red leaves and shallow roots); (iv) *radice buxea* (root resem-

L. Frese · B. Ford-Lloyd (✉)
60 Elm Grove, Bromsgrove, Worcs B61 0DX, UK
e-mail: b.ford-lloyd@bham.ac.uk

E. Biancardi et al. (eds.), *Beta maritima*,
https://doi.org/10.1007/978-3-030-28748-1_6

bling *Buxus sempervirens* L.). Cesalpino also cites the *Plantago* (*Plantago offici-nalis* L.) living in meadows and along roadsides, which is called "*Quinquinervia*" or "*Centinervia*" or "*Beta sylvestris*". The first two terms mean that the leaves bear five or more veins, respectively. The latter name is likely a mistake. In agreement with Gesner (1561), one of the first attempts of taxonomical classification was by Cordus (1551), who cited for *Beta* some unusual German names, such as Beisz-Izol, Romisch-Izol, Rograz and Mangolt. "There are two types of beets differently colored". Coles (1657) listed the following nine sorts of beets together with three sorts of "spinage" (*Spinacia oleracea*): (i) Common white; (ii) common red; (iii) common greene; (iv) Roman red; (v) Italian; (vi) prickly of Candy; (vii) sea; (viii) yellow; and (ix) flat stalked.

Cesalpino's reference to *Beta sylvestris* was corrected by de Tournefort (1700), who best described the flowering features of sea beet, considered by Cesalpino "*sine flore manifesto*" (apparently without flowers). Tournefort, in addition to *Beta sylvestris maritima* (or *Beta sylvestris spontanea marina*), lists the species of cultivated beets: (i) *Beta alba*; (ii) *Beta rubra vulgaris*; (iii) *Beta rubra radice rapae*; (iv) *Beta rubra lato caule*; (v) *Beta pallida virescens*; (vi) *Beta rubra mayor*; (vii) *Beta lutea mayor*; (viii) *Beta costa aurea*; (ix) *Beta foliis et caule flammeis*. In the Appendix of the book, other species are added: *Beta orientalis* and *Beta sylvestris* (also named *Cretica, maritima, foliis crispis*). The species are ranked under their respective "*genera*", an intermediate category between "*familiae*" and "*species*". The *genera*, including *Beta*, were chosen so well that a large proportion of them were adopted by Linnaeus (Jackson 1881). Consequently, the authority for genus *Beta* is also today the abbreviations Tourn. or Tournef. Beets were ranked in the *classis* XV: *De herbis et suffruticibus* (herbs and bushes), *sectio* I, *genus* II *Beta*. This classification was summarized by Valentini (1715) in Fig. 6.1, and was used by Tilli (1723) for the catalogue of the *Hortus Botanicus* of Pisa, Italy.

Ray (1693) divided beet into seven species: *Beta alba, Beta rubra, Beta sylvestris maritima* (*communis* or *viridis*), *Beta rubra radice rapae, Beta lutea mayor, Beta italica*, and *Beta cretica semine aculeato* (see Appendix for translation). He described the characteristics of each species, citing especially Bauhin (1623) and Parkinson. In "*Synopsis methodica stirpium Britannicarum*" (Ray 1690), the drawing of *Beta sylvestris maritima* is shown with the caption "sea-beet", which was used some years earlier by Coles. The description of *Beta cretica* is very detailed and original.

Morison (1715) classified the beets according to their uses and traits (Fig. 6.2). Note that the name of the species *Beta maritima spontanea comminis viridis* "*ad oram*" (until now) has been simplified as *Beta maritima* "*nobis*", that is, with the authority Morison himself. The proposed name "*Beta maritima*" by Morison was used by some later authors. "The Morison's copper plate engravings are very good, although small, but are cumbersome to quote because they are arranged in sections separately numbered, so that three numerals must be used to designate a particular figure" (Jackson 1881). Cupani (1696) mentioned all types of beets known at the time, including *Beta spontanea, Beta maritima, Beta communis, Beta viridis*, and *Beta sylvestris maritima*. The authors of the names of the species are Morison and Bauhin. Cupani also cited some Italian common names: "Gira di spiaggia", "Gira di

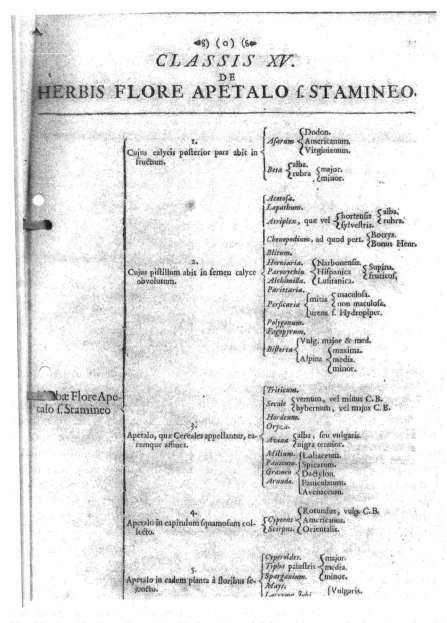

Fig. 6.1 Classification of species with flore "*apetalo*" i.e. without petals. *Beta* (see above) is included among the species "*cuius calycis posterior pars habit in fructus*" (the calyx takes part in the fruit) (Valentini 1715)

Fig. 6.2 Classification of *Beta* according to Morison (1715) see text

ripa di mari". The term "Gira" is not found in references to earlier Italian botanical authors.

In the treatise "*Prodromus theatri botanici*", Bauhin (1622) used, together with the old names "*Beta*" and "*Limonium sylvestris*", the common term "*pyrola*" mentioned by Fuchs. The words "*Beta maritima syl. spontanea*" were used on the posthumous edition of "*Stirpium illustrationes*" edited by Parkinson (1655), which detailed a second type called "*Beta maritima syl. minor*", similar to that above, but smaller in leaf and root development. In the second edition of the "*Pinax theatri botanici*", quoted by de Commerell (1778), Bauhin included "*Beta sylvestris maritima*" in the group *Minores* together with six cultivated species (*Beta alba, Beta communis, Beta rubra vulgaris, Beta rubra radice rapae*, and *Beta lato caule*). In the grouping *Majores*, he included *Beta pallida virens, Beta rubra*, and *Beta lutea* (see Appendix for translation).

Linnaeus, with a more rigorous scientific method, ordered the binomial nomenclature already widely used by botanists beginning with Mattioli, Tara, Bauhin, Pitton, de Tournefort, and so on (Greene 1909). Until the time of Linnaeus, the traditional division was into *herbae, suffrutices, arbores* (herbs, bushes, trees, etc.), or a ranking made according to their use (aromatic, medical, food, etc.). All details of the uses and properties were intentionally ignored by Linnaeus, thus simplifying the nomenclature. He also minimized the number of *genera* and *species* (Greene 1909) and simplified the names of the latter, which were becoming very long and complicated (Jackson 1881). This process of rationalization had already been adopted by Bauhin over a century earlier. Linnaeus also eliminated a large number of synonyms that confounded the precise identification of species.

In the first edition of "*Species plantarum*", Linnaeus (1753) divided the genus *Beta* into four species (*vulgaris, perennis, rubra*, and *cicla*) and eight varieties: Sea beet was named *Beta perennis* var. *sylvestris maritima*. For the remaining varieties, Linnaeus used names introduced by Bauhin (1623). The genus *Beta* was included in the *classis* V *Pentandria* and in the *ordo* II *Digyna*. The use of the term "*vulgaris*" (common) seems to go back to Cesalpino (1583). de Lobel (1576) resumed using the adjective only in the case of *Beta rubra vulgaris non turbinata*.

More complex is the origin of the term "*maritima*" (marine) used by Linnaeus, which was evidently derived from the locations preferred by the species, and partially replacing the various names and adjectives used in the past (Appendix E). The word "*maritima*" appeared in the book "*Pinax Theatri Botanici*" published by Bauhin (1623). In a list of various types of cultivated and wild beets, he used the names "*Beta syl. (sylvestris) maritima*" and "*Beta syl. (sylvestris) spontanea marina*" Lob. ob. Under the heading "*Beta*", de Lobel (1576) listed the forms known at the time, and named the wild plant "*Beta sylvestris spontanea marina*". Shortly after, he also pointed out that the plant grows in "*sabuleti maritimi*" (in sandy seashores). It is likely that Bauhin took this last adjective, which in Latin is equivalent to "*marina*" (marine), already used by Aldrovandi (1551) (cited by Baldacci et al. 1907) (Sect. 1.5). The lack of reference after the name "*Beta syl. maritima*" meant that Bauhin considered himself as the author. The name *Beta sylvestris maritima* followed by the initials of the author, abbreviation of the book, and the page (e.g., *C.B. Pin.* 118) was frequently used until Linnaeus (Dale 1730).

Beta maritima was considered as a separate species in the second edition of "*Species plantarum*" dated 1762. The *genus* was split into two species: the main distinction between *Beta vulgaris* (cultivated beets) and *Beta maritima* (wild or sea beets) was based on the behavior of the seed stalk: erect in *vulgaris* and "*decumbens*" (prostrate) in *maritima* (Fig. 3.18); however, in reality, the stalk is often erect even in *Beta maritima* (Fig. 3.17). The flowers described by Linnaeus "*solitariis aut binis*" (single or double) are actually composed of two or more flowers; only quite rarely they are single. *Beta vulgaris* differs from *maritima* through its biennial cycle "at least in Mediterranean areas" (Greene 1909).

In "*Systema naturae*" on page 276, Linnaeus (1735) split the cultivated species, *Beta vulgaris*, into the sub-species *vulgaris* and *cicla*: the first was grown for the root and the second for the leaves. "*Cicla*" was the ancient Latin name given to the leaf beets. According to Linnaeus, *Beta maritima* differed from *Beta vulgaris* due to the double flowers not being "*congestis*" (numerous) as they were in *vulgaris*, and because it flowered in the first year (annual) rather than in the second (biennial). Between the last two editions, Linnaeus proposed some different classifications for the genus *Beta* (Letschert 1993). Because "*Systema vegetabilium*" was the last book he edited, the latter classification (in which *Beta maritima* is a species and not sub-species) can be considered as definitive. Several authors disagree with this as the final classification. Letschert (1993) gives a detailed review of the taxonomic pre- and post-Linnaean treatment of the genus *Beta*.

6.2   Post-Linnaean and Current Classification

Willdenow (1707) was unwilling to follow the Linnaean system, and divided the genus *Beta* into four species: *Beta vulgaris, Beta patula, Beta cicla,* and *Beta maritima,* each having several sub-species. Stokes (1812) subdivided the genus *Beta* into *Beta esculenta, Beta alba, Greenleaved beet, Reddishleaved beet, Beta rubra, Root of scarcity, Beta rubra radice rapae,* and *Beta lutea mayor,* including about 50 sub-species, but *Beta maritima* was not mentioned. "Root of scarcity" was the initial English name given to the "Mangel Würzel" (fodder beet) based on the literal translation of the first German word (Fig. 6.3) (de Commerell 1778).

After Linnaeus, several modifications of the taxonomy were proposed, among others, by Desfontaines (1789), Kitaibl and Waldstein (1813) (cited by von Lippmann 1925), Hornemann (1813) (cited by von Proskowetz 1896), Marschall (1819), Roxburgh (1832), Mutel (1836), Boissier (1879), and Hohenacker (1838). Desfontaines listed three types of cultivated beets (*Beta vulgaris, Beta rubra vulgaris,* and *Beta rubra*) and two wild types (*Beta maritima* and *Beta sylvestris maritima*). Hornemann reported six species of wild and cultivated beets, and described *Beta maritima,* as having the following characteristics "*floribus geminis, foliis cordatis triangularibus attenuatis*" (double flowers, triangular or heart-shaped leaves). Marschall attributed to genus *Beta* the species *maritima, trigyna,* and *macrorhyza,* and explained that the first species flowered in November, developing inflorescences with 1–4 flowers, bearing *folia subcarnosa* (almost fleshy leaves), and favoring salty water. Mutel (1836) recognized only two species, commune (*Beta vulgaris*) and marine (*Beta maritima*). Lenz (1869) cited Bieberst, who had been referenced by Linnaeus for other sub-species, as author of the name *Beta maritima.*

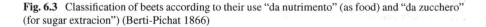

	1. B. campestre	B. *vulgaris* (1).	
	2. B. bianca della	B. *v. pallescens.*	
	3. B. gialla detta	B. *v. lutea.*	
Da nutrimento	4. B. gialla di Germania	B. *v. flava.*	
	5. B. gialla globosa	B. *v. lutea globosa.*	
	6. B. rossa campestre	B. *rubra romana.*	
	7. B. schiacciata	B. *v. compressa.*	

BARBABIETOLE

	8. B. di Slesia	B. *v. saccarifera.*	
	9. B. rossa grossa	B. *v. rubra major.*	
	10. B. di Magdeburg	B. *v. Magdeburg.*	
Da zucchero	11. B. imperiale	B. *imperialis.*	
	12. B. bianca a collare verde	B. *v. viridans.*	
	13. B. id. roseo	B. *v. subrubescens.*	
	14. B. Vilmorin	B. *v. Vilmorin.*	

Fig. 6.3 Classification of beets according to their use "da nutrimento" (as food) and "da zucchero" (for sugar extracion") (Berti-Pichat 1866)

Moquin-Tandon (1840) completely changed the classification of Linnaeus in *"Chenopodearum monographia enumeratio"* in which he brought together the family *Chenopodiaceae*, and split genus *Beta* into eight species (*trigyna*, *longispicata*, *macrorhiza*, *vulgaris*, *orientalis*, *procumbens*, *webbiana*, and *patula*). The species *vulgaris* comprised nine sub-species (*pilosa*, *maritima*, *macrocarpa*, *cicla*, *flavescens*, *purpurescens*, *alba*, *lutea*, and *rubra*), that is, all the cultivated beets and some of the wild beets. *Beta maritima* was described as a plant *"gracilis et glabra. In littoralibus Ocean et Medit."* (delicate and smooth leaved. It lives on Atlantic and Mediterranean shores.)

In *"Prodromus systematis naturalis regni vegetalibus"* published by de Candolle (1849), Moquin-Tandon proposed a new classification, in which the genus *Beta* was divided into ten species. The species *Beta vulgaris* was split into three groups: (i) Bette; (ii) Poirées; and (iii) Bette-raves. The first included the wild species, (α) *pilosa*; (β) *maritima*; (γ) *orientalis*; and (δ) *macrocarpa*. *Beta villosa* (with hairy or velvet leaves) probably corresponded to the above-mentioned *Beta cretica*. It should be noted that the species *Beta villosa* (*cretica*, *pilosa*, etc.), placed by various authors in the Greek islands, Egypt, Corsica, Sicily, and so on disappeared entirely in subsequent classifications. The genus *Beta*, in the Linnaean taxonomy, belonged to the family *"Salsolaceae"* authored Moq. or Moquin.

Another classification was developed by Bertoloni (1837). Although he maintained the Linnaean membership to class *Pentandria*—order *Digyna*, he split the genus *Beta* into three species: *cicla*, *macrocarpa*, and *maritima*. After he listed *maritima* as being found in Italy, he gave a very particular botanical characterization of the species; the flowers were described as "double, rarely triple, and seldom single at the apex of the branches". Berti-Pichat (1866) gave a unique classification of beets based on their two major uses (Fig. 6.4). Gandoger (1910) divided the species *maritima* into two sub-species, *agrigentina* and *atriplicifolia* (leaf similar to *Atriplex* species). The first, given the authority Gdgr. (Gandoger), was declared to be widespread near Agrigento (Sicily), and the second in Spain. The author did not provide any details on other distinctive traits. A somewhat confused classification of genus *Beta* partially taken from Linnaeus was given by Steudel (1871), in which he named *Beta maritima*, *Beta decumbens*, with the authority attributed to Moench without mentioning the written reference.

Some minor changes to classifications within genus *Beta* were made by Joseph Koch (1858), Karl Koch (1839), Ledebour (1846), Heldreich (1877), Boissier (1879), and Radde (1866). Bunge (cited by von Proskowetz 1896) listed under genus *Beta* (Tournef.) 14 species and their respective ranges. Kuntze (1891) divided *Beta maritima* into the following "forms": *macrocarpa*, *orientalis*, *brevibracteolata*, *trigynoides*. The last two were named by Kuntze himself, and were found at Funchal (Madeira) and Garachico (Tenerife), respectively. Gürke (1897) proposed another classification, in which genus *Beta* was divided into seven species; *Beta maritima* was included in the species *vulgaris* together with the sub-species *foliosa*, *pilosa*, *cicla*, and *esculenta*. Some synonyms of *Beta vulgaris maritima* were given, which included *marina*, *deccumbens*, *triflora*, *carnulosa*, *erecta*, and *noëana*. As was proposed by de Wildeman and Durand (1899), *Beta maritima* became the only species of

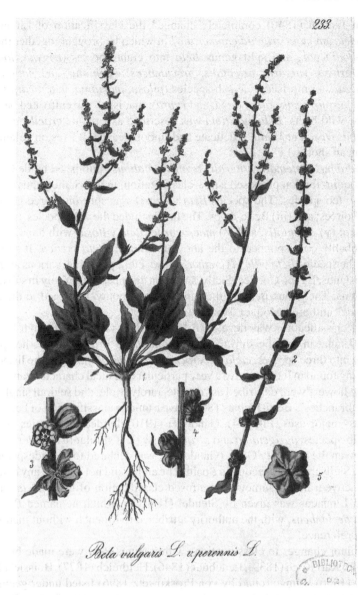

Beta vulgaris L. v. perennis L.

Fig. 6.4 Painting of *Beta perennis* (Reichenbach and Reichenbach 1909)

the sub-family *Betoideae* (family *Chenopodiaceae*) whereas *Beta vulgaris* contained all the cultivated beets.

As we discussed, in the taxonomic evolution of genus *Beta* post 1900, the abbreviations of authorities for genera, species, sub-species, varieties, and so on will be cited only if necessary. As example, the denominations (basionyms) of *Beta maritima* and the respective authors are given (www.tropicos.org).

According to de Vries (1905), "Some authors have distinguished specific types among the wild forms. While the cultivated beets are collected under the heading of *Beta vulgaris*, separate types with more or less woody roots have been described as *Beta maritima* or *Beta patula*". Reichenbach and Reichenbach (1909) classified *Beta maritima* as a "*perennis*" (perennial) variant of *Beta vulgaris* (Fig. 6.5). In this case, genus *Beta* was included in the sub-tribe *Betae*, in the tribe *Chenopodieae*, and in the family *Chenopodiaceae*. Ascherson and Graebner (1919) were quite confused when they subdivided genus (Gesamtart) *Beta* into two species, *Beta trigyna* and *Beta vulgaris*. The wild plants were named *Beta vulgaris perennis*, and under this heading different sub-species and variety synonyms of *Beta maritima* were listed: *perennis, marina, decimbens, triflora, noëana, annua, glabra, pilosa*, and so on, with the respective authorities.

Transchel (1927) divided the genus *Beta* into three undefined "groups": *Vulgares, Corollinae*, and *Patellares* (Coons 1954; de Bock 1986). Ulbrich (1934) called Transchel's groups "sections" and added a fourth section, *Nanae*. He changed the name *Patellares* to *Procumbentes*, a decision supported by Buttler (1977) later. This left genus *Beta* divided into four sections: I *Vulgares*, II *Corollinae*, III *Nanae*, and IV *Procumbentes*. The section *Vulgares* had the widest distribution and was believed to be the primordial species group of the genus *Beta* (Campbell 1984). Ulbrich (1934) divided section *Beta* into two species: *vulgaris* and *macrocarpa*. A cluster analysis based on RFLP DNA fingerprinting showed a higher similarity of *Beta macrocarpa* accession to section *Corollinae* than to section *Beta* accessions (Jung et al. 1993) which supports the proposal of Ulbrich (1934). He considered *Beta maritima* a variety of the species *vulgaris*, which belonged to the sub-species *perennis* along with six other varieties. The division into four sections remained essentially unchanged until recently (Table 6.1).

Coons (1954) adapted Ulbrich's classification, changing the name of section IV back to *Patellares*, and ordering the taxa into sections, species, sub-species, and varieties. As a result, the Latin name of sea beet became "*Beta vulgaris* subsp. *perennis* var. *maritima*". Many other minor changes have been made or proposed by, among others, Komarow (1936) (see Ford-Lloyd 2005), Zossimovitch (1934), Aellen (1938), Ernauld (1945), Helm (1957), Krassochkin (1959) (see de Bock 1986), Mansfeld (1959), Tutin et al. (1964), Davis (1937), Aellen (1967), Buttler (1977) (reviewed in Letschert 1993). We will briefly review the taxonomies developed by Zossimovitch (1934), Burenin and Garvrilynk (1982), and Ford-Lloyd et al. (1975). Zossimovitch subdivided the genus *Beta* into three groups according to their "ecogeographic isolation and the area": (i) eastern (*Beta lomatogona* including *Beta nana, Beta trigyna* (Fig. 6.5), and *Beta macrorhiza*); (ii) central (*Beta vulgaris* with the variety *annua*,

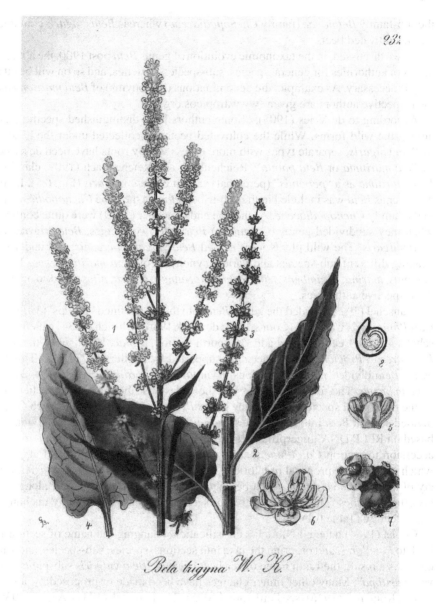

Fig. 6.5 Painting of *Beta trigyna*

Table 6.1 Taxonomy of the genus *Beta* according to Ulbrich (1934)

Genus	Section	Species
Beta	I *Vulgares*	*vulgaris*
		maritima
		macrocarpa
		patula
		atriplicfolia
	II *Corollinae*	*macrorhiza*
		trigyna
		foliosa
		lomatogona
	III *Nanae*	*nana*
	IV *Procumbentes*	*patellaris*
		procumbentes
		webbiana

patula, *macrocarpa*, and *maritima*); (iii) western (*Beta patellaris*, *Beta procumbens*, and *Beta webbiana*).

According to Ford-Lloyd et al. (1975), section *Vulgares*, which subsequently would become section *Beta* (Buttler 1977; Barocka 1985), included only the species *vulgaris*, which was divided into seven sub-species (Table 6.2). The species *maritima* was split into six varieties with the same names as used by Coons (1954). The classification within sections *Nanae* and *Procumbentes* remained the same. In section *Corollinae*, the species named *foliosa* was changed to "*corolliflora*", and was brought into the species *intermedia*. Krasochkin (1959) returned to the classification of *Beta maritima* as species, split into two subspecies: *mediterranea* and *danica*. The former was further subdivided into four varieties: (i) *prostrata*; (ii) *erecta*; (iii) *macrocarpa*; and (iv) *atriplicifolia*.

Another significant revision proposed by Ford-Lloyd and Hawkes (1986) was to divide *Beta* section *Beta* into four sub-species (i) *vulgaris* (including the cultivated beets except leaf beets); (ii) *cicla* (leaf beets); (iii) *maritima* (northern sea beet); (iv) *macrocarpa* (southern sea beets). The International Plant Genetic Resources Institute (IPGRI) in 1993 supported the taxonomy proposed by Ford-Lloyd and Hawkes (1986) with minor changes. *Beta vulgaris* subsp. *maritima*, *Beta prostrata* and *Beta erecta* were no longer considered as separate species. The species *Beta vulgaris* subsp. *vulgaris* was divided into three varieties: *conditiva*, *crassa*, and *altissima* (IPGRI, 1993). The cultivated species were included in *Beta vulgaris* subsp. *cicla* (Swiss chard or leaf beet) and in subsp. *vulgaris* (red beet, fodder beet, and sugar beet).

For the genus *Beta*, including the genus *Patellifolia*, 142 taxon names have been listed of which Hanelt and the Institute of Plant Genetics and Crop Plant Research (2001) only listed 25 names as the accepted taxa (http://mansfeld.ipk-gatersleben. de/apex/f?p=185:145:::NO::P3_BOTNAME:Beta). For *Beta maritima* the USDA-

Table 6.2 Taxonomy of the genus *Beta* according to Ford-Lloyd et al. (1975)

Genus	Section	Species	Sub-species	Variety
Beta	I *Vulgares*	*vulgaris*	*maritima*	*maritima*
				trojana
				macrocarpa
				atriplicifolia
				prostrata
				erecta
			orientalis	
			adanensis	
			cicla	*cicla*
				flavescens
			vulgaris	
			lomatogonoides	
			Patula	
	II *Corollinae*	*macrorhiza*		
		Trigyna		
		Foliosa		
		lomatogona		
	III *Nanae*	*Nana*		
	IV *Procumbentes*	*patellaris*		
		procumbens		
		webbiana		

ARS GRIN taxonomy site (www.usda-grin.gov) and Letschert (1993) listed 25 and 21 synonyms, respectively. Thus, the multiplication of the names of taxa included in the species *Beta vulgaris* continued. This process also involves *Beta maritima*, considered alternatively as species, sub-species, or variety as reviewed by Letschert (1993). According to Letschert et al. (1994), the difficulties of obtaining a satisfactory taxonomic treatment of the genus *Beta* were due, not only to the coexistence of wild and cultivated species, and the difficulties of getting representative samples of all taxa for research, but also to the different professional and cultural background of plant breeders and taxonomists trying to resolve the taxonomic problems.

Today, taxonomists can also rely on molecular marker technologies such as isoenzyme analysis (e.g., van Geyt et al. 1990), various kinds of DNA markers (e.g., Jung et al. 1993; Shen et al. 1996, 1998; Andrello et al. 2017), and comparative genomics (Dohm et al. 2013). The use of genetic markers allows the analysis of the heritable diversity underlying the morphological differentiation of the wild beet taxa. Information on the genetic relatedness has been used to improve the taxonomy of the genus *Beta* (Letschert 1993). However, the variability present within the species still creates problems, and every method applied on a limited number of samples or without a full understanding of the geographic distribution of the species may give varying results.

Based on the literature studies and the analysis of morphological, ecological, and molecular traits, Letschert proposed a revision of the section *Beta* (Letschert 1993; Letschert et al. 1994). According to this revision, the section *Beta* consisted of three species, *vulgaris* (with the sub-species *vulgaris*, *adanensis*, and *maritima*), *macrocarpa*, and *patula*. The Italian scientist Giovanni Arcangeli had also divided the species *vulgaris* into the subsp. *vulgaris* and subsp. *maritima*, as published in the *"Compendium florae italicae"* (Arcangeli 1882). In addition to the usual Linnaean authority for the species (L.), the authority of Arcangeli (Arcang.) was added, resulting in the officially accepted taxon name *Beta vulgaris* (L.) subsp. *maritima* (Arcang.).

The taxonomy proposed by Letschert (1993) for the section *Beta* (syn. *Vulgares*) and for section *Procumbentes* (syn. *Patellares*) (Buttler 1977) seems to have been confirmed by RFLP analysis (Mita et al. 1991). Jung et al. (1993) found a low degree of homology (34%) between the sugar beet and *Patellifolia procumbens* (syn. *Beta procumbens*) after cross-hybridization of sugar beet RFLP probes with *Patellifolia procumbens* probes. In addition, a spinach (*Spinacia olearcea*) sample clustered together with *Patellifolia procumbens* and *Patellifolia webbiana* accession. Scott et al. (1977) suggested ranking section *Procumbentes* as a genus named *Patellifolia*. The cross-hybridization experiment and the high genetic similarity between *Spinacia* and *Patellifolia* accessions further substantiated the proposal.

From Ulbrich (1934) until just a few decades ago, the genus *Beta* was included in the family Chenopodiaceae (Cronquist 1988). Most recently, Kadereit et al. (2006) have suggested the re-introduction of the sub-family Betoideae (excluding Acroglochin), proposed first by Ulbrich (1934), because it resolved as a monophyletic group in molecular analysis (Hohmann et al. 2006) and is morphologically distinct from other sub-families of the Chenopodiaceae/Amaranthaceae alliance. Kadereit et al. (2006) corroborated that the section *Procumbentes* (Ulbrich 1934) should be given the rank of a genus and to keep section *Beta* and *Corollinae* within the genus *Beta* (Table 4.3). This classification also has been supported by the analysis of nuclear ribosomal DNA (Santoni and Bervillè 1992) (Table 6.3).

Following Scott et al. (1977), the genus *Patellifolia* consists of three species, namely *Patellifolia patellaris* (Moq.) A. J. Scott, Ford-Lloyd, and J. T. Williams (syn. *Beta patellaris* Moq.), *Patellifolia procumbens* (C. Sm.) A. J. Scott, Ford-Lloyd, and J. T. Williams (syn. *Beta procumbens* C. Sm.), and *Patellifolia webbiana* (Moq.) A. J. Scott, Ford-Lloyd, and J. T. Williams (syn. *Beta webbiana* Moq.). However, the difficulties in distinguishing the species led some authors to refer an uncertain number of species—two or three (e.g., Wagner et al. 1989; Hohmann et al. 2006; Kadereit et al. 2006) or even only one (Santoni and Bervillè 1992; Thulin et al. 2010). The proposal of Thulin et al. (2010) is based on the analysis of ITS regions of five specimens, namely *Patellifolia procumbens* and *Patellifolia patellaris* from Gran Canaria, *Patellifolia webbiana* and *Patellifolia patellaris* from Tenerife as well as one specimen from *Tetragonia pentrandra* from Socotra (Yemen). While the taxonomic debate continues, the taxonomic system of Scott et al. (1977), which is commonly

Table 6.3 Comparison between the taxonomies of the genus *Beta* proposed by Ford-Lloyd (2005) left; and Kadereit et al. (2006) right

Ford-Lloyd (2005)	Kadereit et al. (2006)
Beta Section *Beta*	*Beta* Section *Beta*
Beta vulgaris L.	*Beta vulgaris* L.
Beta. vulgaris L. subsp. *vulgaris* cultivated form	
Beta vulgaris L. subsp. *maritima* (L.) Arcang.	*Beta vulgaris* L. subsp. *maritima* (L.) Arcang.
Beta vulgaris L. subsp. *adanensis*	*Beta vulgaris* L. subsp. *adanensis*
Beta macrocarpa	*Beta macrocarpa*
Beta patula	
Beta Section *Corollinae*	*Beta* Section *Corollinae*
Beta corolliflora	*Beta corolliflora*
Beta lomatogona	*Beta lomatogona*
Beta intermedia	*Beta trigyna*
Beta trigyna	*Beta nana*
Beta Section *Nanae*	
Beta nana	
Beta Section *Procumbentes*	
Beta procumbens	
Beta patellaris	
Beta webbiana	

used in plant breeding and also applied in threat assessment studies (Bilz et al. 2011), should be used for pragmatic reasons (Frese et al. 2018).

The sections *Beta*, *Corollinae*, and *Nanae* have been differentiated by restriction analyses of the chloroplast DNA (Komarnitsky et al. 1990). Nevertheless, Kadereit et al. (2006) also suggested the elimination of the section *Nanae*, and incorporated *Beta nana* (the lone species in that section) into section *Corollinae*.

Section *Beta* consists of wild and cultivated taxa. The cultivated taxa share a common ancestor with *Beta maritima* as indicated by the RFLP DNA fingerprinting results presented by Jung et al. (1993). Accessions belonging to the genus *Beta* were analyzed using DNA fingerprinting. The results confirmed the taxonomy accepted at this time with the exception that there was too narrow differentiation of *Beta atriplicifolia* and *Beta orientalis* to consider them as distinct species. A high level of similarity was found between Atlantic sea beet populations and cultivated varieties, whereas sugar and leaf beets were widely diverged. Jung et al. (1993) concluded that the hypothesis (Fischer 1989) that sugar beet was derived from an unintentional cross between fodder and leaf beet is unlikely. It also seems probable that there has been more recent gene flow between sugar beet and *Beta maritima* than earlier suspected. Recent research with sequence variations in the ITS1 region of nuclear ribosomal DNA and the molecular structure of the *matK* chloroplast gene has proven useful for phylogenetic discrimination among species within *Beta* (Mglinets 2008; Shen et al. 1998). Ford-Lloyd (2005) revised and updated the current taxonomy taking into account the new research findings. The section *Beta* was modified, as suggested by Lange et al. (1999) and has been accepted for use by the International Database of *Beta* (Germeier and Frese 2004). Compared to the past two centuries of systematic taxonomic investigations, today researchers have access to the full range of genotypes (populations) of the various wild beet taxa and can apply highly sophisticated molecular genetic methods to study the past and ongoing speciation processes generating the interspecific and intraspecific diversity we are trying to classify today. The taxonomy of wild beets will likely be revised again in future when outstanding issues will be resolved. Should *Beta vulgaris* subsp. *adanensis* be treated as species or sub-species? Is the tetraploid form of *Beta macrocarpa* a separate species or a sub-species of *Beta macrocarpa*. Does it make sense to keep *Beta intermedia* and *Beta trigyna* as separate species despite these forms belonging to a highly variable polyploid hybrid complex? Is the genus *Patellifolia* a single, but highly variable species? These are some of the questions still pending (Buttler 1977; Villain 2007; Thulin et al. 2010).

There is often disagreement among taxonomists especially with regard to the classification of the cultivated types (Letschert et al. 1994). Two different approaches have been applied to create a taxonomic system for cultivated beets. In Eastern Europe the cultivated types were given varietas and forma names. Experts familiar with this system immediately know what the material looks like as the name circumscribes specific colors, root, or petiole shapes and further traits of interest to breeders, growers, and consumers. The disadvantage of this so-called "splitter system" is that it does not always allow a clear classification of all types since the trait variation of an outbreeding crop is complex and hampers the unambiguous delineation of forms based on classical taxonomic traits. For this reason Lange et al. (1999) proposed to classify cultivated material according to the use. The resulting "lumper system" is composed of four cultivar groups only and does not require intimate knowledge of the meaning of varietas and forma names, compared to the splitter system names that only provide information on the use type. Therefore, a lumper system should be integrated into an information system that is able to document and provide trait data, that is, trait scores or trait measurements.

Cultivated beets are classified into four cultivar groups or "*culta*" based on their use. The Leaf Beet Group is composed of two types: (i) spinach beet which produces leaves similar to spinach and is used in the same way, and (ii) Swiss chard with developed, white (or colored), and tender petioles and midribs. The unselected root shape has remained similar to that of sea beet.

The Garden Beet Group has a round root more or less flattened, and the skin and flesh often show a deep red color. The beet may also be white or of varying shades and intensities of yellow to orange. The beet (crown, hypocotyl, and taproot) is primarily an enlarged hypocotyl, making up about 85% of the weight. The leaves can be dark-green or red-purple as well. The root contains little fiber and, if harvested at the appropriate time lends itself to be eaten raw or cooked.

The Fodder Beet Group is of any color, shape, and proportion of hypocotyl: taproot. It was developed for easy manual removal from the soil and winter storage. Its high total digestible nutrients make it suitable for feeding all classes of livestock. Beets are very large and can protrude almost completely from the ground.

The Sugar Beet Group has been selected for sucrose production. The roots are ivory white, and cone-shaped, more or less elongated (Fig. 4.1). Root and leaves have uniform characteristics, so that they are not used to distinguish among commercial varieties. The crown protruding from the soil of the taproot is limited. More information regarding cultivated beets is given in Chap. 9.

In 1995, after the publication of the "International Code of Nomenclature for Cultivated Plants" (Trehane et al. 1995), the taxonomy of the section *Beta* was slightly revised (Lange et al. 1999). The changes concerned the species *Beta vulgaris*, which was divided into subsp. *maritima*, subsp. *adanensis*, and subsp. *vulgaris*. *Beta vulgaris* subsp. *vulgaris* was changed to incorporate all beets, including the weedy and wild (feral) beets, which were derived in any way from the cultivated beet crops (Ford-Lloyd 2005). The names indicating the four cultivated groups (*culta*) were slightly modified (Lange et al. 1999). This new approach was endorsed by the World *Beta* Network (WBN), which recommended its use (Frese 2003).

6.3 Phylogeny

Today's distribution area of wild beet species (genus *Patellifolia* and *Beta*) can be divided into two regions of differing phylogenetic significance: the Macaronesian archipelago and the east Mediterranean region. Burenin and Garvrilynk (1982) and Zosimovich (1968, cit. in Burenin and Garvrilynk (1982)) assumed that the genus *Beta* as well as the genus *Patellifolia* (syn. *Beta* section *Procumbentes*) evolved from a hypothetical ancestral form called "Protobeta". Protobeta occurred in the region of the Tethys, an ancient ocean extending from the Caribbean, and the Mediterranean basin to the western shore of Indonesia during the tertiary period 25 million years ago. During the Miocene epoch (late tertiary period), the homogeneity of the Tethys faunas was abruptly disturbed (Hallam 1972). The causes of this phenomenon must have also affected the flora of the Tethys region. It can be assumed that the change from the tropical–subtropical climate of the Miocene and Pliocene epoch to the cool climate of the Pleistocene had a strong impact on the fauna and flora. Along with geological changes induced by the continental plate drift, climate change was also likely to have been an important driver of evolution taking place in the Mediterranean region which is known today as the center of genetic diversity of the genus *Beta* and *Patellifolia*. The climate change was accompanied by the southwards expansion of glaciers in the northern hemisphere. The northwards progression of the deserts in the current Sahara region and the decreasing temperature in the northern hemisphere caused the extinction of many species from the Tethys. Geological and oceanological studies suggest that the continental drift split off the islands Lanzarote and Fuerteventura from North Africa. Thereby parts of the Tethys flora, including prototypes of the genus *Patellifolia*, were on the one hand saved from extinction and, on the other hand, evolved toward geographically isolated relict species (Bramwell and Bramwell 1974) which colonized later sites in South Spain and West Portugal.

The northern movement of the African plate closed not only the northern flank of the Tethys Ocean but also compressed the Eurasian plate leading to the rise of the Alpine orogenetic belt. The orogenetic activities reached a maximum 20 mya, gradually reducing for the past 5 million years and ending with the extinction of volcanic activities during the past million years. During the Alpine orogeny, mountains in Greece, Turkey, and the Caucasus region were formed. The alternating cold and warm periods during the Pleistocene shaped the mountainous areas as we know them today. The last cold period ended only 10,000 years ago.

The progenitors of the section *Corollinae* and section *Beta* presumably shared a number of similar plant traits and formed the evolutionary basis of two phylogenetic lines within the genus *Beta* (Buttler 1977). During the alpidic orogenesis, species of section *Corollinae* may have evolved through adaptation to the harsh environmental conditions of mountains in Greece as well as in the Taurus and Caucasus region. Buttler (1977) realized that the ancestral form of the section *Corollinae* should have had a similar ecological potential comparable to the current species. If only the strong difference in frost tolerance between section *Beta* and *Corollinae* species are taken into account, the existence of two progenitors with very different adaptive potential within the genus *Beta* seems likely. Buttler (1977) regarded *Beta macrorhiza* and *Beta corolliflora* as the primordial taxa with distribution areas in the oriental-turanian region and the east Mediterranean region as well. Both species represent the first phylogenetic line within the section *Corollinae*. *Beta lomatogona* is completely restricted to the oriental-turanian region, specifically adapted to arid habitats, morphologically clearly distinct from *Beta macrorhiza* and *Beta corolliflora* and could be considered the second phylogenetic line. The phylogenetic position of *Beta nana*, a highly specialized species of snow patch vegetation of alpine regions in Greece, has only recently been investigated due to the lack of material.

The progenitor of the section *Beta* occurred in coastal areas of the east Mediterranean basin, spread westwards and finally northwards along the Atlantic coasts when the glaciers withdrew at the end of the last cold period. Populations from the Atlantic part of the distribution area most likely constitute evolutionarily the youngest component of *Beta maritima* (Boughey 1981; Villain 2007).

The hypotheses of Buttler (1977), Burenin and Garvrilynk (1982), Boughey (1981), and Bramwell and Bramwell (1974) based on geology, paleobotany, and geobotany agree with the results of studies using molecular markers. Santoni and Berville (1992) constructed rDNA physical maps of *Beta* section *Beta* and section *Corollinae* as well as *Patellifolia* species. The diversity of restriction sites was higher in sections *Beta* and *Corollinae* when compared to *Patellifolia*. The simple intergenic spacer of *Patellifolia* is likely to be evolutionarily older than the sequences of the *Beta* species. Romeiras et al. (2016) analyzed ITS, matK, trnH-psbA, trnL intron, and rbcL gene sequences to reveal the relationships within the Betoidae sub-family and constructed a molecular clock-dated phylogenetic tree. The genera *Patellifolia* and *Beta* diverged around 25.3 mya (range: 35.9–16.1 mya) which agrees well with the conclusions of Burenin and Garvrilynk (1982) and Bramwell and Bramwell (1974) reached from geological and paleological knowledge. Sections *Corollinae* and *Beta* started to diverge around 7.2 mya (range: 11.5–3.5) which falls into the period of the alpinic orogeny. An Atlantic group (*Beta maritima*, *Beta macrocarpa*) arose 1.4 mya (range: 2.1–0.7) and separated into distinct *Beta maritima* and *Beta macrocarpa* around 0.9 mya (range: 1.9–0.2 mya). The latter taxon is the youngest among the investigated material with an average divergence time of 0.3 mya (range: 0.7–0.1 mya).

The genetic diversity of *Beta maritima* is high compared to *Beta vulgaris* subsp. adanensis, *Beta macrocarpa* (2x and 4x) and *Beta patula* (Letschert 1993; Villain 2007). Letschert (1993) found significantly higher levels of genetic diversity in south eastern and middle Mediterranean populations compared to the Atlantic material. Leys et al. (2014) investigated the spatial distribution of genetic diversity of *Beta maritima* distributed from the Bay of Biscay to the south of Morocco and found a much higher genetic diversity in Morocco as compared to the material sampled north of the Straits of Gibraltar. *Beta maritima* from the Gibraltar region exhibited a particularly high number of private alleles. The results corroborate with Villain (2007), who explained the current spatial distribution patterns of genetic diversity by the existence of three glacial refugia where the species survived cold periods during the Pleistocene. Refugium 1 was located in the northwestern part of the Iberian Peninsula from where sites north of Portugal were recolonized at the end of the Würm glaciations. Refugium 2 existed in Morocco and refugium 3 in the middle to eastern Mediterranean region. Leys et al. (2014) supported this hypothesis and described the Gibraltar region as a refugium and historical buffer zone maintaining genetic diversity of *Beta maritima* without which range expansion after cold periods would not have been possible. The present-day geographic pattern of genetic diversity is the result of a complex microevolutionary process that is not yet fully understood (Villain 2007; Leys et al. 2014; Touzet et al. 2018).

Villain (2007) used polymorphic chloroplastic and nuclear genetic markers to analyze the genetic relationship between *Beta maritima* and *Beta macrocarpa* as well as *Beta maritima* and *Beta vulgaris* subsp. *adanensis*. Within the haplotype network constructed from chloroplastic genetic marker data, *Beta macrocarpa* samples formed a clearly distinct group. Interestingly, a mutation named HK 550 located on the fragment of trnH-psbA distinguishes *Beta vulgaris* from diploid *Beta macrocarpa* and can be used as diagnostic feature. A major haplotype of *Beta macrocarpa* is distributed from the eastern Mediterranean distribution area to the Canary Islands. Villain (2007) concluded from the results that the diploid *Beta macrocarpa* evolved from *Beta maritima* and, in terms of the microevolutionary timescale, spread rapidly. The phylogenic trees constructed from nuclear genetic markers also showed a clear phylogenetic separation between *Beta maritima* and *Beta macrocarpa*. The tetraploid form of *Beta macrocarpa* first detected by Buttler (1977) on the Canary Islands evidently is an allotetraploid species. Lange and de Bock (1989) investigated tetraploid *Beta macrocarpa* and observed a regular diploidized meiosis which is typical for alloploid species. Villain (2007) found in *Beta maritima* and tetraploid *Beta macrocarpa* the same chloroplastic haplotype and concluded that the tetraploid form developed from crosses between *Beta maritima* x *Beta macrocarpa*. All of the diploid *Beta macrocarpa* samples from the Canary Islands proved to be invariable at nine SSR loci while all tetraploid forms, with very few exceptions, showed variation at all of the nine loci. It seems therefore that the time span from the first colonization of the Canary Islands by the diploid *Beta macrocarpa* till today was too short for mutations to have accumulated. The diploid *Beta macrocarpa* on the Canary Islands likely forms the youngest phylogenetic branch of the species and the allotetraploid *Beta macrocarpa* the preliminary end point of speciation within section *Beta*. Self-fertilization is the main mode of reproduction of the diploid and tetraploid *Beta macrocarpa*. Although diploid *Beta macrocarpa* can occur at the same locations, the chance for crosses between *Beta maritima* and *Beta macrocarpa* is low due to the large difference in flowering time. Temporal isolation is a strong reproductive barrier between both taxa and maintains the differences between both species.

The phylogenetic position of *Beta patula* is not yet fully understood. The species is morphologically clearly distinct from all other taxa of the section *Beta*. The substantial morphological divergence and the extremely low allozyme variation was proposed by Letschert (1993) as the signature of an isolated *Beta maritima* population located at the edge of the species' distribution range which adapted rapidly to the environmental conditions of the archipelago of Madeira. As with the diploid *Beta macrocarpa* from the Canary Islands (Villain 2007), the time between the colonization and the present time was probably too short to allow for allozyme polymorphisms to occur (Letschert 1993). This is not the case for the 25 SSR markers applied to study the genetic diversity of *Beta patula*. On average 4.5 alleles per marker locus were observed and only three out of the 25 markers were monomorphic. Compared to the mutation rate for protein coding loci such as allozymes, the mutation rate at SSR loci is higher (Allendorf and Luikart 2007) which may explain the seemingly contradictory results. The principal component analysis grouped *Beta maritima* individuals with those of *Beta patula* from Ilheu Chaos (Madeira) indicating a close genetic relationship between both species (Frese et al. 2012). The results agree with those of Letschert (1993) who calculated the genetic distance between all taxa of the section *Beta* from allozyme data. *Beta patula* was included in the cluster of *Beta vulgaris* sensu lato while diploid *Beta macrocarpa* accessions formed a clearly distinct cluster (Letschert 1993). Andrello et al. (2017) applied 9724 SNP markers to analyze the genetic diversity within a collection of 1512 individuals taken from 1080 Genebank accessions of section *Beta* (cultivated and wild taxa), that is, one to few individuals per accession. The discriminant analysis of principal components (Jombart et al. 2010) was used to identify genetically similar individuals by the k-means algorithm of the ex nihilo cluster method. The single individual of *Beta patula* was assigned to a cluster consisting almost exclusively of *Beta macrocarpa* individuals. In view of the high number of SNP markers used, this finding cannot be interpreted as a random effect caused by the small sample size. It rather shows that the different marker systems detect different kinds of diversity.

In comparison with *Beta macrocarpa*, *Beta vulgaris* subsp. *adanensis* is more difficult to distinguish from *Beta maritima* based on morphological traits. The results of allozyme, nuclear and chloroplastic genetic markers indicate a weak differentiation between *Beta maritima* and *Beta vulgaris* subsp. *adanensis*. The main distribution area of *Beta vulgaris* subsp. *adanensis* encompasses the Aegean Islands, Cyprus and adjacent Turkish coastlines where the two taxa can be found in close proximity. Villain (2007) suggested two hypotheses to explain the lower level of genetic diversity of *Beta vulgaris* subsp. *adanensis* compared to *Beta maritima* and the low level of genetic differentiation between the taxa. The current structure of the *Beta maritima/Beta vulgaris* subsp. *adanensis* complex could be explained as a recent and ongoing speciation process with *Beta maritima* as the progenitor of *Beta vulgaris* subsp. *adanensis*. According to the second hypothesis, the speciation process started from *Beta maritima* populations which survived in isolated East Mediterranean refugia during the past glacial period. However, the microevolutionary time span was not long enough to generate strong reproductive barriers between the autogamous *Beta vulgaris* subsp. *adanensis* and its allogamous ancestor *Beta maritima*. Today, gene flow between the taxa prevents the completion of the speciation process and emergence of a reproductively isolated new species "adanensis".

The ancestor of all cultivated beets, *Beta maritima*, is known to be a diploid species with 2n = 18 chromosomes. Within the section *Beta*, section *Corollinae* and the genus *Patellifolia* ploidy complexes exist. Interspecific hybridization generates novel genetic diversity and can increase the adaptive potential (Castro et al. 2017). The existence of mixed ploidy populations at sites where species grow sympatrically indicates ongoing speciation processes (Castro et al. 2018). Such sites and populations are evolutionary hot spots and should be given high priority by plant genetic resources conservation programs (Frese et al. 2018).

6.4 The Genepool

Crop wild relatives are those genetically related species that can be used to broaden the genetic base of the crop. The phylogenetic distance between the potential donor and the crop species determines the amount of investment into breeding research required to introgress a target trait into the crop's breeding pool and to develop improved varieties. To indicate the degree of relatedness Harlan and de Wet (1971) suggested categorizing species into primary, secondary and tertiary genepools. The primary genepool is composed of all forms of the cultivated species (GP-1A) and closely related wild species (GP-1B). Among the species of GP-1 crossing is easy, the hybrids are fertile and gene segregation is approximately normal. Species of the secondary genepool (GP-2) can be crossed but introgression of a trait into the crop is more difficult due to limited seed set, only partially fertile hybrids, insufficient chromosome pairing or other phenomena. Seed set after crosses between the crop and species of the tertiary genepool (GP-3) is more or less possible but seedlings may show a range of abnormalities for instance lack of root formation.

Researchers have been interested in utilizing the genetic resources of wild species since the emergence of sugar beet breeding programs at the end of the nineteenth century and the beginning of the twentieth century (see Chap. 1). As compared to other wild species of the genus *Beta* and *Patellifolia*, an impressive number of *Beta maritima* population has been sampled in the wild and preserved in Genebanks (Frese 2010). From the plant breeder's perspective, *Beta maritima* is the most valuable species as it not only contains several important resistance genes but can be easily crossed with the sugar beet.

Experimental crosses between species within the genus *Beta* (including *Patellifolia* species) have been conducted for two reasons: (i) to enlarge or replenish the sugar beet breeding pool and (ii) to investigate the genetic and evolutionary relationships between species. Interspecific hybrids have played a key role for enhancing the yield and yield stability of the sugar beet crop. Indeed, wild germplasm has been (and is) used in breeding programs to improve the genetic resistance to sugar beet diseases and pests important for economically and ecologically sound sugar beet production worldwide (see Chap. 8).

Crossing experiments along with cytological studies have been the only way to study the phylogeny of wild beet species until the development of molecular marker technologies. Today, the results of cytological studies can be combined with the results from genetic marker-based phylogenetic studies to describe the position of wild beet species in the crop genepool with greater precision. An enormous amount of detailed information on crossing experiments has been published since then and can be applied to categorize wild beet species according to the genepool concept. A division of wild beet species into the primary, secondary, and tertiary genepool has been suggested by Frese (2010).

Abe and Shimamoto (1989) found no consistent reproductive barriers between *Beta vulgaris* subsp. *vulgaris* and *Beta maritima*. A high percentage of male sterile plants in the backcross generation of *Beta macrocarpa* (A1171) crossed with sugar beet observed by Oldemeyer (1957) is likely the first documented evidence of reproductive barriers between *Beta vulgaris* subsp. *vulgaris* and *Beta macrocarpa*. Abe and Shimamoto (1989) observed pollen and seed abortion in F1 hybrids between *Beta macrocarpa* and *Beta patula*, respectively, with *Beta maritima* and *Beta vulgaris* subsp. *vulgaris* and noted chlorotic plants, dwarf plants, complete male sterile plants and semi-fertile plants in the F2. In addition, significant deviation from the expected F2 segregation ratios of three isozyme loci was observed in the F2 of *Beta macrocarpa* x *Beta vulgaris* subsp. *vulgaris* (Abe et al. 1984) and the reciprocal cross (Abe and Tsuda 1986). By backcrossing the material, Abe and Tsuda (1987) even produced B2F1-plants. Weak reproductive barriers clearly exist but species crosses and backcrosses within section *Beta* are straightforward. Placing *Beta macrocarpa* and *Beta patula* into the GP1 B is therefore justified.

Section *Corollinae* includes *Beta corolliflora*, *Beta macrorhiza*, and *Beta lomatogona*, which Buttler (1977) considers as the base species of derived kinship groups. *Beta nana* and the three base species are sexually isolated and morphologically clearly distinct, populate differing habitats and distribution areas, and therefore match the main criteria for true species (Buttler 1977; Phitos et al. 1995). Filutowicz and Dalke (1976) and Cleij et al. (1976) first introduced genome formulae to designate the genome components in kinship groups and progenies derived from interspecific crosses (see Table 6.4).

Table 6.4 Genome formula of wild beet species according to Cleij et al. (1976), Filutowicz and Dalke (1976), and Dechyeva and Schmidt (2009). *Beta* section *Beta* and section *Corollinae* as well as the genus *Patellifolia* form polyploidy complexes. Not all possible combinations are presented here

Species	Genome formula
Beta vulgaris subsp. *vulgaris* (varieties)	VV, VVV, VVVV
Beta vulgaris subsp. *maritima, Beta vulgaris* subsp. *adanensis, Beta patula*	VV
Beta macrocarpa	VV, VVVV
Beta corolliflora	CCCC
Beta macrorhiza	MM
Beta lomatogona	LL
Beta trigyna	LLCCCC
Beta intermedia	?
Beta nana	NN[a]
Patellifolia procumbens	PP[a]
Patellifolia webbiana	PP[a]
Patellifolia patellaris	PP??

[a]Suggested by the authors of this book chapter

Very comprehensive cytological studies were conducted mainly in Poland and in The Netherlands to investigate the phylogeny of *Corollinae* species and to find ways to tap the secondary genepool for sugar beet breeding. These studies have greatly contributed to our understanding of the relationships between *Corollinae* species and the ploidy complex existing within section *Corollinae*. Dalke (1977) crossed sugar beet (2 = 36) with *Beta corolliflora* and backcrossed the F1 with sugar beet to introgress mosaic virus resistance from the wild species. Of 1570 B2 to B4 plants, 517 virus-resistant plants were selected with chromosome numbers ranging between 18 and 40. Three diploid-resistant B3 plants probably contained only a small wild species chromosome fragment as indicated by the regular course of the meiosis. Vasilchenko and Zhuzhzhalova (2011) paired *Beta vulgaris x Beta corolliflora* and produced triploid hybrid plants. Selection in the progenies gave plants with conical roots, cytoplasmic male sterility, and a high percentage of monogerm plants. The introgression of *Beta corolliflora* genes into the sugar beet breeding lines was evidenced with a *Beta corolliflora*-specific genetic marker. The chromosomes of *Beta corolliflora* can be distinguished clearly from *Beta vulgaris* chromosomes with the genomic in situ hybridization technique (Desel et al. 2002). The authors noticed a weak hybridization of *Beta vulgaris* chromosomes with *Beta corolliflora* DNA and interpreted this observation as hybridization between repetitive sequences that are conserved in both species.

Szota (1995) crossed sugar beet with *Beta macrorhiza, Beta lomatogona*, and *Beta corolliflora* and noticed strong disturbance in meiosis likely caused by a lack of chromosome homology between *Beta vulgaris* subsp. *vulgaris* (sugar beet group) and

Corollinae species. Interspecific hybrids with *Beta corolliflora* or *Beta lomatogona* as male parent were generally partial to fully sterile. Cleij et al. (1976) produced backcross progenies ((sugar beet 2x x *Beta lomatogna* 4x) x sugar beet 2x) and identified plants with amphidiploid genome composition (VVLL). Furthermore, amphihaploid, male sterile offspring of VVLL plants showed some bivalent pairing in meiosis, indicating partial homology between *Beta vulgaris* and *Beta lomatogona* chromosomes. Fertile hybrids were obtained from sugar beet x *Beta macrorhiza* crosses. These hybrids showed regular meiosis and their chromosomes paired in bivalents. Szota (1995) concluded from the experimental results that *Beta macrorhiza* may be phylogenetically closer to *Beta vulgaris* than *Beta corolliflora* and *Beta lomatogona*. The experimental results evidence that introgression of a trait from the base species of section *Corollinae* into the crop species is difficult due to limited seed set, only partially fertile hybrids, insufficient chromosome pairing or other disorders, but it is not impossible. *Beta corolliflora*, *Beta macrorhiza*, and *Beta lomatogona* thus clearly match the criteria of species belonging to the secondary genepool. *Beta nana*, however, has never been used in crossing experiment. Barocka (1985) suggested a close relationship between *Corollinae* species and *Beta nana*. Phylogenetic studies using plastid DNA substantiated this assumption (Fritzsche et al. 1987). Gao et al. (2000) found high homology between the *Beta corolliflora*-specific satellite sequences pBC1279 and pBC1944 with *Beta nana*-specific repeat sequence pRN1 (Kubis et al. 1997). They assumed that the three sequences originated from a common ancestor. Kadereit et al. (2006) finally decided to place *Beta nana* into section *Corollinae* as mentioned earlier in this chapter. Since the criteria of Harlan and de Wet (1971) cannot be applied to the close phylogenetic relationship between *Corollinae* species and *Beta nana* justifies placing this species in the secondary genepool, too.

Two additional accepted taxa of *Corollinae* exist: *Beta trigyna* and *Beta* x *intermedia* (see the USDA-ARS GRIN taxonomy site, www.usda-grin.gov) which Buttler (1977) considers the hybrid species within section *Corollinae*. *Beta trigyna* includes the autotetraploid parental species *Beta corolliflora* and the diploid *Beta lomatogona*. The first indications on the autotetraploid nature of *Beta corolliflora* were given by Ramos-Büttner and Wricke (1993) and this assumption was further substantiated by Gao et al. (2000). Paesold et al. (2012) found two pairs of chromosome 1 in a sample of *Beta corolliflora*, which is further evidence of the autotetraploid nature of the species. Filutowicz and Dalke (1976) concluded from their crossing experiments that *Beta trigyna* (2n = 54, genome formula LLCCCC) stems from crosses between *Beta lomatogona* (2n = 18; genome formula LL) x *Beta corolliflora* (2n = 36, genome formula CCCC). Arapova (1987) crossed diploid *Beta vulgaris* with hexaploid *Beta trigyna*. After pollination of *Beta vulgaris* with *Beta trigyna* only a single hybrid plant was obtained (0.7% of all harvested seeds) while no hybrid plant was found in the reciprocal crosses. Cleij et al. (1968) succeeded to produce F1-hybrid plants and backcross progenies with a tetraploid plant named *Beta intermedia* as pollinator and tetraploid *Beta vulgaris* as seed parent. The hybrid species can be crossed with crop species and backcross progenies can be produced without need for the application of special techniques. The categorization of *Beta trigyna* and *Beta intermedia* as members of the secondary genepool is therefore justified.

Polish researchers paired the base species of section *Corollinae* in different combinations to unravel the phylogeny of the section *Corollinae* species (all possible combinations except for *Beta corolliflora* x *Beta macrorhiza* and *Beta lomatogona* x *Beta corolliflora*). Jassem and Jazdzewska (1980) crossed *Beta macrorhiza* x *Beta lomatogona* (2n = 18) and obtained fertile progeny. Some spontaneous polyploid and fertile apomicts were found in the progeny of *Beta macrorhiza* x *Beta lomatogona* (2n = 18) showing that apomicts can be created by crossing normally sexually reproducing *Corollinae* species. Jassem et al. (1985) said that *Beta macrorhiza* (2n = 18) and *Beta lomatogona* (2n = 18) could be crossed rather easily and gave fertile progeny. Both tended to develop unreduced male gametes, which leads to the development of polyploids. The meiosis of *Beta macrorhiza* (2n = 18) x *Beta lomatogona* (2n = 18) proved to be regular and bivalents were formed (Szota 1995), which indicates a close relationship between both amphimictic species.

Apomixis is strongly expressed in *Beta lomatogona* (2n = 36) and *Beta trigyna* (2n = 54). Chromosome pairing in bivalents and quadrivalents was often observed in offspring of *Beta macrorhiza* (2n = 18) x *Beta trigyna* (2n = 54), whereas chromosome pairing occurred only sporadically in offspring of *Beta lomatogona* (2n = 18) x *Beta trigyna* (2n = 54). The chromosome homology between *Beta macrorhiza* x *Beta trigyna* seems to be stronger as compared to *Beta lomatogona* x *Beta trigyna* indicating that *Beta macrorhiza* is phylogenetically closer to *Beta trigyna* (Szota and Kuzdowicz 1978). Sufficient potential for species crosses within the section *Corollinae*, the tendency to polyploidy combined with facultative apomixis results in the development and fixation of different hybrid forms within the natural habitat and creates a hybrid swarm and a common agamic complex. Jassem (1992) reviewed the extensive crossing experiments and cytological studies of the Polish researchers and concluded from the experimental results that *Beta macrorhiza* is phylogenetically closer to section *Beta* species, *Beta lomatogona* more distant and *Beta corolliflora* maintains an intermediate position. Conclusions with respect to the phylogenetic position of the agamic ploidy complex (*Beta trigyna*, *Beta* x *intermedia*) were not drawn indicating the need for further research.

Patellifolia species have been considered a genetic resource for sugar beet breeding since the beginning of systematic sugar beet breeding (for instance, de Vilmorin 1923). Interestingly, the great potential of *Beta maritima* as gene donor has been overlooked or underestimated and breeding researchers focused efforts on *Beta* section *Corollinae*. Useful pre-breeding material developed from interspecific crosses did not result from the extensive research work (Barocka 1959; Szota 1995; Cleij et al. 1968, 1976). Instead, *Patellifolia* species that proved to be most difficult to handle were the first of the distantly related wild beet species contributing an economically highly valuable trait, the resistance to the beet cyst nematode (*Heterodera schachtii*), to the sugar beet breeding pool.

Szota (1995) conducted crosses among *Patellifolia* species. The F1-generation of *Patellifolia procumbens* 2x x *Patellifolia webbiana* 2x and reciprocal crosses showed a high percentage (>77.9%) of pollen mother cells without any disturbances at anaphase I, II and tetrad stages and a high percentage of viable pollen (87.2–91.1%). When using the self-compatible tetraploid species *Patellifolia patellaris* as pollinator, the percentage of undisturbed PMC at all stages ranged between 14.8 and 42.1% and the pollen viability dropped to between 13.9 and 34.1%. The lack of crossing barriers between *Patellifolia procumbens* and *Patellifolia webbiana* indicates that both outbreeding species are closely related or may even be genetically strongly differentiated forms of the same species. Indeed, there is increasing evidence from diversity studies using isozyme markers (Wagner et al. 1989), RFLP markers (Mita et al. 1991), DNA fingerprinting (Jung et al. 1993), and SSR markers (Frese et al. 2018) that *Patellifolia webbiana* constitutes a spatially isolated population of *Patellifolia procumbens* adapted to a specific habitat. These pieces of evidence have relevance with respect to assumptions on the evolution of *Patellifolia patellaris* which Walia (1971) supposed to be an allotetraploid species. Dechyeva and Schmidt (2009) labeled *Patellifolia procumbens* as well as *Patellifolia patellaris* chromosomes with *Patellifolia procumbens* satellite DNA (clone pTS5) and found in the tetraploid *Patellifolia patellaris* probe the same number of signals as observed in *Patellifolia procumbens*. This genomic in situ hybridization pattern suggests that *Patellifolia procumbens* could be one of the parent species of the likely allopolyploid *Patellifolia patellaris*. Mesbah et al. (1997a, b) characterized monosomic addition lines of sugar beet carrying different chromosomes from *Patellifolia patellaris*. The presence or absence of the *Heterodera schachtii* resistance gene in specific families could best be explained by the existence of two homologous chromosomes 1 in an allotetraploid *Patellifolia patellaris* carrying the resistance gene and two homeologous chromosomes without that gene. Mesbah et al. (1997a, b) found chromosome addition families derived from *Patellifolia patellaris* to be completely resistant to *Polymyxa betae*. However, the introgression of traits suited to enhance the resistance of sugar beet to pests and diseases is very difficult. Desel et al. (2002) used genomic DNA of *Patellifolia procumbens* to detect chromatin of *Patellifolia procumbens* in nematode resistant *Beta vulgaris* introgression lines. They observed, unlike the weak labeling of *vulgaris* chromosomes with *Beta corolliflora* DNA, no labeling of the *Beta vulgaris* chromosomes with *Patellifolia procumbens* DNA. Their observation further underpins that the phylogenetic difference between *Patellifolia* species and *Beta vulgaris* is greater than the difference between section *Corollinae* species and *Beta vulgaris*.

In hybrids between sugar beet and *Patellifolia* species almost no chromosome pairing occurs in meiosis causing development and fertility distortions. Speckmann and de Bock (1982) and Löptien (1984) reported the main problems impeding the introgression of traits from the *Patellifolia* species into the sugar beet. *Beta vulgaris* x *Beta procumbens* F1 and BC1 plants lacked functioning roots and seedlings need to be grafted to *Beta vulgaris* hypocotyls to produce flowering plants. F1-plants show high degrees of sterility and even plants of backcross generations do not thrive well. Only after extensive crossing and screening work Savitsky (1978) succeeded to select introgression lines carrying the resistance to *Heterodera schachtii* from *Patellifolia procumbens*. There is therefore sufficient experimental evidence that

Patellifolia species match the criteria of Harlan and de Wet (1971) for species of the third genepool.

References

Abe J, Shimamoto Y (1989) Evolutionary aspects and species relationships. In: IBPGR. International crop network series. 3. Report of an international workshop on *Beta* genetic resources. International board for plant genetic resources, Rome, pp 71–79

Abe J, Tsuda C (1986) Genetic behaviour of wild chromosomes in the species hybrids made between sugar beet and *B. macrocarpa*. Proc Sugar Beet Res Asso Jpn 28:25–30

Abe J, Tsuda C (1987) Genetic analysis for isozyme variation in the section *Vulgares*, genus *Beta*. Jpn J Breed 37:253–261

Abe J, Yoshikawa H, Chikahiro T (1984) Isozyme variation used in selecting cultivated type plants out of the segregated population of the interspecific hybrids between sugar beet and the wild relative, *B. macrocarpa*. Proc Sugar Beet Res Asso Jpn 26:37–43

Aellen P (1938) Die orientalische *Beta* Arten. Ber Schweiz Bot Ges 48:408–479

Aellen P (1967) Flora of Turkey and the East Aegean Islands. Univ Press, Edinbourgh, UK, pp 296–299

Allendorf FW, Luikart G (2007) Conservation and the genetics of populations. Blackwell Publishing Ltd., Maldon, USA

Andrello M, Henry K, Devaux P, Verdelet D, Desprez B, Manel S (2017) Insights into the genetic relationships among plants of *Beta* section *Beta* using SNP markers. Theor Appl Genet 130:1857–1866

Arapova TS (1987) Hybrids between cultivated beet and the wild apomictic species *Beta corolliflora* Zoss. and *B. trigyna* W. et Kit. In: Problemy apomiksisa i otdalennoi gibridizatsii. Nauka (Publisher), pp 146–150

Arcangeli G (1882) Compendio della flora italiana. Ermanno Loescher, Torino, Italy

Ascherson P, Graebner P (1919) Synopsis der mitteleuropäischen Flora. Verlag von Gebrüder Borntraeger, Leipzig, Germany

Baldacci A, de Toni E, Frati L, Ghigi A, Gortani M, Morini F, Ridolfi AC, Sorbelli A (1907) Intorno alla vita ed alle opere di Ulisse Aldrovandi. Libreria Treves di L. Beltrami, Bologna, Italy

Barocka KH (1959) Die einzelfrüchtigen Arten der Gattung *Beta* L. in Hinblick auf ihre mögliche Verwendung zur Einkreuzung in *B. vulgaris* L. subsp. *vulgaris* (Zucker- und Futterrübe). Züchter 29:193–203

Barocka KH (1985) Zucker und Fütterrüben (*Beta vulgaris* L.). In: Hoffmann W, Mudra A, Plarre W (eds) Lehrbuch der Züchtung landwirtschaftlicher Kulturpflanze, vol 2. Verlag Paul Parey, Berlin, Germany, pp 245–247

Bauhin G (1622) *Catalogus plantarum circa Basileam sponte nascentium … etc.* Basel, Switzerland

Bauhin G (1623) *Pinax theatri botanici … etc.* Basel, Switzerland

Berti-Pichat C (1866) Corso teorico e pratico di agricoltura. Unione Tipografico-Editrice, Torino, Italy

Bertoloni A (1837) *Flora italica*, vol 3. Typographeo Richardi Morii, Bologna, Italy

Bilz M, Kell S, Maxted N, Lansdown RV (2011) European red list of vascular plants. Pub. Office of the European Union, Luxembourg

Boissier E (1879) *Flora orientalis sive enumeratio plantarum in Oriente a Graecia et Aegipto ad Indiae fines hucusque observatarum.* Apud H Georg Bibliopolam, Lugduni (Lyon), France

Boughey CL (1981) Evolutionary and taxonomic studies in wild and cultivated beets. PhD thesis, University of Birmingham, UK

Bramwell D, Bramwell Z (1974) Wild flowers of the Canary Islands. Stanley Thornes (Publishers) Ltd., Cheltenham, UK

Burenin VI, Garvrilynk IP (1982) Systematics and phylogeny of the genus *Beta* L. Trudy po Prikladnoi Botanike, Genetike i Selektsii 72:3–12

Buttler KP (1977) Revision von *Beta* Sektion *Corollinae* (*Chenopodiaceae*) I. Selbststerile Basisarten. Mitt Bot München 255–336

Campbell GKG (1984) Sugar beet. In: Simmonds NW (ed) Evolution of crop plants. Longmann, London, UK

Castro S, Loureiro J, Iriondo J, Rubio Teso ML, Duarte MC, Romeiras MM, Pinheiro de Carvalho MAA, Santos Guerra A, Rey E, Frese L (2017) Cytogenetic diversity of *Patellifolia* species. In: Poster presented at 6th Global Botanic Gardens Congress, Botanic Gardens Conservation International, Conservatory and Botanical Garden of the City of Geneva, Switzerland, 26–30 June 2017. http://www.6gbgc.org/en/sample-page/

Castro M, Castro S, Figueiredo A, Husband B, Loureiro J (2018) Complex cytogeographical patterns reveal a dynamic tetraploid–octoploid contact zone. AoB PLANTS 10(2). https://doi.org/10.1093/aobpla/ply012

Cesalpino A (1583) *De plantis libri* XVI. Apud Georgium Marescottum, Florence, Italy

Cleij G, de Bock TSM, Lekkerkerker B (1968) Crosses between *Beta intermedia* Bunge and *Beta vulgaris* L. Euphytica 17:11–20

Cleij G, de Bock TSM, Lekkerkerker B (1976) Crosses between *Beta vulgaris* L. and *Beta lomatogona* F. et M. Euphytica 25:539–547

Coles W (1657) Adam in Eden or natures paradise. Printed by F Streater, London, UK

Coons GH (1954) The wild species of *Beta*. Proc Am Soc Sugar Beet Technol 8:142–147

Cordus V (1551) *Adinotationes in Dioscoridis de medica material libros*. Apud Buil. Morelium, Paris, France

Cronquist A (1988) The evolution and classification of flowering plants. Thomas Nelson & Sons Ltd., London, UK

Cupani F (1696) *Hortus catholicus* etc. Apud Franciscum Benzi, Naples, Italy

Dale S (1730) The history and antiquities of Harwich and Dovercourt. Printed for C Davis and T Green, London, UK

Dalke L (1977) Interspecific hybrids between sugar beet and *Beta corolliflora* of the *Corollinae* section. In: Interspecific hybridization in plant breeding. Proceedings of the 8th Eucarpia Congress, Madrid, Spain, pp 113–118

Davis P (1937) Flora of Turkey. Edinburgh Univ Press, Edinburgh, UK

de Bock TSM (1986) The genus *Beta*: domestication, taxonomy and interspecific hybridization for plant breeding. Acta Hort 182:335–343

de Candolle A (1849) *Prodromus systematis naturalis regni vegetali*. Librarierie de Victor Masson, Paris, France

Dechyeva D, Schmidt T (2009) Molecular cytogenetic mapping of chromosomal fragments and immunostaining of kinetochore proteins in *Beta*. Int J Plant Genomics. https://doi.org/10.1155/2009/721091

de Commerell A (1778) Of the culture and use of Mangel Wurzel, or root of scarcity. Printed for Charles Dilly, London, UK

de Lobel M (1576) *Plantarum seu stirpium historia* … etc. Anterwep, Belgium

Desel C, Jansen R, Dedong G, Schmidt T (2002) Painting of parental chromatin in *Beta* hybrids by multi-colour fluorescent in situ hybridization. Ann Bot 89:171–181

Desfontaines R (1789) *Flora atlantica*. Blanchon, Paris, France

de Vilmorin MJL (1923) L'hérédité chez la betterave cultivée. Thèse de Doctorat, Gauthier-Villars et Cie, Paris, France

de Vries U (1905) Species and varieties. Open Court Publishing Co, Chicago IL, USA

de Tournefort JP (1700) *Institutiones rei herbariae*. Thypographia Regia, Paris, France

de Wildeman E, Durand T (1899) Prodrome de Flore Belge. Alfred Castaigne Editeur, Brussels, Belgium

Dohm JC, Minoche AE, Holtgräwe D, Capella-Gutiérrez S, Zakrzewski F, Tafer H, Rupp O, Sörensen TR, Stracke R, Reinhardt R, Goesmann A, Kraft T, Schulz B, Stadler PF, Schmidt

T, Gabaldón T, Lehrach H, Weisshaar B, Himmelbauer H (2013) The genome of the recently domesticated crop plant sugar beet (*Beta vulgaris*). Nature 505:546–549. https://doi.org/10.1038/nature12817

Ernauld L (1945) Les espèces botanique du genre *Beta*. Publ IRBAB 13:219–254

Filutowicz A, Dalke L (1976) Interspecific hybrids in the *Corollinae* section, genus *Beta*. Hodowla Roślin, Aklimatyzacja i Nasiennictwo 20(1):1–17

Fischer HE (1989) Origin of the 'Weisse Schlesische Rübe' (white Silesian beet) and resynthesis of sugar beet. Euphytica 41:75–80

Ford-Lloyd BV (2005) Taxonomy. In: Biancardi E, Campbell LG, Skaracis GN, de Biaggi M (eds) Genetics and breeding of sugar beet. Science Publishers, Enfield NH, USA

Ford-Lloyd BV, Hawkes JG (1986) Weed beets, their origin and classification. Acta Hort 82:399–404

Ford-Lloyd BV, Williams ALS, Williams JT (1975) A revision of *Beta* section *Vulgares* (*Chenopodiaceae*), with new light on the origin of cultivated beets. Bot J Linn Soc 71:89–102

Frese L (2003) Sugar beets and related wild species—from collecting to utilisation. In: Knüpffer H, Ochsmann J (eds) Schriften zu Genetischen Ressourcen, vol 22. Zentralstelle für Agrardokumentation und-information (ZADI), Bonn, Germany

Frese L (2010) Conservation and access to sugar beet germplasm. Sugar Tech 12:207–219

Frese L, Nachtigall M, Enders M, Pinheiro de Carvalho MÂA (2012) *Beta patula* Aiton: genetic diversity analysis. In: Maxted N, Dulloo ME, Ford-Lloyd BV, Frese L, Iriondo JM, Pinheiro de Carvalho MÂA (eds) Agrobiodiversity conservation: securing the diversity of crop wild relatives and landraces. CABI Publishing, Wallingford, UK

Frese L, Nachtigall N, Iriondo JM, Rubio Teso ML, Duarte MC, Pinheiro de Carvalho MÂA (2018) Genetic diversity and differentiation in *Patellifolia* (Amaranthaceae) in the Macaronesian archipelagos and the Iberian Peninsula and implications for genetic conservation programmes. Genet Resour Crop Evol

Fritzsche K, Metzlaff M, Melzer R, Hagemann R (1987) Comparative restriction endonuclease analysis and molecular cloning of plastid DNAs from wild species and cultivated varieties of the genus *Beta* (L.). Theor Appl Genet 74:589–594

Gandoger M (1910) *Novus conspectus florae Europeae*. Hermann et fils, Paris, France

Gao D, Schmidt T, Jung C (2000) Molecular characterization and chromosomal distribution of species-specific repetitive DNA sequences from *Beta corolliflora*, a wild relative of sugar beet. Genome 43:1073–1080

Germeier CU, Frese L (2004) The international database for *Beta*. In: Frese L, Germeier CU, Lipman E, Maggioni L (eds) Report of a working group on *Beta* and world *Beta* network. Second joint meeting, 23–26 October 2002, Bologna, Italy. International Plant Genetic Resources Institute, Rome, Italy, pp 84–102

Gesner K (1561) *Simesusij Annotationes in Pedacij Dioscoridis. Anazarbei De medica materia Libros V*: etc. Argentorati Excudibat Iosias Rihelius

Greene EL (1909) Linnaeus as an evolutionist. Proc Washington Acad Sci 9:17–26

Gürke M (1897) *Plantae europeae*. Paris, France & Lipsia, Germany

Hallam A (1972) Continental drift and the fossil record. Sci Am 227:56–66

Hanelt P, Institute of Plant Genetics and Crop Plant Research (eds) (2001) Mansfeld's encyclopedia of agricultural and horticultural crops (except Ornamentals). Springer, Berlin, Germany

Harlan J, de Wet J (1971) Towards a rational classification of cultivated plants. Taxon 20:509–517

Heldreich J (1877) Pflanzen der Attische Ebene. Engelmann, Lipsia, Germany

Helm J (1957) Versuche einer morphologisch-systematischer Gliederung der Art *Beta vulgaris* L. Züchter 27:203–222

Hohenacker M (1838) Pflanzen der Provinz Talish. Soc. Imp. Naturalists, Moskow, Russia

Hohmann S, Kadereit JW, Kadereit G (2006) Understanding Mediterranean-Californian disjunctions: molecular evidence from *Chenopodiaceae-Betoideae*. Taxon 55:67–78

Jackson BD (1881) Guide to the literature of botany; Being classified selection of botanical works. Longmans & Green, London, UK

Jassem B (1992) Species relationship in the genus *Beta* as revealed by crossing experiments. In: Frese L (ed) International *Beta* genetic resources network. A report on the 2nd international *Beta* genetic resources workshop held at the Institute for Crop Science and Plant Breeding, Braunschweig, Germany, 24–28 June 1991. IBPGR, Rome, Italy, pp 49–54

Jassem B, Jazdzewska E (1980) Cytoembryological investigation of triploid F2 hybrids of *Beta macrorhiza* Stev. X *Beta lomatogona* f. 2x F. et M. (Badania cytoembriologiczne nad triploidalnymi mieszancami pokolenia F2 *Beta macrorhiza* Stev. X *Beta lomatogona* f. 2x F. et M.). Hodowla Roslin, Aklimatyzacja i Nasiennictwo 2:707–715

Jassem B, Jazdzewska E, Szota M (1985) Badania nad Filogeneza Dzikich Gatunkow Sekcji *Corollinae* Rodzaju *Beta* (Investigations on phylogenesis of wild species of the *Beta* genus within the *Corollinae* section). Hodowla Roslin, Aklimatyzacja i Nasiennictwo 29:1–10

Jombart T, Devillard S, Balloux F (2010) Discriminant analysis of principal components: a new method for the analysis of genetically structured populations. BMC Genet 11:94

Jung C, Pillen K, Frese L, Fähr S, Melchinger AE (1993) Phylogenetic relationships between cultivated and wild species of the genus *Beta* revealed by DNA "fingerprinting". Theor Appl Genet 86:449–457

Kadereit G, Hohmann S, Kadereit JW (2006) A synopsis of *Chenopodiaceae* subfam. *Betoideae* and notes on the taxonomy of *Beta*. Willdenowia 36:9–19

Koch K (1839) Das natürliche System des Pflanzenreichs. Hofhausen, Ben Karl, Jena, Germany

Koch J (1858) *Synopsis florae Germanicae et Helveticae*, 3rd edn. Sumptibus Gerhard et Reisland, Lipsia, Germany

Komarnitsky IK, Samoylov AM, Red'ko VV, Peretyayko VG, Gleba YuYu (1990) Intraspecific diversity of sugar beet (*Beta vulgaris*) mitochondrial DNA. Theor Appl Genet 80:253–257

Krasochkin VT (1959) Review of the species of the genus *Beta*. Trudy Po Prikladnoi Botanike. Genetik i Selektsii 32:3–35

Kubis S, Heslop-Harrison JS, Schmidt T (1997) A family of differentially amplified repetitive DNA sequences in ther genus *Beta* reveals genetic variation in *Beta vulgaris* subspecies and cultivars. J Mol Evol 44:310–320

Kuntze O (1891) *Revisio plantarum*. Lipsia, Germany

Lange W, de Bock TSM (1989) The Diploidised Meiosis of Tetraploid *Beta macrocarpa* and its Possible Application in Breeding Sugar Beet. Plant Breed 103:196–206

Lange W, Brandenburg WA, de Bock TSM (1999) Taxonomy and cultonomy of beet (*Beta vulgaris* L.). Bot J Linn Soc 130:81–96

Ledebour CF (1846) Flora rossica, sive enummeratio plantarum etc. Sumptibus Librariae E Schweizerbart, Stuttgart, Germany

Lenz HO (1869) Botanik der alten Griechen und Römer. Verlag von Thienemann, Gotha, Germany

Letschert JPW (1993) *Beta* section *Beta*: biogeographical patterns of variation, and taxonomy. Dissertation, Wageningen Agricultural University Papers 93-1, Wageningen, The Netherlands

Letschert JPW, Lange W, Frese L, van Der Berg RG (1994) Taxonomy of *Beta* selection *Beta*. J Sugar Beet Res 31:69–85

Leys M, Petit EJ, El-Bahloul Y, Liso C, Fournet S, Arnaud JF (2014) Spatial genetic structure in *Beta vulgaris* subsp. *maritima* and *Beta macrocarpa* reveals the effect of contrasting mating system, influence of marine currents, and footprints of postglacial recolonization routes. Ecol Evol 4(10):1828–1852. https://doi.org/10.1002/ece3.1061

Linnaeus (1735) *Systema Naturae*. Typie et Sumptibus Io Iac Curt, Halae Magdeburgicae (Halle), Germany

Linnaeus (1753) *Species plantarium exhibentes plantas rite cognitas ... etc.*, 1st edn. Stockholm, Sweden

Löptien H (1984) Breeding nematode-resistant beets. 1. Development of resistant alien additions by crosses between *Beta vulgaris* L. and wild-species of the section *Patellares*. Z Pflanzenzüchtg 92:208–220

Mansfeld R (1959) Vorläufiges Verzeichnis landwirtschaftlich oder gartnerish Pflanzenarten. Kulturpflanze 2:38–45

Marschall P (1819) *Flora taurico-caucasica*, vol 1. Charkov, Leipzig, Germany

Mesbah M, de Bock TSM, Sandbrink JM, Klein-Lankhorst RM, Lange W (1997a) Molecular and morphological characterisation of monosomic additions in *Beta vulgaris*, carrying extra chromosomes of *B. procumbens* or *B. patellaris*. Mol Breed 3:147–157

Mesbah M, Scholten OE, de Bock TSM, Lange W (1997b) Chromosome localisation of genes for resistance to *Heterodera schachtii*, *Cercospora beticola* and *Polymyxa betae* using sets of *Beta procumbens* and *B. patellaris* derived monosomic additions in *B. vulgaris*. Euphytica 97(1):117–127

Mglinets AV (2008) Phylogenetic relationships of genus *Beta* species based on the chloroplast *trnK* (*matK*) gene intron sequence information. Dokl Biochem Biophys 420:135–138

Mita G, Dani M, Casciari P, Pasquali A, Selva E, Minganti C, Piccardi P (1991) Assessment of the degree of genetic variation in beet based on RFLP analysis and the taxonomy of *Beta*. Euphytica 55:1–6

Moquin-Tandon A (1840) *Chenopodearum monographica enumeratio*. J-P Loss, Paris, France

Morison R (1715) *Plantarum historiae universalis oxoniensis*. Apud Pl Vaillant, Oxford, UK

Mutel A (1836) Flore française destinée aux herborizations ou descriptions des plantes. Levrault, Paris, France

Oldemeyer RK (1957) Sugar beet male sterility. J Am Soc Sugar Beet Tech 9:381–386

Paesold S, Borchardt D, Schmidt T, Dechyeva D (2012) A sugar beet (*Beta vulgaris* L.) reference FISH karyotype for chromosome and chromosome-arm identification, integration of genetic linakge groups and analysis of major repeat family distribution. Plant J. https://doi.org/10.1111/j.1366-313x.2012.05102.x

Parkinson J (1655) *Matthiae de L'Obel stirpium illustrationes*. Warren, London, UK

Phitos D, Strid A, Snogerup S, Greuter W (eds) (1995) The Red Data Book of rare and threatened plants of Greece. WWF-Greece, Athens

Radde A (1866) Reisen an der Persisch-russischen Grenze. Engelmann, Leipzig, Germany

Ramos-Büttner SM, Wricke G (1993) Evidence of tetrasomic inheritance in *Beta corolliflora*. J Sugar Bett Res 30:321–327

Ray J (1690) *Synopsis methodica stirpium Britannicarum … etc*. Apud Samuel Smith, London, UK

Ray J (1693) *Historia plantarum generalis*. Smith & Walford, London, UK

Reichenbach L, Reichenbach HG (1909) *Icones florae Germanicae et Helveticae*. Sumptibus Federici de Zezschwitz, Leipzig, Germany

Romeiras MM, Vieira A, Silva DN, Moura M, Santos-Guerra A, Batista D, Duarte MC, Paulo OS (2016) Evolutionary and biogeographic insights on the Macaronesian Beta-Patellifolia species (Amaranthaceae) from a time-scaled molecular phylogeny. PLoS ONE 11(3):e0152456. https://doi.org/10.1371/journal.pone.0152456

Roxburgh W (1832) Flora indica or description of Indian plants. Printed for W Thacker andCo, Calcutta, India

Santoni S, Bervillè A (1992) Two different satellite DNAs in *Beta vulgaris* L.: evolution, quantification and distribution in the genus. Theor Appl Genet 84:1009–1016

Savitsky H (1978) Nematode (*Heterodera schachtii*) resistance and meiosis in diploid plants from interspecific *Beta vulgaris* x *B. procumbens* hybrids. Can J Genet Cytol 20:177–186

Scott AJ, Ford Lloyd BV, Williams JT (1977) *Patellifolia*, nomen novum (Chenopodiaceae). Taxon 26:284

Shen Y, Newbury HJ, Ford-Lloyd BV (1996) The taxonomic characterisation of annual *Beta* germplasm in a genetic resources collection using RAPD markers. Euphytica 91:205–212

Shen Y, Newbury HJ, Ford-Lloyd BV (1998) Identification of Taxa in the genus *Beta* using ITS1 Sequence Information. Plant Mol Bio Rep 16:147–155

Speckmann GJ, de Bock TSM (1982) The production of alien monosomic additions in *Beta vulgaris* as a source for the introgression of resistance to beet root nematode (*Heterodera schachtii*) from *Beta* species of the section *Patellares*. Euphytica 31:313–323

Steudel E (1871) *Nomenclator Botanicus … etc*. Sumptibus LG Cottae, Stuttgart and Tübingen, Germany

Stokes J (1812) A botanical materia medica etc. J Johnson & Co., London, UK

Szota M (1995) Meiosis in species of sections *Procumbentes* and *Corollinae* and in interspecific hybrids within these sections and with sugarbeet (*Beta vulgaris* L.) (Mejoza gatunkow sekcji *Procumbentes* i *Corollinae* oraz miedzygatunkowych mieszancow w obrebie tych sekcji i z burakiem cukrowym (*Beta vulgaris* L.) Hodowla Roslin, Aklimatyzacja i Nasiennictwo 39(4):3–59

Szota M, Kuzdowicz A (1978) Meiosis in *Beta macrorhiza* 2x X *B. trigyna* 6x and *B. lomatogona* 2x X *B. trigyna* 6x F1 and F2 hybrids (Mejoza mieszancow F1 i F2 *Beta macrorhiza* 2x X *Beta trigyna* 6x oraz F1 i F2 *Beta lomatogona* 2x X *Beta trigyna* 6x). Hodowla Roslin, Aklimatyzacja i Nasiennictwo 22:279–285

Thulin M, Rydberg A, Thiede J (2010) Identity of *Tetragonia pentandra* and taxonomy and distribution of *Patellifolia* (Chenopodiaceae) Willdenowia 40:5–11

Tilli MA (1723) Catalogus plantarum horti pisani. Tartino & Franchi, Florence, Italy

Touzet P, Villain S, Buret L, Martin H, Holl A-C, Poux C, Cuguen J (2018) Chloroplastic and nuclear diversity of wild beets at a large geographical scale: Insights into the evolutionary history of the *Beta* section. https://doi.org/10.1002/ece3.3774

Tranzshel VA (1927) Obzor vida roda *Beta*. Trudy prikl. Bot Genet Selek 17:203–223

Trehane P, Brickell CD, Baum BR, Hetterscheid WLA, Leslie AC, McNeil J, Spongberg SA, Vrugtman F (1995) International code of nomenclature for cultivated plants—1995 (ICNCP or Cultivated plant code). Quarterjack Publishing, Wimbourne, UK

Tutin TG, Heywood VH, Burgess NA, Valentine DH, Walters SM, Webb DA (1964) Flora Europaea. Cambridge Univ Press, Cambridge, UK

Ulbrich E (1934) *Chenopodiaceae*. In: Engler A, Harms H (eds) Die Natürlichen Pflanzenfamilien. Wilhelm Engelmann, Leipzig, pp 375–584

Valentini CB (1715) Tournefortius contractus sub forma tabularum. Frankfurt, Germany

van Geyt JPC, Lange W, Oléo M, de Bock TSM (1990) Natural variation within the genus *Beta* and its possible use for breeding sugar beet: a review. Euphytica 49:57–76

Vasilchenko EN, Zhuzhzhalova TP (2011) Variability of morphological and biochemical traits of beet interspecific hybrids. Sakharnaya Svekla 1:18–21

Villain S (2007) Histoire evolutive de la section *Beta*. Dissertation, Université des Sciences et Technologies de Lille, France

von Lippmann EO (1925) Geschichte der Rübe (*Beta*) als Kulturpflanze. Verlag Julius Springer, Berlin, Germany

von Proskowetz E (1896) Über die Culturversuche mit *Beta* im Jahre 1895, und über Beobachtungen an Wildformen auf natürlichen Standorten. Österreiche-Ungarische Zeitschrift für Zuckerindustrie und Landwirtschaft 33:711–766

Wagner H, Gimbel E-M, Wricke G (1989) Are *Beta procumbens* Chr. Sm. and *B. webbiana* Moq. different species? Plant Breeding 102:17–21

Walia K (1971) Meiotic prophase in the genus *Beta* (*B. vulgaris* 2x and 4x, – *B. webbiana* and *B. patellaris*). Z. Pflanzenzüchtg. 65:141–150

Willdenow KL (1707) *Species plantarum exibentes plantas rite cognitas* … etc. Nauck, Berlin, Germany

Zossimovitch V (1934) Wild species of beets in Transcaucasia. VNIS2-3

Chapter 7
Uses

Enrico Biancardi

Abstract The many uses of the different parts of *Beta maritima* harvested in the wild are described. Although eaten as a potherb before recorded history, most of our information about the uses of sea beet, and beets in general, are as a medicinal herb because this was the interest of most of the ancient authors who wrote about plants. Many of these medicinal uses have lost their importance with the advances of medical science. Nonetheless, sea beets (and other beets) are still used in homeopathic and "natural" remedies and have a number of useful qualities, both for the smooth function of the digestive tract, and to prevent diseases.

Keywords *Beta maritima* · Digestive aid · Beet juice · Beet fiber · Betacyanin

7.1 Medicinal Uses

Leaves and roots of sea beet have been used since prehistory against several ailments and diseases (Fig. 7.1). Some important applications are recognized by current medicine as well. (http://www.celtnet.org.uk/recipes/miscellaneous/fetch-recipe.php?rid=misc-sea-beet-quiche, http://www.magicgardenseeds.com/BET05). The roots are described as more medicinally effective than the leaves and sea beet as more active than the cultivated beets, stated by Galen around 100 AD. When cooked, the beet loses part of its properties, because the main part of the vegetation matters (Galen 1833).

It was claimed that the Babylonians were relatively immune to leprosy because they frequently ate beets cooked in different ways (Anonymous 2011). According to Theophrastus (400 BC) and Hippocrates (around 460 BC), the raw leaves are good material for binding wounds, whereas the boiled leaves relieve skin burns. Some properties of wild beet juice were listed in the *Herbarium* of Crateuas (around 300 BC): including (i) clears the head; (ii) reduces ear pain if infused in the nose mixed with honey; (iii) fights dandruff; and (iv) mollifies the chilblains. Moreover,

E. Biancardi (✉)
Stazione Sperimentale di Bieticoltura, Viale Amendola 82, 45100 Rovigo, Italy
e-mail: enrico.biancardi@alice.it

This is a U.S. government work and not under copyright protection in the U.S.; foreign copyright protection may apply 2020
E. Biancardi et al. (eds.), *Beta maritima*,
https://doi.org/10.1007/978-3-030-28748-1_7

Fig. 7.1 Drawing of an old "Pharmacia" (1450?), Biblioteca Ariostea, Ferrara, Italy

the leaves used as a poultice heal leprosy, the itching caused by alopecia, and skin sores (Biancardi 2005).

According to Pliny, *Beta candida* possesses purgative properties, whereas the *Beta nigra* was rather astringent. Some digestive properties were listed by Vigier (1718), including the efficacy against the intestine worms. He also noted that the leaves were used to treat burns, and the powder obtained by grinding the seed was useful to relieve dysentery. Dioscorides had the same opinion: the decoction of *Beta candida* softened the intestines, and *Beta nigra* cured diarrhea (Kühn 1829). Beet juice is introduced into nostrils "*expurgat caput*" (lightens the head). The same means and methods were advised by Coles (1657) "against Head-ach and Swimmings therein, and turnings of the Braine." The decoction made using roots and leaves reduced dandruff (Dioscorides 89 mAD); the leaves applied to the skin healed wounds and ulcerations; however, if eaten in excessive quantity, beet increases the evil humors.

In "*Tractatus de virtutibus herbarum,*" (da Villanova 1509) it is possible to find a long list of applications taken, in part, from the Arab physicians Avicenna and Serapion (Fig. 7.2). According to da Villanova (1509), beet juice was useful against San Antony's fire (herpes zoster), infected wounds, and mouth ulcerations. If put in the ears, the juice relieved earaches. Dioscorides wrote that when beet is cooked with vinegar and mustard, it was effective against several diseases of the liver and spleen. Mixed with eggs, it reduced the effects of herpes zoster and skin burns.

Platina (1529) recommended drinking beet juice for reducing garlic breath. The same was advised by Cato (1583). Moreover, it reduces the consequences of summer heat and "*nutrientes foeminas plutimo lacte implet et sedat menstrua*" (brings plenty of milk to nursing women and cures menstrual pains). The sea beet, having a "hot nature similar to saltbush (*Atriplex* spp.) but less humid," causes weakness and slowness (Averroes cited by Bruhnfels 1531). Simone Sethi, cited by Fuchs (1551), together with the recipes of the classic authors, confirmed that the beet juice as being hot in nature (see Galen 1833), "*ventrem constringit et sitim affert*" (it blocks the intestine and makes thirsty).

Jean Ruel in the book "*Diosciridae pharmacorum*" (Ruel 1552) many times referred to *Beta sylvestris*, *Beta agrestis*, as well as to sea beet by its old names, *limonion* and *neuroides* noted and infusion made with leaves was useful against colic (Ruel 1552). The same recipe is mentioned by Ibn Sina (900) in a manuscript in Latin translation "*Liber canonis medicinae*" (Fig. 7.3), and by Ritze (1599). If the leaves of wild beet are chewed, a disease of the eyes named "mal del piombo" in old Italian (likely glaucoma), could be reduced (Durante 1635). The poultice obtained from roots boiled in vinegar relieved a toothache if taken bound into the hands. The same, put under food, cured sciatica. If applied around the wrists, it afforded recovery from scabies. Finally, the juice was effective against the bite of a wolf (Durante 1635).

Bruhnfels (1531), who took references from some Arab authors (Serapion, Averroes, Zacharia, and so on), asserted that sea beet juice is effective for ulcerations of the nostrils, hair loss, lice, and reduced dandruff as well. Dorsten (1540), confirmed that "*Betae omnes frigidam et umidam naturam habent*" (Beets behave a cool and wet nature), and also stated that "*Radix decocta, si inde tres vel quatuor calidae*

Fig. 7.2 Harvest of leaf beets represented on "*Tacuinum sanitatis*" of uncertain origin (1100?) and reporting recipes of Dioscorides and Arab writers

guttae auribus instillentur, tollit dolorem earum" (three or four drops of hot root decoction put into the ears reduce the ache of them).

Parkinson (1629) cited the use of enemas prepared with water used from boiling beet leaves as an effective laxative; "The leaves are much used to mollify and open the belly being used in the decoction of blisters." In "*Stirpium illustrations*" Parkinson (1655) mentioned that *Beta maritima sylvestris minor* due to "... *gustu salso & nitroso commendatur ad hydripicorum aquas educendas*" (the salty and nitric taste of *Beta maritima sylvestris minor* is recommended in edema for reducing the liquid inside the tissues). The same was confirmed by Magnol (1636), who added that sea beet also "*calefacit & siccat*" (heats and dries) owing to its "*nitrositatem*" (high content of nitrogen). If drunk, the decoction improved the function of the spleen and,

Fig. 7.3 School of medicine (Ibn Sina 900?) from Wikipedia

according to Mattioli (1557), relieved itching. Like other vegetables, beet nourishes little, but benefitted the liver, especially if eaten seasoned with mustard and vinegar (Mattioli 1571).

Squalermo (1561) wrote that the cooked roots "conferiscono molto agli appetiti di Venere" (much improved sexual energy). This effect was confirmed by http://www.godchecker.com/. Beets generated good blood, removed stains from the face, and reduced hair loss (de Crescenzi 1605). For Ray, in "Synopsis" (1690), *Beta sylvestris maritima* was a laxative. The pulverized root snuffed up the nose caused sneezing, diminished the bad humors of the brain, and cured headache even if chronic. Meyrick (1790) confirmed the efficacy of this system "in order to provoke the discharge of humors from the head and parts adjacent." The roots lightly boiled and mixed with vinegar improved the appeal of foods and liver activity. Finally, *"veteres tamen fatuitatem iis exprobant"* (they help recovering the memory of aged people) (Ray 1738).

Ficino (1576) argued that the beet soup, if eaten frequently, is a valid means of protection against the plague. Tanara (1674) quoted the Latin proverb: *"Ventosam betis si sapis adde fabam"* (for reducing the flatulence caused by beet, eat it mixed with broad beans, *Vicia faba* L.). The same author mentioned that pieces of root could be used as a suppository and that the leaves cooked under ashes are effective against burns. Among the negative effects, Tanara, cited Pythagoras, and wrote that the misappropriate consumption of beet may cause excessive amount of fluids in the circulatory system and a disease called "hydropsy" (edema) in tissues.

Dodoens (1586) in "A new herbal or historie of plants," along with the common uses, asserted that the juice of beet "put into the ears takes away the pains in the same, and also reduces the singing or the humming noise." "Beets make the belly soluble and cleanse the stomach," whereas the juice is "a good antioetalgic being poured into the ears; and opens the opulations of the liver and spleen" (K'Eogh 1775). In agreement with Dodoens, the leaves used as impiaster (poultice), reduced the severe effects of choler, and "the roots put as a suppository into the fundament soften the belly."

Quite widespread was the use of beet leaves for binding wounds (Clark 2011). Culpeper (1653) recommended the use of beet juice for reducing headache, vertigo, and all the brain diseases. According to Hill (1820), the white beet juice also was a useful drug for toothache. It promoted sneezing if inhaled through the nose. The red beet root had the same uses but was less effective than the white one and much less still than sea beet. The beneficial action of the juice introduced into the nostrils against headache, even if chronic, was confirmed by Blackwell (1765). Salmon (1710) listed in the "English herbal" both the virtues of the different types of beet, and the different means for using them: (i) liquid juice; (ii) inspissate (thick) juice; (iii) essence; (iv) decoction; (v) cataplasm (poultice); and (vi) saline tincture.

More recently, beet juice (red beet in particular) is considered an effective means for reducing blood pressure (http://en.wikipedia.org/wiki/Beet). Red beet also has been recommended for the prevention of intestinal tumors and the seed boiled in water is said to be effective against the same disease affecting the genital organs (http://dukeandthedoctor.com). Moreover, red beet juice regularly consumed is said to (i) keep the elasticity of arteries; (ii) drop the risk of defects in newborns, because it contains folic acid; (iii) stimulate the function of the liver; (iv) relieve constipation, etc. Beet juice and water boiled with the seed has been said to have therapeutic value against several diseases including cancer (Allioni 1785) and leukemia (Duke and Atchley 1984; http://www.life-enthusiast.com/; http://www.pfaf.org/user/default.aspx).

Another current use of beets is the root fiber, which has higher water- and fat-holding capacity than other dietary fibers. Therefore, for several years, beet fiber (by-product of sugar beet factories, adequately processed) is finding an important use to promote regular bowel movement and as blood detoxifier (http://www.whfoods.

com/). The seed of sea beet, called "silaijah" or "silaigah," has been sold commonly in the Indian and Iranian bazaars for different medical uses (Hooper 1937). The decoction of leaves is used in South Africa as purgative and against hemorrhoids.

An updated list of pharmacological activities of sea- and sugar-beet roots can be found in Jasmatha et al. (2018).

7.2 Food Uses

The beginning of the use of *Beta maritima* as a potherb is lost in the prehistory. At least at the beginning, it likely was limited to the leaves because the roots, woody, fangy, and deep in the soil were not suited for human consumption being too hard for chewing (Fig. 7.4). Among other things, harvest would have been quite difficult. Thus, the root was used only as a botanical drug because of the smaller quantities needed. To make the roots more suited for food, a long selection process to improve the shape, weight, and reduce woodiness was necessary.

The various recipes for preparing the leaves do not always specify whether they are intended for the wild or cultivated plants. However, according to opinions of many, the wild beets are always tastier and more appreciated. Pliny (75 A.D.) reported that the leaves were prepared together with beans, lentils, and mustard, to eliminate their insipidity (Giacosa 1992). With a lightweight placed over the leaves at an early stage, beet develops a broad blanched head "more than two feet" much appreciated by the Romans. This practice also was widespread in Greece (Lindley and Moore 1866).

Some recipes using leaves were given by von Megenberg (1348) in "Das Buch der Natur," which was is believed to be among the earliest printed books. It was explained (in old German) that beet leaves became a good dish especially if mixed with parsley (*Petroselinum* subsp.). In the earlier cited "English herbal," Salmon (1710) wrote that "Beets are used (I mean the root) as a sallet, and to adorn and furnish out dishes of meat withall, being as sweet and good as any carrot. If boiled as carrots, and eaten with butter, vinegar, salt, and pepper, it makes a most admirable dish, and very agreeable with the stomach."

Evelin (1740), after citing some epigrams of Martial, wrote "the rib of the white beet leaves were boiled melts and can be eaten like the marrow (*Cucurbita pepo* L.). But there is a beet growing near the sea called *Beta maritima sylvestris*, which is most delicate of all." The young leaves, collected in winter or early spring, are boiled and in this way become a good wholesome dish (Taylor 1875; Thornton 1812). If harvested later, the leaves taste bitter (www.wildmanwildfood.co.uk). In France, the leaves were often mixed with sorrel (*Rumex acetosa* L.) to lessen the acidity of the latter (Lindley and Moore 1866).

In Ireland, sea beet is well known to people living on the coast, who call it cliff-spinach or perpetual spinach (Fig. 7.4), and frequently cultivate it in their gardens using seed collected on the wild plants (Sturtevant 1919; Henreitte's 2011). The same is done in England. "This form has been ennobled by careful culture, continued until a mangold was obtained" (Sturtevant 1919; Burton and Castle 1838) refers that *Beta*

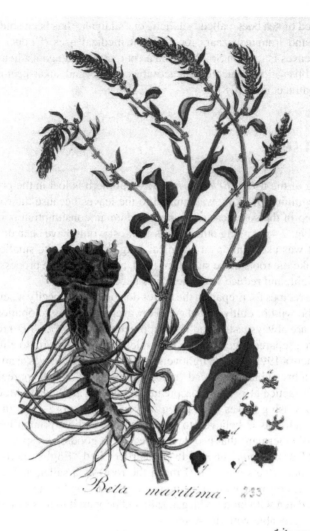

Beta maritima

Fig. 7.4 Drawing of aged plants of *Beta maritima* (Kops et al. 1865)

maritima is extensively used "as a pikle and salad, preserved as a confiture, made a substitute of coffee, and yielding a beautiful varnish." The following information was given by Williams (1857), "sea beet, which frequently grows in great abundance on the sea-beach, the salt marshes, and all about the cliffs, is very useful and is as good as the cultivated spinach. As an edible vegetable it is often cultivated on the coast of Cork."

In Italy, where collection of sea beet is still widespread (Ghiradini et al. 2007) and some attempts of cultivation using wild seed have been made (Branca 2001), the leaves are mixed with fresh cheese in order to prepare a specialty sort of "tortellini." In another popular recipe, the leaves, boiled briefly, are cooked together with scrambled eggs. Rivera et al. (2006) reported the recipes of two popular dishes from Sardinia (Italy) and Valencia–Alicante (Spain). The first is named "minestra delle 18 erbe" (18 greens soup), and prepared with a mix of *Borrago officinalis, Silene vulgaris, Beta maritima, Carduus spp, Sonchus arvensis, Papaver rhoeas,* etc. The second, named "Cocas" or "Mintxos," is a sort of pizza filled with fish and wild greens (*Sonchus* spp, *Reichardia* spp, *Beta maritima,* etc.). On the island of Cyprus, the leaves of 11 wild herbs, including sea beet, *Papaver rhoeas,* etc., are used as main ingredient of the traditional pie named "pittes" (Della et al. 2006).

In the kitchen, the young leaves of cultivated types of beet have the same use as spinach (*Spinacia oleracea* L.), which also belongs to the family of Chenopodiaceae. The boiled leaves of beet are perhaps more appreciated than spinach because they are not astringent in taste and "are quite as good" (Johns 1870). Seed of *Beta maritima* to be used in gardens for leaf production is currently sold by some firms such as Magic Garden Seeds (Regensburg, Germany).

According to Pratt (1856) "Of all our sea-side plants, boiled for table vegetables," the one which seemed to the writer of these pages most to deserve commendation for the purpose is the sea beet (*Beta maritima*). Unlike the silvery glaucous foliage of the orache and goosefoot, the leaves of this plant are of a deep rich green color, very succulent and wavy at the edges. This seaside spinach is certainly very wholesome, and if it were not a wild plant would be in much request. The roots of all the beets contain much saccharine matter, and the well-known experiments of the French on another species, the red beet, for the purpose of obtaining sugar, need not be referred to. No such quantity of sugary substance is yielded by other European esculents as by this. This plant is also common as a culinary root and is also frequently used for salads. On some parts of the coast, it is gathered from the cliff or the muddy shore for food, yet it is often left unnoticed. The English proverb, which our old writer, Fuller, so often quotes, "Fetched far, and cost dear, is fit for ladies," applies, seemingly, as well to the other portion of humanity as to the fair sex."

There are countless methods and recipes for cooking the roots. In this case, the type used most often is the red or garden beet. Apicius (35 BC?) provided several methods for cooking beet roots. In a more recent edition of "*Ars Coquinaria*" (Lister 1709), that book was integrated with recipes from other authors like Humelbergius, Barthius, Reinesius, van der Linden, etc. A number of recipes including those of beets are cited in English by Henriette (2011). Atheneus reported that the roots of sea beet have "a sweet taste and grateful, much better than cabbage." According to Ray (1738). "*Beta estur ut olus, eaque nihil in culina usitatius*" (beet is as spinach, and nothing is more used in the kitchen).

The following recipe is given in "The young housewife's daily assistant" (Anonymous 1864), "Wash off the mould, being careful to not to rub the skin; place the beetroot in a moderate oven and bake about two hours. When cold, take off the skin and use the beetroot as may be required. It is very good dressed as cucumber and served with fish and cold meat thus: cut the beetroot into thin slices, sprinkle over a salt spoonful of pepper, the same of salt, two tablespoonfuls of oil, and one of vinegar." At the site, http://www.guardian.co.uk/, sea beet is described as follows: "dark green, robust, spinach-like leaves, wild chervil, the perfect accompaniment to salmon, sea purslane, delicate, salty, succulent pods that explode on the tongue, and of course no end of chanterelles, morels, ceps and other wild fungi that inhabit our meadows and woods." Countless recipes are available on the WEB for cooking the roots (http://recipes.wikia.com/wiki/Sea_beet; http://www.celtnet.org.uk/recipes/miscellaneous/fetch-recipe.php?rid=misc-sea-beet-quiche; http://www.celtnet.org.uk/recipes/miscellaneous/fetch-recipe.php?rid=misc-sea-beet-quiche.

The fibrous matter extracted from beets added in proper proportion to different foods has the following properties: (i) will keep bread soft for longer time; (ii) improves the action of dough; (iii) reduces grilling losses in hamburger steaks; (iv) fried croquette scarcely burst, and so on (Dillard and Bruce German 2000).

A good beer and a pleasant wine may be made from the fermented roots (Burton and Castle 1838). After acetic fermentation, the sliced root is the main ingredient in the dish named "barszcz" in Poland and "borscht" in the Balkan countries (Chaumeton 1815). Beet soup is listed among the foods of propitious omen to be eaten by the Jewish people on the first day of the year (www.jewishencyclopedia.com). Betacyanin, the main pigment of red beet, may cause red urine in organisms unable to break down it (http://en.wikipedia.org/wiki/Beeturia).

The dried root was used as a substitute for coffee (Miller 1768). During the last world war, the beets were considered one of the better vegetables suited to be canned for the Allied soldiers (http://aggie-horticulture.tamu.edu). In 1975, a sort of beet purée was served on board of Soyuz 19 shuttle during the meeting with the Apollo 18 astronauts. The food was canned in tubes like toothpaste and it was squeezed in the mouth (http://www.healthdiaries.com/eatthis/25-facts-about-beets.html). The very latest citation of sea beet as food is described on the application "Ultimate SAS Survival Guide" downloadable on mobile phones and similar devices. Here, *Beta maritima* is listed along with the edible plants available in case of emergence along the European seashores (Wiesemann 2010).

7.3   Other Uses

When stored wine has the flavor of cabbage, it can be remedied by soaking in beet leaves, and the water utilized for boiling beet roots removed stains from fabrics, parchment, and clothes (Pliny 75 A.D.). The decoction also removed lice from hair (Bruhnfels 1531), whereas the beet juice was useful for polishing gold and silver (Berthelot and Ruelle 1888). A beauty mask prepared with a mixture of gridded beet root and milk cream was said to be very effective for delaying the signs of the age from the face (Messegué 1979).

Sea beet has a high salt-removing capacity, which is helpful where the soil salinity is high (Aksoy et al. 2003). Trist (1960) asserts that *Agropyron pungens* is the best grass for sea walls. *Beta maritima* is considered a particularly damaging weed because its deep roots can make conditions favorable for erosion. Moreover, roots create holes in the dams, through which water under pressure can easily penetrate easily enlarging the hole. The pathogens of several diseases, including beet yellows virus (BYV), beet mosaic virus (BMV), the causal agents of rust (*Uromyces betae*) and downy mildew (*Peronospora schachtii*), respectively, were found to be common in sea beet growing on the seashores of southern Wales and southern England. In early spring, the viruses infecting the overwintering beets are easily transmitted by aphids into the cultivated beet fields (Gibbs 1960; Sorensen and Marcussen 1996).

References

Aksoy U, Kaykcoglu H, Kukul YS, Hepaksoy S, Can HZ, Balc B (2003) An environmentally friendly technique to control salination: salt removing crops. Acta Hort 593:137–142

Allioni C (1785) Flora pedemontana. Excudebat Iohannes, Turin, Italy

Anonymous (1864) The young housewife's daily assistant. Simpkin, Marshall, and Co., London, UK

Anonymous. Jewish encyclopedia. http://www.jewishencyclopedia.com/

Berthelot M, Ruelle C (1888) Collection des alchimistes Grecs. France, Paris

Biancardi E (2005) Brief history of sugar beet cultivation. In: Biancardi E, Campbell LG, Skaracis GN, de Biaggi M (eds) Genetics and breeding of sugar beet. Science Publishers Inc, Enfield (NH), USA, pp 3–9

Blackwell E (1765) Sammlung der Gewachse. de Launoy, Nürnberg, Germany

Branca F (2001) Prove di coltivazione di specie spontanee utilizzate in Sicilia per scopi alimentari. Italus Hortus 8:22–26

Bruhnfels O (1531) In hoc volumine contenitur insignium medicorum … etc. Strasbourg, France

Burton BH, Castle T (1838) British flora medica. Cox, London, UK

Cato H (1583) L'agricoltura et casa di villa … etc. Appresso GB Ratteri, Turin, Italy

Chaumeton FP (1815) Flore medicale, vol 2. Panckoucke Editeur, Paris, France

Clark PA (2011) A cretan healer's handbook. Ashgate, Burlington VT, USA

Coles W (1657) Adam in Eden or natures paradise. Printed by F Streater, London, UK

Culpeper T (1653) Complete herbal. Culpeper's complete herbal. Evans, Richard, London, UK

da Villanova A (1509) Tractatus de virtutibus herbarum. Johannes Rubeus, Venice, Italy

de Crescenzi P (1605) Trattato dell'agricoltura, Florence, Italy

Della A, Pareskeva-hadjichambi D, Hadjichambis AC (2006) A ethnobotanical survey on wild edible plant of Paphos a countryside of Cyprus. J Ethnobiol Ethnomed 34:1–10

Dillard CJ, Bruce German J (2000) Phytochemicals: nutraceuticals and human health. J Sci Food Agric 80:1744–1756

Dodoens R (1586) A new harbal or histoire of plants, London, UK

Dorsten T (1540) Botanicon, continens herbarum aliorumque simlicium, Frankfurth, Germany

Duke JA, Atchley AA (1984) Proximate analysis. In: Cristie BR (ed) The handbook of plant sciences in agriculture. CRC Press Inc., Boca Raton, FL, USA

Durante C (1635) Herbario nuovo. Jacomo Bericchi et Jacomo Ternierij, Rome, Italy

Evelin J (1740) Acetaria or a discourse of sallets, London, UK

Ficino M (1576) Contro alla peste. Giunti, Florence, Italy

Fuchs L (1551) De historia stirpium commentarii insignes. Arnolletum, Lyon, France

Galen C (1833) De almentorum facultatibus. In: Kühn CG (ed) Medicorum graecorum opera. Officina Libraria Caroli Cnoblochii, Lipsia, Germany

Ghiradini MP, Carli M, del Vecchio N et al (2007) The importance of taste. A comparative study on wild food plant consumption in twenty-one local communities in Italy. J Ethnobiol Ethnomed 22:1–14

Giacosa IG (1992) A taste of ancient Rome. Chicago Univ Press, Chicago, IL, USA

Gibbs AJ (1960) Studies on the importance of wild beet as a source of pathogens for the sugar beet ceop. Ann Appl Biol 48:771–779

Henreitte's (2011) Henriette's Herbal Homepage. http://www.henriettesherbal.com/

Hill J (1820) The family herbal. In: Brightley G, Kinnersley T, Bungay, UK

Hooper D (1937) Useful plants and drugs of Iran and Iraq. Field Museum of Natural History, Chicago, MI, USA

Jasmatha SK, Shenon A, Hedge K (2018) A review on *Beta vulgaris* (Beet-roots). Int J Pharm Chem 4:136–140

Johns CA (1870) Flowers in the field … etc., 12 edn. George Routledge & Sons, London, UK

K'Eogh J (1775) Botanologia universalis hibernica or a general Irish herbal. Lane, UK

Kops J, Hail HC, Trappen JE (1865) *Flora Batava*. Sepp JC, Amsterdam, The Netherland

Kühn CG (1829) *Medicorum graecorum opera quae exstant*. Pendanium Dioscoridem Anazarbeum, Lipsia, Germany

Lindley J, Moore T (1866) The treasure of botany … etc. Longmans, Greene, and Co., London, UK

Lister M (1709) Apicius (35 BC?) De arte coquinaria. Reprinted in: Lister M (1709) Apicii Coelii De opsoniis et condimentis., 2nd edn. Apud Jannsonio-Waesbergios, Amsterdam, the Netherlands

Magnol P (1636) *Botanicum* Montspelliense. Ex Officina Danielis Pech, Montpellier, France

Mattioli PA (1557) I discorsi di Pietro Andrea Mattioli, medico senese. Venice, Italy

Mattioli PA (1571) *De simplicium*. Apud Gulielmum Rouillium (sub scuto Veneto), Lyon, France

Messegué M (1979) Ha ragione la natura. Italy, Milan

Meyrick W (1790) The new family herbal. Thomas Pearson, Birmingham, UK

Miller P (1768) Gardener's dictionary. Printed by Francis Rivington et al, London, UK

Parkinson J (1629) Paradisi in sole paradisus terrestris, or a garden of all sorts of pleasant flowers. Printed by Humfrey Lownes and Robert Young, London, UK

Parkinson J (1655) Matthiae de L'Obel stirpium illustrationes. Warren, London, UK

Platina B (1529) De honesta voluptate etc. Ex Oficina Eucharn, Colonia, Germany

Pliny TE (75 A.D.) Historia naturalis. In: Giulio Einaudi Editore (ed) Storia naturale. Milan, Italy

Pratt A (1856) Common thing on the sea-coasts. Sea side plants. Society for Promoting Christian Knowledge, London, UK

Ray J (1690) Synopsis methodica stirpium Britannicarum … etc. Apud Samuel Smith, London, UK

Ray J (1738) Travels through the low-countries, 2nd edn, Germany, Italy, and France, London, UK

Ritze V (1599) *Dispensatorium* … etc. Apud Theobaldum Paganus, Lyon, France

Rivera D, Obón C, Heinrich M, Inocencio C, Verde A, Farajado J (2006) Gathered mediterranean food plants—ethanobotanical investigators and historical development. In: Heinrich M, Müller WE, Galli C (eds) Local mediterranean food plants and nutraceuticals. Forum Nutr., Karger, sorensen J (1529) *Dioscoridae pharmacorum simplicium*. Ioh. Schortum, Lyon, France

Ruel J (1552) *Pedanii Dioscoridis Anazarbei, de medicinali materia*. Apud Balthazarem Arnolletum, Lyon, France

Salmon W (1710) The English herbal, Daves, London, UK

Sorensen F, Marcussen C (1996) Rust *Uromyces betae* in Denmark inoculums sources and effect on sugar beet yield. In: 59th congres institute international de recherche betteravieres, Bruxelles, Belgium

Squalermo L (1561) liber de simplicibus … etc. Valgrisi, Venice, Italy

Sturtevant J (1919) Notes on edible plants. JB Lyon and Co., Albany, New York, USA

Tanara V (1674) Economia del cittadino in villa, Curti, Stefano, Venice, Italy

Taylor JE (1875) Science-gossip, an illustrated medium for interchange and gossip, London, UK

Thornton RJ (1812) Elements of botany. J. Whiting, London, UK

Trist JPO (1960) Protective flora of sea walls. Agriculture 67:228–231

Vigier J (1718) Historia das plantas da Europa e das mais usada. Anisson & Posuel, Lyon, France

von Megenberg K (1348) Puch der Natur. Stuttgart, Germany
Wiesemann JL (2010) Ultimate SAS survival (Donwnloaded by I-Pad). Harper Collins Publishers, London, UK
Williams C (1857) Picking on the sea-shore. Judd and Glass, London, UK

Chapter 8
Source of Useful Traits

Leonard W. Panella, Piergiorgio Stevanato, Ourania Pavli
and George Skaracis

Abstract   In the late 1800s, there already was speculation that *Beta maritima* might provide a reservoir of resistance genes that could be utilized in sugar beet breeding. European researchers had crossed *Beta maritima* and sugar beet and observed many traits in the hybrid progeny. It is impossible to estimate how widely *Beta maritima* was used in the production of commercial varieties, because most of the germplasm exchanges were informal and are difficult to document. Often these crosses of sugar beet with sea beet germplasm contained undesirable traits, e.g., annualism, elongated crowns, fangy roots, high fiber, red pigment (in root, leaf, or petiole) and much lower sucrose production. It is believed that lack of acceptance of *Beta maritima* as a reservoir of genes was because most of the evaluations of the progeny were done in early generations: The reactions of the hybrids *vulgaris* × *maritima* were not impressive, and it is clear now that they were not adequately studied in the later generations.

Keywords   Disease resistance · Rhizomania · Cercospora · Nematodes · Drought · Salt stress · Root rot · Curly top · Virus yellows · Powdery mildew · *Polymyxa betae*

Contrary to other species of the genus *Beta*, the evolutionary proximity between the sea beet and the cultivated types favors casual crosses (Hjerdin et al. 1994). Important characters of resistance to diseases, currently present in cultivated varieties, have been isolated from wild material (Table 8.1). According to several authors, *Beta maritima* is also an important means to increase the genetic diversity of cultivated types, now rather narrow from a domestication bottleneck and continuous selection for improvement of production and quality traits (Bosemark 1979; de Bock 1986; Doney 1998;

L. W. Panella (✉)
Colorado State University, 3944 Century Dr., Fort Collins, CO 80526, USA
e-mail: lee.panella@colostate.edu

P. Stevanato
DAFNAE, University of Padua, Padova, Italy

O. Pavli · G. Skaracis
Agricultural University of Athens, Athina, Greece

E. Biancardi et al. (eds.), *Beta maritima*,
https://doi.org/10.1007/978-3-030-28748-1_8

Table 8.1 Useful traits in the Genus Beta (Frese 2011, personal communication)

Beta and *Patellifolia* Taxa

TRAIT	1	2	3	4	5	6	7	8	9	10	11	12	13	14	15	16	17	18
Annual life cycle	■	■	■	■	■	■	■	■	■									
Monogermity					■	■	■			■	■	■	■		■	■	■	■
Hard seedeness										■	■			■	■		■	■
Seed shattering										■	■	■	■	■	■		■	■
CMS	■																	
Genetic male sterility							■											
Salt tolerance						■				■	■	■	■	■				
Frost tolerance										■	■	■	■	■	■			
Curly Top						■	■					■	■			■		
Yellowing viruses BYV						■						■	■					
Beet mild yellowing virus BMYV								■				■						
Beet mosaic virus BMV						■	■			■		■				■		
Beet necrotic yellow vein virus BNYVV																		■
Yellow wilt				■														
Peronospora farinosa				■	■		■			■			■					
Erysiphe betae																		■
Rhizoctonia solani			■															
Cercospora beticola			■										■	■		■	■	
Polymyxa betae																		
Black leg disease				■														
Erwinia subsp.																■	■	■
Heterodera schachtii						■										■	■	
Heterodera trifolii						■										■	■	
Meloidogyne hapla						■										■	■	
Meloidogyne incognita																		
Meloidogyne javanica																		
Meloidogyne arenaria																		
Myzus persicae						■					■							
Pegomya spp.											■							

1. *Beta vulgaris* subsp. *vulgaris* (*Bv*), 2. *Bv* leaf beet group, 3. *Bv* garden beet group, 4. *Bv* fodder beet, group, 5. *Bv* sugar beet group, 6. *Beta vulgaris* subsp. *maritima*, 7. *Bv* subsp. *adanensis*, 8. *Beta* (*Beta*) *macrocarpa*, 9. *Beta patulaI*, 10. *Beta corolliflora*, 11. *Beta macrorhiza*, 12. *Beta lomatogona*, 13. *Beta intermedia*, 14. *Beta trigyna*, 15. *Beta nana*, 16. *Patellifolia* (*Patellifolia*) *procumbens*, 17. *Patellifolia webbiana*, 18. *Patellifolia patellaris*

Jung et al. 1993; McGrath et al. 1999). This is especially true of sugar beet varieties, due to the common origin from the White Silesian Beet (Achard 1803; Fischer 1989), whose variability, according to Evans and Weir (1981), could have been enhanced by crosses with North Atlantic sea beet. Moreover, this narrowing of genetic diversity was increased through the widespread use both of Owen's cytoplasmic genetic male sterility (CMS) and the monogermy trait transferred to the current varieties by means of inbred lines (Jung et al. 1993; Owen 1945; Savitsky 1952). The attempts to transfer useful traits from sea beet are still underway. In a recent paper, Campbell (2010) described the performance of four crosses between *Beta maritima* and commercial varieties, which performed quite well, both in yield and resistance to some diseases (Rhizoctonia root and crown rot, rhizomania, powdery mildew, Cercospora leaf spot, Aphanomyces root rot, and Fusarium yellows).

However, the association of negative characters with the traits to be transferred often has made the improvement of cultivated genotypes difficult (Coons 1975; Mita et al. 1991). The major problems associated with such hybridizations are (1) the dominance of the annual life cycle in some wild forms, (2) the very bad shape of the root, (3) woodiness of roots, (4) elongated and multiple crowns, (5) low sugar content, (6) poor root yield, (7) low processing quality (Oltmann et al. 1984), (8) growth habit of the seed stalk, (9) prostrate seed stalk, (10) early seed shattering, etc. (Rasmussen 1932; van Geyt et al. 1990). Similar problems also arise when crossing sea beet with fodder, leaf, and garden beets. Several backcrosses and repeated selection cycles are necessary before such hybrids can acquire a satisfactory morphology and sufficient agronomic qualities (de Bock 1986; Munerati 1932).

The ancestors of the modern crops are defined as "crop wild relatives" (CWR), which also include other species closely related to them (Hajjar and Hodgkin 2007). Their commercial worth is invaluable (www.biodiversityinterna-tional.org). Many wild species, including *Beta maritima*, are threatened through reduction, degradation, or fragmentation of their habitat. Therefore, we need to identify not only the species to be protected in their respective areas but also the facilities for their in situ and ex situ conservation (Frese and Germeier 2009). Maxted et al. (2006) subdivided the species of the genus *Beta* into gene pools (GP) (Harlan and de Wet 1971) according to the difficulty of using the pool as a source of traits for the beet crops: (1) primary gene pool includes the cultivated forms (GP-1A) and the wild or weedy forms of the crop (GP-1B); (2) secondary gene pool (GP-2) includes the less closely related species from which gene transfer to the crop is difficult, but possible, using conventional breeding techniques; and (3) tertiary gene pool (GP-3) includes the species from which gene transfer to the crop is impossible or requires sophisticated techniques. Consequently, *Beta maritima* was classified as explained in Table 6.2. A PGR Forum was organized both to better define CWR and to compile a list of the more endangered species (Ford-Lloyd et al. 2009).

8.1 Resistances to Biotic Stresses

Most of the breeding work with *Beta maritima* has been to use it as a source of resistance to varied pests and diseases. Lewellen (1992) theorized that because the sugar beet and the white Silesian fodder beet source were developed and produced in the temperate climate of Northern Europe, there was less pressure to maintain plant resistance to biotic stress because of the mild disease incidence and "As a consequence, this narrowly based germplasm may never have had or may have lost significant levels of genetic variability for disease resistance or the factors that condition disease resistance occur in the germplasm at low frequencies" (Lewellen 1992). However, once sugar beet production moved out of Northern Europe, east into Russia and Asia, south into Mediterranean Europe and North Africa, and west into England and North and South America, many new diseases endemic to these areas limited production of sugar beet (Lewellen 1992).

The first documented instance of successfully transferring disease resistance from sea beet to sugar beet was by Munerati using sea beet growing in the Po Delta as a source of resistance to Cercospora leaf spot (Munerati et al. 1913a). Following Munerati's success, other European researchers began working with *Beta maritima* as a source of disease resistance (Margara and Touvin 1955; Schlösser 1957; Zossimovich 1939; Asher et al. 2001a). Nonetheless, for many of the reasons enumerated by Coons (1975), it is unlikely that much of this effort resulted in commercial varieties with sea beet in their genetic background, and due to the proprietary status of commercial germplasm, this information has not found its way into the literature.

8.1.1 Yellowing Viruses

Virus yellows (VY) is an important disease of sugar beet (Fig. 8.1). It is most severe and persistent in mild maritime climates such as Pacific coastal states of the USA, Western Europe, and Chile. These climates provide a long season for sugar beet for both root and seed crops, give a potentially continuous reservoir of virus–host sources, and favor the overwinter survival of the aphid species that transmit the viruses. VY is caused by the closterovirus *Beet yellows virus* (BYV), and the poleroviruses *Beet western yellows virus* (BWYV), *Beet chlorosis virus* (BChV) (Duffus and Liu 1991; Liu et al. 1999), and *Beet mild yellows virus* (BMYV). The principal aphid vector is the green peach aphid (*Myzus persicae* Sulzer) (Watson 1940) but many other species are known to vector one or more of these viruses. BMYV, BChV, and BYV can decrease sugar yield by at least 30%, 24%, and 49%, respectively (Smith and Hallsworth 1990; Stevens et al. 2004). Breeding for resistance in sugar beet started in Europe in 1948 and in 1957 in the USA (Bennett 1960; de Biaggi 2005; Duffus 1973; Duffus and Ruppel 1993; Hauser et al. 2000; Luterbacher et al. 2004; McFarlane and Bennett 1963; Rietberg and Hijner 1956; Stevens et al. 2004, 2005, 2006).

Fig. 8.1 Vein of beet yellows virus on sugar beet

Likely, the agents that cause VY have coevolved with *Beta* spp. It would seem then that a desirable place to search for high host–plant resistance to one or more of the viruses would be in the primary and secondary germplasms (Luterbacher et al. 2004; Panella and Lewellen 2007). Conventional breeding for resistance to VY has been moderately successful within sugar beet, but most sources of resistance are quantitatively inherited and have low heritabilities. This makes transfer from exotic sources to elite breeding lines and parents of hybrids very difficult. Other than the cultivated beet crops, *Beta maritima* would be the most logical place to find the desired genetic variability. However, little known research has been done within *Beta maritima* for VY resistance.

Grimmer et al. (2008a) reported that resistance to BMYV was identified in wild accessions and successfully transferred to early generation backcrosses with sugar beet. Luterbacher et al. (2004) assessed resistance to BYV in 597 *Beta* accessions collected worldwide and identified highly resistant individual accessions. Resistant individual plants were crossed with sugar beet plants to generate populations for mapping (Francis and Luterbacher 2003). The results from mapping these populations were reported by Grimmer et al. (2008b). Using AFLP and SNP markers, a locus controlling vein-clearing (Fig. 8.2) or mottling symptoms caused by incipient BYV infection was mapped to chromosome IV and given the name *Vc1*. Three BYV resistance QTLs were identified and mapped to chromosomes III, V, and VI. QTLs on chromosomes III and V acted only in plants showing mottled symptoms. Vein-clearing symptoms were controlled only in plants with allele *Vc1* on Chromosome VI. These results and concurrently run ELISA tests for BYV suggest that BYV resistance breeding can be facilitated by employing molecular marker techniques (Grimmer et al. 2008b) but the inheritance of resistance is still rather complex with unknown outcomes in the field.

Breeding for VY resistance at Salinas, CA had been one of the long-term objectives of the sugar beet breeding program starting in 1957 for BYV (McFarlane and Bennett

Fig. 8.2 Virus yellows inoculated trials at Salinas

1963), then changing to BWYV (Lewellen and Skoyen 1984), and then to BChV (Lewellen et al. 1999). Despite preliminary tests with wild beet species that suggested "It seems unlikely that any of the wild species tested will be of value in the program of breeding for resistance to beet yellows" (McFarlane and Bennett 1963), it seemed important to determine if higher, more heritable resistance could be found in *Beta maritima*. Several lines with resistance have been released from this later work, including C927-4 (Lewellen 2004d).

The development and traits of line R22 also called C50 and C51 (Lewellen 2000b) are discussed in Sects. 8.1.3 and 8.1.11.1. Other populations, for example, C26 and C27, containing *Beta maritima* germplasm also were developed (Lewellen 2000b). One of the objectives in breeding R22, C26, and C27 was to find higher resistance to VY from *Beta maritima*. Advanced cycle synthetics of R22 were further backcrossed into sugar beet and reselected for VY resistance (Lewellen 2004c). Spaced plants grown in the field were inoculated with BYV, BWYV, and/or BChV and selected on the basis of individual sugar yield and freedom from yellowing symptoms.

Trials in the UK with BChV were run to show that BChV caused significant losses (Stevens and Hallsworth 2003). At Salinas, compared to susceptible, unselected sugar beet, germplasm lines with *Beta maritima* had reduced losses to BChV (Table 8.2). However, in developing R22 and its backcrosses, moderately VY-resistant/tolerant sugar beet parents were used that showed similar responses to VY. It is unclear if any additional genetic variation for resistance was introduced from the *Beta* maritima sources. These tests did suggest, however, that mass selection for VY resistance based on components of sugar yield lead to higher sugar yield and percentage sugar performance than what might be expected for lines with up to 50% of their germplasm from *Beta maritima*.

Table 8.2 One component of virus yellows is *Beet chlorosis virus* (BChV). Comparison of breeding lines under BChV inoculated and non-inoculated conditions at Salinas, CA, including lines with germplasm from *Beta maritima*

Variety	References	Description	BChV Inoculated		% Loss[2]	Yellows score[3]
			SY[1] (kg/ha)	% Sugar		
Susceptible checks						
SP6322-0	Coe and Hogaboam (1971)	Selected without exposure to VY[4]	9860	14.3	36	6.9
US 75	McFarlane and Price (1952)	Selected from US 22	11,100	13.1	28	5.2
Virus yellows selected starting 1957						
C37	Lewellen et al. (1985)	VY selected from US 75	17,200	16.1	7	2.7
C31/6	Lewellen (PI 590799)	VY selected from US × European VY selections	16,200	15.4	7	2.9
C76-89-5	Lewellen (1998)	Full-sib family from C31/6	17,900	16.3	1	2.0
C69/2	Lewellen (2004a, b, c, d)	VY selected composite of all VY selections	19,000	17.0	6	3.5
Lines with Germplasm from *Beta maritima*						
C67/2	Lewellen (2004a, b, c, d)	10% *Beta maritima* through R22 (C51)	18,000	16.5	6	3.5
C26 × C27	Lewellen (2000b)	50% *Beta maritima* C37 × Atlantic *Beta maritima*	17,000	16.2	2	3.1
LSD(0.05)			1700	0.9		0.4

[1]SY is gross sugar yield (root yield × % sugar). Field trial area fumigated with methyl bromide in 2000 to reduce the effects of soilborne diseases and pests

[2]Relative % loss due to BChV calculated from variety means from adjacent companion tests planted on February 27, 2002, BChV inoculated on May 9, 2002, and harvested on October 15, 2002

[3]Virus yellows foliar symptoms scored every 3 weeks during chronic infection from late June to mid-August on a scale of 1–9, where 9 = 100% yellowed canopy. $r = 0.81**$ for % loss × VY scores

[4]VY = BYV, BWYV, and BChV in the USA

8.1.2 Beet Mosaic Virus

Infection by *Beet mosaic virus* (BtMV) is one of the most common diseases of sugar beet and other cultivated beets (Lewellen and Biancardi 2005). In California, it is almost always found in weed and wild beets of various origins growing near the Pacific coast in a perennial manner. The virus is transmitted nonpersistently by aphids including the green peach aphid (*Myzus persicae* Sulzer), often in association with VYs and is easily mechanically transmitted (Dusi and Peters 1999). It is common where cultivated beet is grown as a winter crop or overwintered for seed production (Shepherd et al. 1964). The damage caused by BtMV is small compared to that caused by VYs (Shepherd et al. 1964).

Because damage from most BtMV infections is modest, it has received low priority or no interest from breeders and seed companies. Major gene resistance was not known in sugar beet. However, in a self-fertile (*Sf*), annual (*BB*) line of sugar beet developed by Owen (1942) from Munerati germplasm (Abegg 1936), Lewellen (1973) identified an incompletely dominant gene that conditions resistance. He named this gene *Bm*. In both classical linkage and molecular marker research, this gene was found to be linked to the locus for genetic male sterility (*A1*) on Chromosome 1 (Friesen et al. 2006). The *Bm* allele was also backcrossed into biennial (*bb*) sugar beet backgrounds and evaluated under artificially inoculated conditions in replicated field trials (Lewellen et al. 1982). When all plants were inoculated in the four- to six-leaf stage, *BmBm/Bmbm* plants expressed high resistance, whereas the susceptible *bmbm* recurrent parents showed sugar yield losses that ranged from 8 to 22%. In singly and dually inoculated treatments with components of VYs, the damage caused by BtMV was additive as previously shown by Shepherd et al. (1964). BtMV-resistant breeding lines were released as C32 (PI 590675), C43 (PI 590680), and C719 (PI 590761) (Lewellen et al. 1982).

The *Bm* factor for resistance to BtMV was not found in *Beta maritima* directly, but in a sugar beet annual that likely had a *Beta maritima* source from Munerati's annual (Owen 1942). This suggests that even when not done intentionally, over time useful genes and traits from *Beta maritima* have probably enriched sugar beet germplasm.

8.1.3 Rhizomania

Rhizomania, caused by *Beet necrotic yellow vein virus* (BNYVV), is one of the most destructive diseases of sugar beet (Biancardi et al. 2002; Tamada and Baba 1973). BNYVV is transmitted by the obligate root parasite *Polymyxa betae* Keskin (Fujisawa 1976). Rhizomania was initially found in Italy (Fig. 8.3), then Japan, and it gradually spread over most sugar beet-growing areas worldwide (Biancardi et al. 2002; Brunt and Richards 1989; Scholten and Lange 2000). *Polymyxa betae* is distributed more widely than the BNYVV (Brunt and Richards 1989). Rhizomania is a disease, but its control is well reviewed by Biancardi and Tamada (2016).

Fig. 8.3 Roots severely
diseased by rhizomania
(above) and by cyst
nematodes (below). (Donà
dalle Rose 1951)

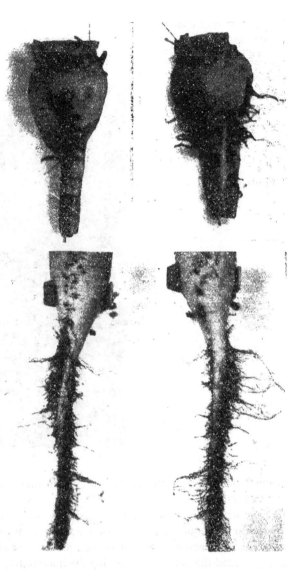

The first assessments of commercial varieties in rhizomania-infested fields began
in 1958 (Bongiovanni 1964), i.e., before the discovery of the disease's causal agent,
attributed to Canova (1966).[1] Results from early field tests (Fig. 8.4), along with data
from trials of seed companies from 1966 onward (Gentili and Poggi 1986), showed
clearly that Alba P and some other similar multigerm diploid varieties of Italian

---

[1]Canova used the Italian term "rizomania" for the disease, which had been introduced around
50 years earlier by Munerati (Munerati and Zapparoli 1915). According to Biancardi et al. (2010),
this term and not "rhizomania" should be employed for the disease.

Fig. 8.4 Susceptible variety sown between "Alba"-resistant multigerm families (San Pietro in Casale, Italy, 1979)

origin were the most productive varieties in rhizomania-infested soils (Biancardi et al. 2002).

The varieties in question also possessed good Cercospora Leaf Spot (CLS) resistance as a consequence of their parentage from Munerati's genotypes, from which the CLS resistance was obtained (Sect. 8.1.7). It is likely that these old genotypes also provided the genes conditioning the quantitative resistance to rhizomania carried by the variety Alba P (Biancardi et al. 2002; Lewellen and Biancardi 1990). It has been ascertained that the resistance of "Alba type" is governed by genes with additive effects (Biancardi et al. 2002; Frese 2010; Lewellen and Biancardi 1990). In the period from 1980 to 1985, the variety Rizor was bred at the SES-Italy breeding station, carrying a gene for qualitative rhizomania resistance (Fig. 8.5). The variety was much more productive than the varieties with quantitative resistance cultivated at the time (de Biaggi 1987). Additional information regarding the Alba and Rizor resistances is given in step 11, Sect. 1.7.

In 1983, rhizomania was first found in North America in a field located in California on the USDA-ARS station, Salinas, CA by R. T. Lewellen and confirmed to be BNYVV (Duffus et al. 1984). Individual beets, exhibiting symptoms of both necrotic yellow veins and root bearding, were found in a field where beet cyst nematode (*Heterodera schachtii* Schmidt) trials had been conducted. In order to enrich the nematode inoculum, soil had been incorporated from several commercial sugar beet fields reported to be infested with beet cyst nematode (McFarlane et al. 1982). It may be that the root damage on nematode-resistant genotypes, owing to the *Patellifolia procumbens* resistance, was not due to sensitivity to cyst nematode infection, as reported by McFarlane et al. (1982), but instead was due to BNYVV.

Following the initial reports on rhizomania to the sugar beet industry in 1983, suspicious fields were further reported in several locations. One of these was the variety trial field of Holly Sugar's breeding program at Tracy, CA, where severe damage was observed by Erichsen on all entries except for one series of experimental three-way hybrids. The researchers at Salinas were asked by Erichsen to visit the trial (Fig. 8.6). It was determined that BNYVV rather than cyst nematode likely caused this differential reaction (Biancardi et al. 2002) (Fig. 8.7).

Plants from Holly experimental hybrids were crossed to susceptible sugar beet, and the F1 plants were selfed. In a field test at Salinas under rhizomania conditions, 13-week-old individual S_1 families were either homozygous susceptible or segregated approximately 3 resistant:1 susceptible, thus supporting the hypothesis that resistance was controlled by a single dominant gene (Lewellen et al. 1987) (Fig. 8.8). Individually and collectively, the segregating S_1 families fitted the expected 3:1 (resistant:susceptible) ratio (Fig. 8.9).

Fig. 8.5 Rhizomania diseased field at Phitiviers, France, showing the resistant plot (1983)

Fig. 8.6 Rhizomania diseased field at Tracy, CA (1983)

Fig. 8.7 Susceptible variety USH11. Non-fumigated (left) and fumigated soil

The gene for resistance, unofficially called the "Holly" gene, initially was named *Rz* (subsequently referred to as *Rz1*) (Lewellen 1988). The source of *Rz1* could not be determined by pedigree and breeding records (Erichsen, personal communication, 1987), but it is thought that it likely arose from unknown or unintended outcrosses to *Beta maritima*, as no other similar gene could be found within cultivated beets (Biancardi et al. 2002). This gene provided high-level resistance to BNYVV. The resistance found in the commercial cultivar "Rizor" (developed by SES in Italy) (Biancardi et al. 2002; de Biaggi 1987; de Biaggi et al. 2003) and *Rz1* are the only major resistance genes found in the commercial sugar beet gene pool (Biancardi et al. 2002; Scholten and Lange 2000). The origin of the quantitative resistance to

Fig. 8.8 S1 families under rhizomania at 10 weeks, Salinas CA, 1986

Fig. 8.9 Roots showing segregation within S1 family at 13 weeks, Salinas CA, 1986

rhizomania "type Alba" and qualitative (type "Rizor" and "Holly") is attributable to materials derived from crosses with *Beta maritima* and obtained from Munerati (Biancardi et al. 2002). More recently, using molecular tools, it was confirmed that the resistance found in Rizor and the Holly material did not come from separate genetic sources (Stevanato et al. 2015). This evidence is indicative of the fact that the SES pollinator used most likely originated from the Ro 281 family (from Munerati's work) or a similar germplasm, which had been probably bred in public and private programs and then found its way to Holly Sugar through typical exchanges of germplasm (Panella and Biancardi 2016).

Once rhizomania was recognized in California, an extensive program to find host resistance by screening *Beta* genetic resources (cultivated and wild) was initiated by the USDA-ARS at Salinas. The identified resistance sources were incorporated into elite sugar beet germplasm (Biancardi et al. 2002). The *Rz1* allele proved to be handled easily in breeding programs. Resistance breeding to rhizomania has deployed the *Rz1* gene in elite germplasm worldwide (Amiri et al. 2009; Azorova and Subikova 1996; Barzen et al. 1997; Lewellen et al. 1987; Nouhi et al. 2008; Thomas et al. 1993; Whitney 1989b). However, as single dominant resistance genes often are eventually overcome by mutations in a variable pathogen gene pool, additional sources of resistance were sought by breeding programs worldwide. Since no additional resistant sources were found in the cultivated sugar beet gene pool, various genetic resources, especially *Beta maritima* accessions, were screened for rhizomania resistance (Francis and Luterbacher 2003; Geyl et al. 1995; Panella and Lewellen 2007).

The USDA-ARS germplasm improvement program used two different breeding approaches. The first breeding method focused on major gene resistance. When discovered, genes were backcrossed into elite sugar beet germplasm. Lewellen and coworkers identified several BNYVV-resistant *Beta maritima* accessions (Lewellen 1995a, 1997a), using field resistance and levels of virus titer (by ELISA) as preliminary evaluation assays (Whitney 1989b). A resistant accession from Denmark, WB42, was crossed with sugar beet parental line C37 (Lewellen et al. 1985) and was released as germplasm C48 and C79-3 (Lewellen 1997a; Lewellen and Whitney 1993). This resistance was shown to be different from *Rz1*. In growth chamber tests, it conferred higher resistance than *Rz1* and was designated as *Rz2* (Scholten et al. 1996, 1999). Thus far, there are five sources of resistance conditioned by a single gene from *Beta maritima*, although most sources have been shown to be either *Rz1* or *Rz2* (Biancardi et al. 2002; Panella and Lewellen 2007). *Rz3*, which maps to chromosome III, has been shown to be linked to *Rz1* and *Rz2* (Gidner et al. 2005). The source of *Rz3* is a *Beta maritima* accession, WB41 (Denmark). There is a variable BNYVV-resistant expression in the heterozygote in the genetic background in which it has been evaluated.

Nonetheless, sugar beets with the combination of *Rz1* and *Rz2* or *Rz3* (in the heterozygous state) showed a lower virus titer than *Rz1* alone (Gidner et al. 2005). Using R36 (Lewellen and Whitney 1993), a composite population of many *Beta maritima* accessions, Grimmer et al. (2007) identified a major QTL, named *Rz4*, that appeared to be different from *Rz1*, *Rz2*, or *Rz3* and also located on chromosome III. Using a mapping population, based on C79-11 as the resistance donor, another potential resistance gene, referred to as *Rz5*, was identified (Grimmer et al. 2008c). The resistance in C79-11 (Lewellen and Whitney 1993) was from *Beta maritima* accession, WB258 (step 12, Sect. 1.7). *Rz4* and *Rz5* map close to *Rz1* and each other, thus raising the possibility of belonging to an allelic series.

In the Imperial Valley (IV) of California (near the border with Mexico) in 2003, resistant hybrids, winter beet cultivars carrying the *Rz1* gene, showed rhizomania symptoms in a few fields. Over the next couple of years, laboratory, greenhouse, and field tests at Salinas confirmed that *Rz1* resistance gene had been overcome (Liu et al. 2005; Rush et al. 2006). Since then, resistance-breaking strains have been found in major growing regions, including Colorado, Idaho, Minnesota, Nebraska, and Oregon (Liu and Lewellen 2007). Only partial resistance to these strains of BNYVV is conferred by *Rz2* and *Rz3* from *Beta maritima*, although combinations of *Rz1* and *Rz2* appear to condition more resistance than either alone. Encouragingly, progeny families of C79-9 (resistance from *Beta maritima* accession WB 151–PI 546397) appeared to have higher levels of resistance to resistance-breaking strains of BNYVV (Lewellen 1997a; Panella and Lewellen 2007).

The emergence of resistance-breaking strains of BNYVV rekindled the interest in the C79 populations with multiple, different sources of rhizomania resistance backcrossed to C37, created by Lewellen at Salinas (Lewellen et al. 1985; Lewellen 1997a, b). The 11 germplasms in the C79 series were from different genetic sources of resistance to BNYVV. They had been backcrossed 1 to 6 times with C37 (Lewellen 1997a, b). The seed from these sources had been poly-crossed in the field at Salinas and, following selection, was designated as R740 and placed in storage (Panella et al. 2018). With the renewed interest in other sources of genetic resistance, this seed was sent to the USDA breeding program at Fort Collins, Colorado. SNP markers, which were linked to *Rz1* and *Rz2* (Stevanato et al. 2012, 2014a; Panella et al. 2015a, b), were used to select individual plants. Two germplasms were released from this project: FC1740 was selected as homozygous resistant to SNP markers linked to both *Rz1* and *Rz2* resistance genes (inferred genotype—*Rz1Rz1Rz2Rz2*), and FC1741 was selected as homozygous to the marker linked to the *Rz2* gene for resistance and homozygous susceptible for *rz1* (inferred genotype—*rz1rz1Rz2Rz2*) (Panella et al. 2018). There is a possibility that other resistance genes may also be present in these germplasms but there were no SNP markers publicly available to ascertain this at the time of their release.

The second breeding method involved individual screening of *Beta maritima* populations and pooling the selected resistant plants—a composite approach (Doney 1993). The pooled plants were increased in mass, and there was no effort to classify the resistance sources as *Rz1*, *Rz2*, etc., or other factors. Several breeding populations were developed using this method and have been released as C26, C27, C51, R21, C67, R23, R23B, and R20 (Lewellen 2000b, 2004b). Although there are most likely major genes in these populations, the existence of additional minor resistance genes may eventually lead to a more durable resistance.

In an attempt to discover novel sources of quantitative multigene resistance, Richardson et al. (2019) conducted a thorough screening of available *Beta maritima* germplasm collection under field and greenhouse conditions using both resistance-breaking and nonresistant-breaking strains of BNYVV. Overall findings from field and greenhouse assays pointed to the superiority of accessions from Denmark in combating BNYVV as well as resistant breaking strains of BNYVV, thus providing evidence for their possible exploitation as pre-breeding donor material in future efforts aiming at the development of rhizomania-resistant varieties.

Recently, the University of Padua, Italy, through a sponsored research project, has collected seeds of 35 populations of *Beta maritima* along the Italian and Croatian coasts of Adriatic Sea. Representative seed samples from each population were planted the year after collection both in the field and glasshouse. Molecular analyses were performed in order to examine the presence of the *Rz1* source of resistance. Preliminary results showed that the frequency of the *Rz1* allele was significantly higher in sea beet populations collected on the Italian Adriatic coast. This would provide additional genetic proof about the speculated origin of *Rz1* from the Italian sea beet gene pool (Stevanato, personal communication). In a collaborative project between the University of Padua and the USDA Fort Collins program, 24 individuals from 64 populations were screened with markers for *Rz1* and *Rz2*. Many populations contained the *Rz1* SNP marker, while there were areas where the *Rz2* marker was present (unpublished data). A big future challenge is to determine the allelic diversity within these populations and to gain insight into its effect in relation to the level of resistance.

8.1.4 Beet Curly Top Virus

Curly top in beets is caused by a mixture of at least three closely related Curtoviruses in the family *Geminiviridae*: *Beet curly top virus* (BCTV), *Beet mild curly top virus* (BMCTV), and *Beet severe curly top virus* (BSCTV) (Strausbaugh et al. 2008). They are all transmitted by the beet leafhopper, *Circulifer tenellus* Baker (Fig. 8.10), which attacks sugar beet and many other crops cultivated in semi-arid areas (Western USA, Mexico, Turkey, and Iran) (Bennett 1971; Bennett and Tanrisever 1958; Briddon et al. 1998; Duffus and Ruppel 1993; Panella 2005b). Similar viruses occur in Argentina, Uruguay, and Bolivia (Bennett 1971).

Almost as soon as the sugar beet industry was established in the Western United States, BCTV severely impacted yields (Bennett 1971; Carsner 1933; Murphy 1946). Production in California was begun in 1870, and shortly thereafter BCTV symptoms were observed on beets grown there, and by the 1920s, it was clear the sugar beet industry required varieties with resistance to BCTV to survive (Bennett 1971; Bennett and Leach 1971; Carsner 1933; Coons 1953; Murphy 1946) (Fig. 8.11). The early breeding efforts resulted in the release of US 1, a curly top-resistant open-pollinated variety that was a huge step forward (Carsner 1933). At the time of its release, researchers already were looking at *Beta maritima* as a potential source of resistance to BCTV (Coons et al. 1931), which probably is why Coons was commissioned in 1925 to collect *Beta maritima* in Europe (Coons et al. 1955). Further increases in resistance to BCTV were achieved with US 33 and US 34 selected from heavily curly top infested fields of US 1, and eventually they were superseded by US 12 and US 22, which were further improved in US 22/2 and US 22/3 (Coons et al. 1955). However, as stated by Coons et al. (1955): "Hybridizations [of *Beta maritima*] with sugar beets were made and the segregating generations were selected for both leaf spot resistance and curly top resistance. The outlook of obtaining resistant strains in this way was promising but not more so than from the selections made from the sugar beet itself. Since breeding work with the sugar beet did not present the problems of ridding the progenies of multicrowns and rootiness, the emphasis on wild hybrids gradually dwindled."

Fig. 8.10 Leafhopper
(*Circulifer tenellus*)

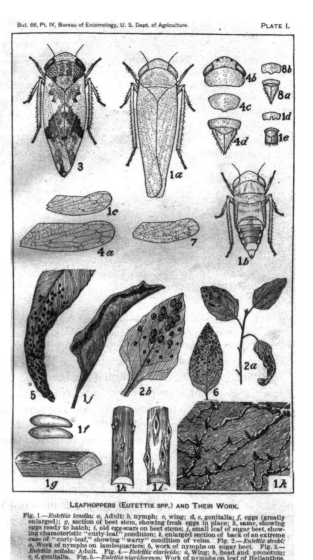

Bul. 66, Pt. IV, Bureau of Entomology, U. S. Dept. of Agriculture.   PLATE I.

LEAFHOPPERS (EUTETTIX SPP.) AND THEIR WORK.

Fig. 1.—*Eutettix tenella:* a, Adult; b, nymph; c, wing; d, e, genitalia; f, eggs (greatly
enlarged); g, section of beet stem, showing fresh eggs in place; h, same, showing
eggs ready to hatch; i, old egg-scars on beet stems; j, small leaf of sugar beet, showing characteristic "curly-leaf" condition; k, enlarged section of back of an extreme
case of "curly-leaf," showing "warty" condition of veins. Fig. 2.—*Eutettix strobi:*
a, Work of nymphs on lambsquarters; b, work of nymphs on sugar beet. Fig. 3.—
Eutettix scitula: Adult. Fig. 4.—*Eutettix clarivida:* a, Wing; b, head and pronotum;
c, d, genitalia.   Fig. 5.—*Eutettix nigridorsum:* Work of nymphs on leaf of Helianthus.
Fig. 6.—*Eutettix straminea:* Work of nymphs on leaf of another Helianthus. Fig.
7.—*Eutettix insana:* Wing.

Despite what Coons states, Owen speculated that his source of extreme resistance
to BCTV, which he called "strain 286", was most likely a chance hybridization with
a "wild beet" in California (Owen et al. 1939). We know that wild beets in California
encompass introductions of *Beta macrocarpa* and *Beta maritima* from Europe, and
may include feral domestic beets (chard, table beet, sugar beet) (Bartsch et al. 1999;
Carsner 1928; McFarlane 1975). Owen also declared "However, some accidental
hybridization of parental strains of US 1 and progenies comparable in origin with
286 is now suspected." Certainly, the spangled roots of early 286 progeny in the

Fig. 8.11 Beets diseased by BCTV (left)

photograph in the 1946 Proceedings of the ASSBT (Owen et al. 1946) resemble progeny of sugar beet crossed with a sea beet. It is during the development of US 1 that Carsner comments on the wild beets in southern California (Carsner 1928), which lends credence to Owen's remarks. The performance of 286 showed extreme resistance to curly top (Carsner 1926; Owen et al. 1946). CT9 and later, C569, which were widely used in the Western USA as components of curly top-resistant hybrids, were derived from this line (McFarlane et al. 1971; Owen et al. 1946). This example of *Beta maritima* being a largely unrecognized source of resistance and yet being characterized by Coons as difficult to work with when other sources were present in the sugar beet germplasm typifies the attitude of many of the commercial breeders who made little use of sea beet germplasm during the first 60 years of the last century (Lewellen 1992). Most of the beet curly top-resistant material in use today stems from this gene pool, which was widely used by USDA-ARS plant breeders and provided sources of strong resistance to curly top and may have been a source of resistance to other diseases. Nonetheless, there is continued screening of sea beet for resistance to all of the curly top viruses in a cooperative curly top nursery managed by the Beet Sugar Development Foundation and USDA-ARS planted in Kimberly, Idaho (Doney 1998; Hanson and Panella 2002b, 2003b, 2004a; Panella 1998b, 1999a, 2000b; Panella and Hanson 2001b; Panella and Strausbaugh 2011a, b, 2013; Strausbaugh and Panella 2014, 2015, 2016, 2017). In a recent search of the USDA-ARS National Plant Germplasm System's (NPGS) Germplasm Resources Information Network (GRIN) Database, there are two *Beta maritima* accessions that had better resistance than intermediate (rating of <5; 0 to 9 scale; immune to dead) to beet curly top (PI 518338 and PI 504185) (USDA-ARS 2011a).

8.1.5 Powdery Mildew

Damage from powdery mildew caused by *Erysiphe polygoni* DC (syn. *E. betae* Weltzien) is common almost everywhere sugar beet is grown. Major gene resistance has not been found in sugar beet germplasm; however, quantitatively conditioned tolerance is known and widely used in commercial varieties (Lewellen 1995b; Whitney et al. 1983). In an initial screen of *Beta maritima* accessions at Salinas in field plots in the late 1970s and early 1980s, resistance to powdery mildew was identified in several accessions. In greenhouse tests on seedlings plants, Whitney (1989a) confirmed that high resistance segregated among these accessions.

Two accessions (WB97 and WB242) that showed high resistance were chosen as sources of resistance in a program to determine the inheritance of resistance and transfer this resistance to sugar beet (Lewellen 2000a). WB97 (PI 546394) was in the Salinas collection assembled and evaluated by McFarlane. WB97 was sent to Salinas from the Japan Sugar Beet Improvement Foundation in 1968 and identified as *Beta patula* WB46 from the Wageningen collection. If WB97 (WB46) is *Beta patula*, then it would have been collected from dos Embarcaderos near Madeira (Lange et al. 1999). McFarlane noted that WB97 was variable and did not have typical *Beta patula* characteristics and was more likely *Beta maritima* or crosses between *Beta patula* and *Beta vulgaris/Beta maritima*. Resistance to powdery mildew was transferred from WB97 to sugar beet, and a series of germplasm releases identified as CP01, CP03, CP05, and CP07 were made (Lewellen 2000a, 2004a, b). Resistance is conditioned by one dominant gene (Lewellen and Schrandt 2001) (Figs. 8.12 and 8.13).

WB242 (PI 546413) was obtained for the Salinas collection from Rietberg, Bergen op Zoom, the Netherlands in May 1974. It was reported to have been collected from the Loire River Estuary, France, and to have reduced nematode cyst counts in tests

Fig. 8.12 Segregation for reaction to *Erysiphe polygoni* within plot of CP04 with WB242 source

Fig. 8.13 Adjacent 5-monts-old plants segregating for reaction to powdery mildew

at IRS, Bergen op Zoom. It is probably similar to other accessions obtained from the Netherlands including one called Le Pouliguen Group 2 (PI 198758–59) received from Boss in 1987. Germplasm developed from the introgression of powdery mildew resistance into sugar beet from WB242 has been more extensively studied than that from WB97. Sequential backcrosses and improvements were released as germplasm lines CP02, CP04, CP06, CP08, and CP09CT (Lewellen 2000a, 2004a, b).

Resistance to powdery mildew from WB242 is conditioned by one major gene named *Pm* (Lewellen and Schrandt 2001). Molecular markers to this resistance factor were identified (Janssen et al. 2003; Weiland and Lewellen 1999). WB242 is susceptible to rhizomania and backcrosses to introgress *Pm* into sugar beet utilized recurrent sugar beet lines that had resistance to rhizomania (*Rz1*). During field tests under both rhizomania and powdery mildew conditions, it was observed that derivatives from line CP02 also carried resistance/tolerance to sugar beet cyst nematode. Population CN12 was released as a source for resistance genes for powdery mildew (*Pm*), rhizomania (*Rz1*), and sugar beet cyst nematode in a background with adaptation for the Western USA (Lewellen 2006b). Other releases have included CN12-446, CN12-751, CN12-770, CN12-8-407, CN07-410, CN07-413, and CN18-438 (Lewellen, unpublished). Although resistance to downy mildew caused by *Peronospora farinosa* (Fr.) Fr. f.sp. *betae Byford* (syn. *Peronospora schachtii* Fckl.) has been reported (Dale et al. 1985), we are not aware of any breeding programs using this source for commercial varieties.

8.1.6 Root Rots

Rhizoctonia crown and root rot of sugar beet (caused by *Rhizoctonia solani* Kühn) affects or threatens sugar beet-growing areas worldwide (Ahmadinejad and Okhovat 1976; Büttner et al. 2003; Herr 1996; Ogata et al. 2000; Panella 2005c; Windels et al. 2009). In the USA, where it is registered for use, Quadris™ (an azoxystrobin fungicide) effectively controls this disease; however, the timing of application is critical (Stump et al. 2004). As crop rotations are shortened in the USA, Europe, and worldwide, this disease is becoming an increasing problem. Rhizoctonia root rot is best managed through an integrated program, based on resistant germplasm using good cultural practices and timely fungicide application (Herr 1996).

In the 1950s, Gaskill (USDA-ARS at Fort Collins, Colorado) began a Rhizoctonia crown and root rot resistance breeding program primarily based on the Great Western Sugar Co. (GWS) sugar beet germplasm (Lewellen 1992; Panella 1998a). Schneider and Gaskill (1962) also were looking at introduced germplasm at that time. Although in their report most everything is described as *Beta vulgaris* (Schneider and Gaskill 1962), they comment that much of the material is annual, which suggests that if it is not *Beta maritima*, it had most likely hybridized with it at some point. Some of this *Beta maritima* germplasm made its way into SP5831, released for resistance to Aphanomyces black root (Doney 1995). This source, as well as other sources of *Beta maritima*, was incorporated into some of the early Rhizoctonia-resistant releases. These included FC706 (Hecker and Ruppel 1979), FC708 (Hecker and Ruppel 1981), and FC710 (Hecker and Ruppel 1991; Panella 1998a, 2005c). Although commercial sugar beet breeding companies used and exchanged this germplasm, much of this activity was informal and it is not easy to document the use of *Beta maritima* (Lewellen 1992).

Since the 1980s, efforts to screen *Beta maritima* for new sources of resistance to *R. solani* have increased (Asher et al. 2001b; Burenin 2001; Luterbacher et al. 2000, 2005; Panella and Frese 2003; Panella and Lewellen 2007). Most of the Rhizoctonia-resistant germplasm (commercial and public) can trace its parentage to the USDA-ARS program at Fort Collins, Colorado, started by Gaskill (Panella 2005c). This program continues to screen *Beta maritima* for resistance to *Rhizoctonia solani* and to incorporate resistant accessions into enhanced germplasm for release (Hanson and Panella 2002c, 2003c, 2004b, 2005, 2006, 2007; Panella 1999b, 2000c; Panella et al. 2008, 2010, 2011b, 2012, 2013, 2014, 2015a, 2016; Panella and Hanson 2001c; Panella and Ruppel 1998).

Fusarium yellows is an important soilborne disease found in sugar beet (*Beta vulgaris* L.) production areas throughout sugar beet-growing areas worldwide (Christ and Varrelmann 2010; Panella and Lewellen 2005; Hanson et al. 2018). Many *Fusarium* species have been reported to cause Fusarium yellows (Hanson 2006; Hanson and Hill 2004; Hanson and Lewellen 2007; Ruppel 1991; Windels et al. 2009); however, the primary causal agent in sugar beet is *Fusarium oxysporum* Schlechtend. Fr. f. sp. *betae* (Stewart) Snyd & Hans. (Stewart 1931). The severity of *Fusarium* yellows is influenced by temperature, inoculum dose, and presence of sugar beet cyst nematode (*Heterodera schachtii* Schm.) (Gao et al. 2008; Hanson et al. 2009a, b; Landa et al. 2001). When conditions favor its occurrence, yield losses can be devastating (Hanson et al. 2009a, b).

Unfortunately, *F. oxysporum* f. sp. *betae* is highly variable in its morphology, pathogenicity, and genetic structure (Harveson and Rush 1997; Hanson et al. 2018; Hill et al. 2011; Ruppel 1991). Other species of *Fusarium* also have been shown to cause yellowing-like symptoms on sugar beet (Burlakoti et al. 2012; Hanson and Hill 2004). Research to date has identified resistant commercial cultivars and a high degree of variability in virulence (Hanson et al. 2009a, b). Management of this disease is heavily dependent on the use of resistant hybrid cultivars (Franc et al. 2002; Hill et al. 2011). In sugar beet, *F. oxysporum*-resistant lines are known, but the genetic system that controls *Fusarium* diseases is still unclear (de Lucchi et al. 2017). Some public breeding has been done, and *Beta maritima* accessions do have resistance (Panella et al. 2015b). Currently, germplasms containing *Beta maritima* germplasm are being screened by the USDA sugar beet breeding program in Fort Collins, Colorado, and field resistance is correlated to molecular markers (unpublished data) linked with resistance to *F. oxysporum* f. sp. *betae* (de Lucchi et al. 2017).

8.1.7 Cercospora Leaf Spot

Cercospora leaf spot (CLS) caused by the fungus *Cercospora beticola* Sacc. is the main fungal disease of beet-growing areas in temperate and humid environments (Fig. 8.14) and affects approximately one-quarter of the cultivated acreage (Holtschulte 2000; Jacobsen and Franc 2009). Pioneering studies on genetic resistance to CLS began in the late 1800s, but only in the early 1900s did the efforts in hybridization and selection made by Munerati achieve the first results. No other source of resistance has been isolated against this disease and incorporated into sugar beet cultivars, except for the "C2 form", which was active only against rarely distributed strains (Lewellen and Whitney 1976). Therefore, the CLS-resistant varieties currently used are derived from crosses with *Beta maritima* obtained by Munerati (de Bock 1986). Mass selections on sea beet began on plants sown in cultivated soil, followed by inbreeding, with the main objective being to fix enough bienniality (Munerati et al. 1913b). Crosses with the sea beet were begun, first using predominantly biennial lines, followed by a number of backcrosses to eliminate the negative traits of the wild parents (fangy and fibrous roots, tendency toward bolting, etc.). Further selections improved bolting resistance and, after 10 years, led to the release of the line RO581, which was considered the first substantially improved CLS-resistant line (Coons et al. 1955). The line was distributed to public and private breeding stations. The American variety US201 is cited as one of the oldest derived lines, together with the Italian Cesena R and Mezzano 71, the Polish Buszczynski CLR, the French Desprez RC2, and the Dutch Vanderhaven AC (Bongiovanni et al. 1958). The increased effort of the breeding companies has produced an improvement in sugar

Fig. 8.14 Drawing of beet moderately diseased by CLS (KWS Cercospora Tafel)

yield and bolting resistance, which had been the main negative traits of the CLS-resistant varieties. With the recent breeding progress, sugar yield is today at similar levels to that of the susceptible varieties (Panella and Lewellen 2007) (Fig. 8.15).

It has been estimated that a severe epidemic in the USA can cause up to a 42% loss of gross sugar (Smith and Ruppel 1973), or up to a 43% relative dollar loss (Shane and Teng 1992). In the USA, initial breeding efforts were based on inbred germplasm developed from Pritchard's (1916) lines and other European lines (Coons 1936) along with germplasm selected by American Crystal in the Arkansas Valley of Colorado (Skuderna 1925). However, as this breeding effort was getting underway, there was another source of Cercospora resistance brought into the USA from Europe (Coons et al. 1955). This material had been seen by Coons in 1925 when it still had many of the undesirable traits from *Beta maritima.* It had been further developed by Italian breeders, and by the time Coons saw it again in 1935, it had been greatly improved (Coons et al. 1955).

The Italian germplasm was incorporated into Great Western Sugar Company varieties GW 304 and GW 359 (source Cesena) and the USDA-ARS researchers also used "Mezzano 71" (Coons et al. 1955; Lewellen 1992). Brewbaker et al. (1950) also referred to breeding lines from some other crosses with European *Beta maritima*, as well as wild beet (most likely *Beta maritima*) out of California. Although it is not known if US 201 (PI 590678) developed from Mezzano 71 was ever used in a commercial hybrid (Lewellen 1992), it found its way into many of the ARS breeding programs (Panella 1998a). It is these early CLS-resistant germplasm pools

Fig. 8.15 Performance of CN12 progenies under severe nematode conditions, Imperial Valley, May 2007. Individual plants from CN12 were selfed and the S1 progeny evaluated under severe nematode conditions in overwintered Imperial Valley. This picture contrasts the differences in reaction to SBCN under these conditions among sets of S3 lines that had been selected for NR (foreground) and nematode susceptibility (background)

that formed the basis of Cercospora resistance breeding in the USA, and much of that resistance came from the *Beta maritima* sources out of Munerati's program and, later, from the curly top germplasm that was added to the Cercospora breeding pools to incorporate resistance to these two important diseases. Further efforts at breeding for resistance in ARS to CLS were focused on combining CLS resistance with other disease resistances, mainly through inbreeding (Panella 1998a). These early breeding efforts have been reviewed in several publications (Coons 1975; Coons et al. 1955; Lewellen 1992; Panella and McGrath 2010; Skaracis and Biancardi 2000). In the last 40 years, because of the renewed interest in using *Beta maritima* as a genetic resource in sugar beet breeding, developing new sources of resistance to CLS has become an important goal. Efforts in the 1980s by the USDA-ARS Sugar Beet Crop Advisory Committee (now Crop Germplasm Committee—CGC) focused on evaluations of sea beet for resistance to CLS as one of the most important goals (Doney 1998). In Europe, innovative methods to introgress genes from sea beet into sugar beet were developed by Bosemark (1969, 1971, 1989), which lead to the efforts of the Genetics and Breeding Work Group of the IIRB to develop "buffer populations" for CLS resistance, as described by Frese et al. (2001) in an example for rhizomania. Efforts in evaluating *Beta maritima* in Europe and the United States were intensified, and some of this germplasm with CLS resistance was discovered (Panella and Frese 2000). In the USA, sea beet germplasm has been screened by the Sugar Beet CGC since 1986 (Hanson et al. 2009a, b, 2010, 2011; Hanson and Panella 2002a, 2003a; Panella 1999c, 2000a; Panella and Hanson 2001a; Panella et al. 1998), and

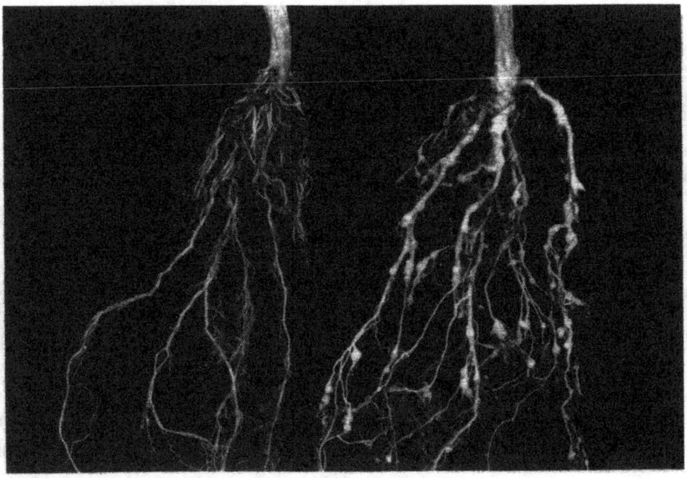

Fig. 8.16 Nematode resistance in commercial hybrids derived from *Beta maritima*. In this picture, two commercial hybrids (SBCN susceptible on left, partially resistant on right) are shown in an Idaho, USA field under SBCN conditions (courtesy Betaseed, Inc). Hybrids with partial resistance to SBCN are now being commercially grown across the northern growing areas of USA. Unlike the *Beta procumbens* resistance, yield drag does not occur in the absence of *Heterodera schachtii*

there are now 123 accessions in GRIN of *Beta maritima* that have been screened for resistance to CLS. Of these, 13 were rated as very resistant (3 < on a scale of 1 = no disease to 9 = dead) (USDA-ARS 2011b) (Figs. 8.16). The GENRES CT95 42 project in Europe evaluated 82 *Beta maritima* accessions, 10 of which were scored very resistant (3<; same scale) (Frese 2004a). Many of these accessions have been incorporated into breeding programs, which are being released to increase the genetic base of the CLS-resistant commercial varieties (Panella and Lewellen 2007; Panella et al. 2015b).

8.1.8 *Polymyxa Betae*

Polymyxa betae (Fig. 8.17) is the vector of numerous soilborne viruses of sugar beet (Abe and Tamada 1986; Kaufmann et al. 1992; Liu and Lewellen 2008; Wisler et al. 1994), including *Beet necrotic yellow vein virus* (BNYVV), the cause of rhizomania (Tamada and Baba 1973). BNYVV is transmitted by viruliferous zoospores of this plasmodiophorid protozoan. *Polymyxa betae* is an obligate parasite and is found in almost every soil in which sugar beet is grown (Liu and Lewellen 2007). Beet is infected by anterior bi-flagellate zoospores. *Polymyxa betae* forms long-living resting spores clustered together to form cystosori. Viruliferous cystosori can survive many decades in the field. The life cycle, ecology, and infection process have been well documented (Keskin 1964; Tamada and Asher 2016a, b). As a parasite per se,

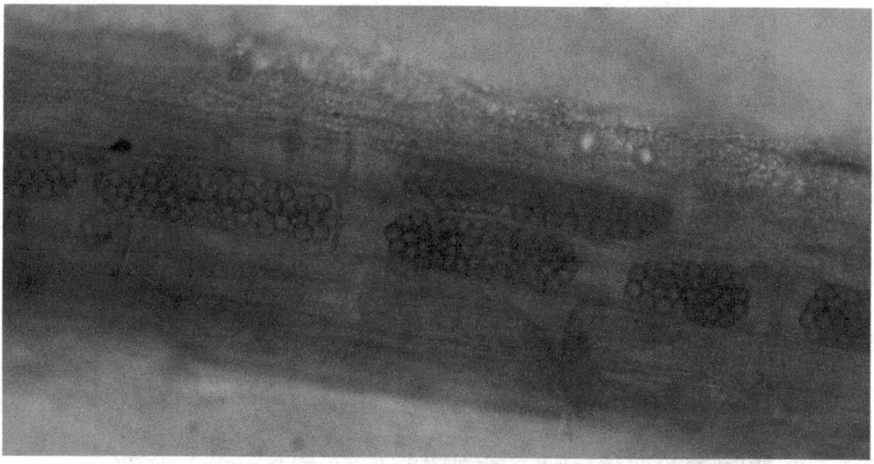

Fig. 8.17 *Polymixa betae* is the vector of BNYVV. Shown here are *Patellifolia betae* cystosori in sugar beet root cells (courtesy John Sears)

Polymyxa betae is usually not considered to cause measurable damage. However, in well-designed and controlled tests, it has been shown to cause reductions in yield (Liu and Lewellen 2008; Wisler et al. 2003).

To quantify the level of *Polymyxa betae* in sugar beet roots, in addition to microscopic techniques, end-point PCR methods were developed (Mutasa et al. 1993, 1995, 1996). However, these methods only indicate *Polymyxa betae's* presence or absence at one specific time. Moreover, the presence of DNA from non-infecting or dead zoospores attached to roots can give misleading results. Kingsnorth et al. (2003) developed protocols for both sequence-independent and hybridization probe real-time PCR for the detection of *Polymyxa betae* glutathione-S-transferase (GST) in infected sugar beet roots. They also demonstrated that real-time PCR analyses of both serially diluted zoospore suspensions and infected root material provided a close relationship between the threshold cycle and the amount of *Polymyxa betae*.

One strategy for breeding more durable resistance to BNYVV is to combine virus resistance genes (e.g., *Rz1*, *Rz2*) (Sect. 8.1.3) with resistance to the vector, *Polymyxa betae* (Asher et al. 2009; Barr et al. 1995; Pavli et al. 2011). A two-gene system (*Pb1/Pb2*) conferring resistance against *Polymyxa betae* has been identified and mapped (Asher et al. 2009). The resistance to the vector is simply inherited and acts additively to the *Rz1* resistance against BNYVV, while it also confers protection comparable to *Rz1* in individuals lacking this gene.

In research at Salinas by Liu and Sears, Kingsnorth's methods were modified to screen *Beta* germplasm for possible resistance to *Polymyxa betae* (Liu, personal communication 2010). In a screen of germplasm, 38 materials were tested including accessions of *Patellifolia procumbens*, *Patellifolia webbiana*, and *Patellifolia patellaris*. Four commercial hybrids received from KWS and Betaseed, Inc. ("Roberta" (*rzrz*), "Beta4430R" (*Rz1*), "Angelina" (*Rz1Rz2*), and "BetaG017R" (*Rz2*)), which have been extensively used in rhizomania research at Salinas (Liu and Lewellen 2007, 2008; Liu et al. 2005), were used as checks. The remaining 31 entries represented a broad germplasm base from the breeding program at Salinas and included rhizomania-resistant and rhizomania-susceptible sugar beet inbreds, populations, and open-pollinated lines. Many of the Salinas entries had germplasm from *Beta maritima* in their background. Based on the GST copy number, where lower values indicated more resistance or lower incidence of *Patellifolia betae*, there was a range from 9 to 881,000 copies. *Patellifolia patellaris*, *Patellifolia procumbens*, and *Patellifolia webbiana* were highly resistant to *Polymyxa betae* with an average of 52 copies. This agrees with previous findings (Paul et al. 1992, 1994). The four commercial hybrids ranged from 48,000 to 881,000 copies with "Angelina" being most susceptible. This result was supported by microscopic examinations, in which "Angelina" had the most cystosori. Except for three entries, the sugar beet lines fit in the same range of susceptibility. The exceptions were monogerm C790-15 (PI 564758) (Lewellen 1994), CP04 (PI 632285) (Lewellen 2004a), and monogerm C812-41 (PI 651522). C790-15 and CP04 were identical to *Patellifolia* accessions for copy number suggesting high resistance. C812-41 had ten times more copies and although partially inbred would likely segregate at most loci. These results need to be confirmed but suggest that high resistance may occur within sugar beet. C790-15 does not have known *Beta maritima* germplasm and is susceptible to rhizomania although in the field at Salinas showed tolerance (Lewellen, unpublished). C790-15 was selected in an S_1 progeny, recurrent selection program that may have favored selection for resistance to *Polymyxa betae*, if genetic variability occurred. CP04 and C812-41 have germplasm from *Beta maritima* and resistance to rhizomania, *Rz1* and *Rz2* or *Rz3*, respectively. WB242 was the *Beta maritima* line used to breed CP04 (Sects. 8.1.5 and 8.1.11.1). C812-41 has WB41 and WB42 *Beta maritima* germplasm through C48 (PI 538251) (Lewellen and Whitney 1993) collected from Denmark and the source of the *Rz2* and *Rz3* resistance to BNYVV (Sect. 8.1.3). It is not known if this putative *Polymyxa betae* resistance came from *sea beet* or not. For C812-41, C790-15-type germplasm was used as the final sugar beet recurrent parent.

8.1.9 Black Root

Aphanomyces root rot or black root and Aphanomyces damping-off are caused by the oomycete, *Aphanomyces cochlioides* Drechs (Buchholtz and Meredith 1944; Drechsler 1929). Black root is a chronic rot of the mature root, which can be a component of a root rotting complex, often including Fusarium yellows and Rhizoctonia crown and root rot (Harveson and Rush 2002). Aphanomyces root rot has been reported in Canada, Chile, Eastern Europe, France, Germany, Hungary, Japan, Russia (and the former Soviet Union), the UK, and the USA (Asher and Hanson 2006; Panella 2005a; Windels and Harveson 2009).

Early *Aphanomyces* resistance breeding programs were centered in the Red River Valley (Minnesota and North Dakota, USA) and with the USDA-ARS stations at

Beltsville, MD and East Lansing, MI. Progress was slow until a greenhouse screening method was developed by Coe and Schneider (Coe and Schneider 1966; Doxtator and Downie 1948; Doxtator and Finkner 1954; Schneider 1954). In the early generations of testing, curly top and leaf spot-resistant material found its way into this program, some of which contained a significant contribution from *Beta maritima* germplasm (this chapter). Schneider and Gaskill (1962) tested a number of foreign accessions (including some *Beta maritima*) for resistance. It is unknown how much of a contribution was made by resistance genes from sea beet.

More recently, evaluations by the USDA-ARS Sugar Beet Crop Germplasm Committee (CGC) and the European GENRES project ("*Evaluation and enhancement of Beta collections for the extensification of agricultural production*"— GENRES-CT95-42) have screened sea beet germplasm for resistance to *Aphanomyces* (Asher et al. 2001a; Doney 1998; Panella and Frese 2003). In the European evaluations of 159 accessions of *Beta maritima*, 5 had high resistance to *Aphanomyces cochlioides* (Luterbacher et al. 2005), and of the 87 screened by the USDA-ARS, 11 had high resistance to this disease (USDA-ARS 2011c). The USDA-ARS breeding program at East Lansing, MI, continues developing *Aphanomyces*-resistant germplasm and studying its inheritance (McGrath 2006; Yu 2004).

8.1.10 Minor Fungal Diseases

High resistance to blackleg disease caused by *Pleospora bjoerlingii* Byford (*Phoma betae* Frank) was observed on fodder beets and on hybrids with *Beta maritima* (Burenin and Timoshenko 1985; Kazantseva 1975). Under severe attack of rust (*Uromyces betae*), Coons (1975) identified some *Beta maritima* population free from infection.

8.1.11 Nematodes

8.1.11.1 Cyst Nematodes

Sugar beet cyst nematode (SBCN) (*Heterodera schachtii* Schm.) is among the most damaging pests known on sugar beet worldwide. Major gene resistance has not been found in sugar beet germplasm (Doney and Whitney 1969). However, high resistance is well known in the Genus *Patellifolia* (formerly Section *Procumbentes* of Genus *Beta*) Ulbrich (Schneider 1937). Resistance from *Patellifolia procumbens* was transferred by Helen Savitsky to sugar beet as a 19-chromosome alien addition line reduced to 18 chromosomes containing a translocated fragment (Savitsky 1975, 1978) (Fig. 1.41). Similar interspecific hybrids have been made and advanced many times since (Jung et al. 1994). This nematode resistance was named *Hs1pro-1* and has been cloned (Cai et al. 1997). The literature on nematode resistance from *Patellifolia procumbens* has been reviewed (Jung et al. 1994; Panella and Lewellen 2007; Yu 2005). Commercial varieties using *Hs1pro-1* have been developed by commercial seed companies but show a yield penalty under most cultural conditions (Lewellen and Pakish 2005). Resistance to nematode, in which there is no yield drag, remains needed.

Among the *Beta maritima* accessions assembled at Salinas by McFarlane were several that had been reported to be partially resistant to SBCN or have reduced numbers of cysts (although we will refer to this as a "partial resistance" to SBCN, it is often referred to as tolerance rather than resistance) (Heijbroek et al. 1977). Among these was accession WB242 (PI 546413) (Sect. 8.1.5) that had been provided by Rietberg, IRS, Bergen op Zoom, the Netherlands in May 1974 and stated to be an accession collected from Loire River Estuary in France. The accessions with partial resistance were crossed with about 60 other individual sea beet accessions to sugar beet (Lewellen and Whitney 1993). The bulked F_2s were placed in the USDA-ARS NPGS *Beta* collection (NSSL serial no. 206290). The F_2s also were mass selected at Salinas under rhizomania conditions to produce a broadly based sugar beet × sea beet population called R22. R22 was released as C50 (PI 564243) (Lewellen and Whitney 1993). After five cycles of recurrent phenotypic selection, an improved R22 line released as C51 (PI 593694) was produced (Lewellen 2000b). The primary emphasis was selection for resistance to rhizomania and Virus Yellows.

In 1995, an experimental hybrid with R22 was grown in an Imperial Valley of California test under rhizomania conditions in comparison to "Rhizosen" (*Rz1* Holly Hybrids cultivar) and a rhizomania-susceptible commercial cultivar "HH41" that had been grown widely in Imperial Valley (Lewellen and Wrona 1997). As had been observed previously for R22 and R22 hybrids at Salinas, R22 and R22 hybrids seemed to express greater resistance to rhizomania than that conditioned solely by *Rz1*. It was unclear whether this greater resistance was due to improved resistance to rhizomania or resistance to some other pest or disease present in the field. Resistance to beet cyst nematode was suspected by JR Stander and RT Lewellen because most of the rhizomania trial areas also were infested with cyst nematode. Despite its 12.5% *Beta maritima* germplasm, the R22 hybrid had significantly higher sugar yield than Rhizosen (Lewellen and Wrona 1997). A field trial area was established on the Brawley Station, Imperial Valley of California (IV) for evaluation of reaction to rhizomania. Later, it became evident that the cyst nematode population also had increased and had become the predominant disease factor in this trial area (Becker et al. 1996). Since 1995, an expanded area has been successfully used to screen and select *Beta* germplasm resources and breeding lines for resistance to SBCN.

During the later stages of development of C51, R22 was being backcrossed into self-sterile sugar beet breeding lines such as C78 (Lewellen 1997b). In the same 1995 trial with R22, some of these backcross-derived lines also were superior to lines with only *Rz1*, suggesting that the factor from R22 for enhanced performance or disease resistance had been further introgressed into sugar beet and was highly heritable and efficacious. Line C67/2 (PI 628750) (about 6% *Beta maritima*) (Lewellen 2004c) and C72 (PI 599342) (about 3% *Beta maritima*) were as resistant as R22. Based upon subsequent greenhouse tests, it was shown that cyst counts were highly correlated

Table 8.3 Performance of a C927-4 experimental hybrid under non-diseased and severe sugar beet cyst nematode (SBCN) conditions in the Imperial Valley of California in comparison to commercial hybrids

Variety	Rz1, Rz1, R22 (Bvm)	Severe SBCN		Non-SBCN
		SY[a] (kg/ha)	Appearance[b]	SY (kg/ha)
US H11		3800	3.3	
Beta 4430R	Rz1	7800	3.1	15,200
Phoenix	Rz1	6300	3.8	14,300
C927-4H5	Rz1, R22 (Bvm)	11,200	1.8	13,200
LSD$_{(0.05)}$		1900	0.7	1800

[a]SY is refined white sugar yield
[b]Appearance is scored from 1 (healthy) to 5 (dead)

with canopy appearance scores in the IV (higher scores for greater canopy loss); sugar yield was significantly, inversely correlated with canopy scores and cyst counts (Lewellen and Pakish 2005). From these tests, it was determined that the superior performance of R22 and populations extracted from it was due to partial resistance to *Heterodera schachtii* and that this differential canopy response gave a reliable way to identify and discriminate SBCN resistance from susceptibility.

Crosses and backcrosses from R22 to C931 (Lewellen 2006a) to produce a self-fertile Doggett-type population were made to transfer *Beta maritima*-derived rhizomania resistance to sugar beet. Large numbers of individual plants were selfed to produce selfed progeny lines for evaluation. One of the specific lines with enhanced performance was released as C927-4 (PI 640421) (Lewellen 2004d). Subsequent tests in Imperial Valley and at Salinas in the field and greenhouse showed that C927-4 performance has been due in part to resistance to SBCN (Table 8.3).

From C927-4, a series of selfed progeny lines were developed and tested for resistance to SBCN. Based on nematode tests under field and greenhouse conditions, CN927-202 (PI 640420) was selected from C927-4 and released (Lewellen 2007). From other backcrosses to sugar beet populations derived from R22, another selfed progeny line was found that had partial resistance to SBCN. This line was ultimately released as CN926-11-3-22 (PI 640421) (2% *Beta maritima*) after two additional cycles of selfing and reselection for resistance to SBCN (Fig. 1.44) (Lewellen 2007). From two different sugar beet × *Beta maritima* broadly based populations called C26 (PI 610488) and C27 (PI 610489) (Lewellen 2000b), a selfed progeny line from a backcross to C931 was identified that appeared to be resistant to SBCN. This nematode-tolerant line was the only one identified from this material and was released as CN921-306 (PI 640422) (25% *Beta maritima*) (Lewellen 2007).

The specific accession(s) among the Salinas collection of sea beet lines that contributed the resistance gene(s) for cyst nematode resistance to R22 was not known. One of the logical candidates was WB242, which was being used concurrently in the powdery mildew (Sect. 8.1.5) resistance genetics and breeding program (Lewellen 2000a; Lewellen and Schrandt 2001). For the powdery mildew research, WB242 and WB97 (PI 546394) were crossed and backcrossed to sugar beet to set up a Doggett population. When individual plants of this population were examined and selected, it was observed that in addition to segregation for reaction to powdery mildew (*Pm_:pmpm*), some root systems were heavily infested with SBCN cysts, whereas intermingled roots from some adjacent plants were completely free of visible cysts. As mother roots and stecklings were being advanced from sequential backcrosses to sugar beet for resistance to *Erysiphe polygoni* and rhizomania, the root system of each plant was also examined and, where possible, preference was given for seed production to ones without nematode cysts. Within the population that became P912, there appeared to be a low frequency of SBCN-resistant plants. Similar selections originating with WB242 lead to CP04, CP06, CP07, and CP08 (Lewellen 2004a, b). When evaluated under the Imperial Valley conditions, these progressions of backcross lines from WB242 germplasm showed similar performance for resistance to SBCN as R22- and R22-derived material (Lewellen and Pakish 2005). P912 was released as CN12 (Lewellen 2006a, b). From CN12, individual selfed progeny lines were evaluated and selected (Fig. 8.16). Some of these have been released as CN12-446 (PI 657939) and CN12-770 (PI 657940).

In an informal exchange of breeding lines for disease resistance, an accession of *Beta maritima* was received from IRS, the Netherlands in 1987. This accession was reported to be Le Pouliguen Group 2 PI 198758–59. Le Pouliguen Group 2 had been selected for low SBCN cyst counts from *Beta maritima* collected from Le Pouliguen, Brittany, France by Cleij and coworkers at IRS, Bergen op Zoom and SVP, Wageningen (Hijner 1951; Lange and de Bock 1994). These materials were shown to have partial resistance to SBCN but initially thought not to be useful in sugar beet breeding (Heijbroek 1977; Heijbroek et al. 1977). Repeated selection was carried out, and rather high levels of resistance were achieved (Mesken and Lekkerkerker 1988). In 1990, several of the selected stocks were released to the European breeding companies (Lange and de Bock 1994). In tests at Wageningen by Lange and de Bock (1994), it was found that the resistant selections from this *Beta maritima* reduced the number of cysts by about two-thirds. In addition, it was shown that the *Beta maritima* resistance resulted in many of the cysts being much smaller than those on the susceptible control varieties. These smaller cysts contained fewer eggs and reduce the multiplication rate of the nematodes even further. Greenhouse tests at Salinas showed Le Pouliguen Group 2 to have reduced cyst counts as compared to susceptible sugar beet. Although Le Pouliguen Group 2 did not enter the breeding program at Salinas, it was believed to be similar to WB242 and corroborated the value of partial resistance in *Beta maritima*. Eight years later, similar *Beta maritima* material called accession N499 (PI 599349) at Salinas was obtained from KWS seed company. After initial tests in the field at Salinas and Brawley, CA under SBCN conditions, this weedy appearing annual sea beet was backcrossed into sugar beet population C931. An improved population was released as CN72 (PI 636339) (Lewellen 2006b). From CN72, individual selfed progeny families were evaluated at Salinas and Brawley and one line was released as CN72-652 (PI 657938). The SBCN partial resistance from this *Beta maritima* source from Le Pouliguen, France progressed to commercial

Fig. 8.18 Field trials in Imperial Valley of California are used to select and evaluate reactions to cyst nematode

usage in hybrids developed by KWS and Betaseed, Inc. to ameliorate the damage caused by *Heterodera schachtii* (Fig. 8.18).

The genetic relationship for resistance to SBCN from *Beta maritima* among R22 populations, WB242, CN12, Le Pouliguen Group 2, and N499 (CN72) is now known (Stevanato et al. 2014b). Because most of these lines and sources have been derived from the Loire River Estuary in France, all seem have the same gene for SBCN resistance. Nonetheless, WB242 has high resistance to *Erysiphe polygoni* (*Pm*) and has a compact, dark green canopy with slow bolting tendency that distinguishes it from the SBCN resistance from the other sources, particularly N499. In Imperial Valley tests, it appears that partial resistance to *Empoasca* sp. also may occur in WB242-derived material. The SBCN resistance derived from *Beta maritima* is not immunity, but conditions lowered reproduction of cyst nematode (Lange and de Bock 1994; Lewellen and Pakish 2005) and greatly reduces the losses caused by *Heterodera schachtii* under field conditions (Lewellen and Pakish 2005). Similar resistance from *Beta maritima* has been advanced by the commercial seed companies into commercial hybrids and shows equally favorable resistance without sugar yield drag associated with the *Beta procumbens* source under commercial sugar beet production.

Many technologies have been developed to very quickly genotype large numbers of SNPs in DNA samples (Stevanato et al. 2014a). SNP markers linked to the nematode tolerance were developed using the WB242 source. A segregating F_2 population, developed from WB242 as pollinator was crossed to a male sterile line was used for bulked segregant analysis to develop an SNP marker linked to the gene for sugar beet nematode tolerance, named HsBvm-1 (Pegadaraju et al. 2013; Stevanato et al. 2014b). This marker was able to select among a set of 13 tolerant (heterozygous for the marker) and 13 susceptible commercial (homozygous susceptible) as well as the homozygous-resistant F_2 plants (Stevanato et al. 2014b). These results have been confirmed in another segregating F_2 population with WB242 as the resistance donor parent (unpublished data).

8.1.11.2   Root-Knot Nematodes

Damage from root-knot nematode (RKN) caused by numerous species of *Meloidogyne* is common where sugar beet is grown in a subtropical or warm temperate climate. Resistance to RKN could not be found in cultivated *Beta vulgaris* in a screen of 190 accessions (Yu 1995) (Fig. 1.45). In an initial search of 113 *Beta maritima* accessions, resistance was identified in WB66 (PI 546387). The original source of WB66 is unknown but likely was found within a collection from Wageningen (WB37) in 1963 by way of the Japan Sugar Beet Improvement Foundation in 1968. Resistance from WB66 has been transferred to sugar beet (Yu 1996, 2001; Yu et al. 1999, 2001; Yu and Lewellen 2004). An isozyme marker was identified for RKN resistance (Yu et al. 2001).

Beet germplasm with resistance initially was released and registered as germplasm line M66 (Yu 1996). A molecular marker was identified, and the inheritance of resistance was shown to be conditioned by a single dominant gene named *R6m-1* (Weiland and Yu 2003). Subsequently, resistant beet germplasm from backcrosses to sugar beet was released as M6-1 (Yu 2001). An additional release was made following the fifth backcross to sugar beet after homozygous-resistant plants were selected (Yu and Lewellen 2004). The *R6m-1* gene in lines M66, M6-1, and M6-2 has been shown to condition resistance to at least six species of *Meloidogyne* (Yu et al. 1999; Yu and Roberts 2002).

Resistance to RKN was also discovered in WB258 (PI 546426) (Yu 1997, 2002a, b). WB258 was collected by de Biaggi and Biancardi in the Po Delta in 1979 and sent to McFarlane at Salinas (step 12, Sect. 1.7). WB258 was also shown to have resistance to rhizomania (Lewellen 1995a, 1997a; Whitney 1989c) (Sect. 8.1.3). Root-knot nematode resistance from WB258 is near immunity and conditions resistance to all *Meloidogyne* species tested (Yu et al. 1999). Resistance from WB258 and WB66 may or may not be the same, whereas resistance from WB66 is marked by an isozyme (Yu et al. 2001), which from WB258 is not (Yu 2002b). This difference suggests that WB66 and WB258 were collected from different locations and populations. Resistance to root-knot nematode may be essential in the development of sugar beet for subtropical areas, where *Meloidogyne* spp. cause severe losses.

8.1.12 Insects

In *Beta maritima*, some degree of resistance has been found to bean aphid (*Aphis fabae*) colonization (Dale et al. 1985) and to the multiplication rate of green peach aphid (*Myzus persicae*) (Lehmann et al. 1983). Lowe and Russell (1969) ascertained that the resistance to aphids is inherited in pattern suggesting a trait under polygenic control. These findings have not led to any practical application.

8.1.13 Multiple Resistances

The diseases of beet crops may appear alone or, more frequently, associated with one another. In this case, genotypes endowed with multiple resistances would be useful (McFarlane 1971), and, indeed many hybrids are multiple disease resistances. Many recent public germplasm releases, multigerm, monogerm, and O-type lines, have multiple disease resistances (e.g., Lewellen 2006b; Panella and Lewellen 2005; Panella et al. 2011a, 2015). These materials were crossed with genotypes bearing the monogenic resistances to rhizomania taken from *Beta maritima*. Luterbacher et al. (2005, 2004) published the results of a large survey including cultivated and wild germplasm belonging to the genus *Beta*. Between 580 and 700 accessions were evaluated in several European countries in the presence of three foliar diseases (VYs, powdery mildew, Cercospora leaf spot). The assessment of resistances was performed both in field and glasshouse conditions. In taxa within section *Beta*, there were some cases of multiple resistances identified in *Beta maritima*. The rate of entries displaying more than one resistance was higher in the genus *Patellifolia* and section *Corollinae*. Regarding the soilborne diseases caused by *Aphanomyces cochlioides*, *Pytium ultimum, Rhizoctonia solani*, and BNYVV, *Beta maritima* showed the highest number of accessions endowed with multiple resistances. By this term, Scholten et al. (1999) also mean the combination in the same genotype of different types of resistance to the single disease. The combination of diverse resistances increases the plant's ability to combat the effects of the disease with complementary reaction mechanisms (Lewellen and Biancardi 1990). This synergy is currently employed for contrasting the yield reduction in severe rhizomania diseased fields (Sect. 8.1.3).

8.2 Resistances to Abiotic Stresses

Surveys conducted on commercial varieties of sugar beet have shown the existence of a reduced genetic variability for tolerance to water stress. The physiological basis of salt resistance in *Beta maritima* has been explored by Koyro (2000) and Bor et al. (2003). The habitat of *Beta maritima* requires resistance to abiotic stresses caused by both salinity and drought (Shaw et al. 2002). These traits are ones that have been sought in sugar beet for many years (reviewed by van Geyt et al. 1990), especially in climates where sugar beet cultivation is rain-fed. The effect of climatic and precipitation patterns on rain-fed sugar beet production areas in Europe has been studied (Pidgeon et al. 2001), and there is concern on the effect that global climate change will have on continued production (Jones et al. 2003; Pidgeon et al. 2004).

8.2.1   Drought and Heat Tolerance

Drought tolerance has long been of interest to sugar beet breeders (van Geyt et al. 1990) and is one of the often-mentioned rationales for conserving and using *Beta maritima* as a genetic resource of sugar beet (Doney and Whitney 1990; Frese 2003, 2004b; Stevanato et al. 2004). Because of the variability of rainfall in the UK, researchers there have long been interested in drought tolerance in sugar beet and *Beta maritima* germplasm, and in developing assays to determine drought tolerance (Thomas et al. 1993). The GENRES CT95 42 project in Europe evaluated 155. *Beta maritima* accessions (Frese 2004a). In this test, a standard was used, the cultivar "Saxon", and data from all accessions that were significantly different in weight than Saxon were normalized to Saxon and the deviation from the mean for individual accessions was divided into a 1 to 9 scale with 1 as the most tolerant (Frese 2004a). Five of the seven most drought-tolerant accessions (scored 1) were *Beta maritima* as were three with a drought stress score of 2 (Fig. 8.19). The drought screening was done at Broom's Barn Research Station in the UK and much of the subsequent investigations and reporting out of these results have been done by scientists located there (Ober et al. 2004a, b, 2005; Ober and Rajabi 2010; Ober and Luterbacher 2002).

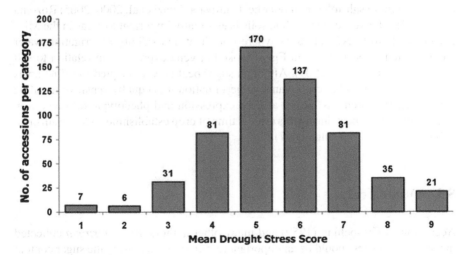

Fig. 8.19 Drought stress tolerance frequency distribution (Frese 2004a)

Some researchers working with *Beta maritima* are approaching the issue from examining the life history traits of the sea beet and how these traits, including resistance to drought, have evolved over time as important survival traits (Hautekèete et al. 2002, 2009; Wagmann et al. 2010). Although many of the countries, which grow winter beet in the Mediterranean and other heat and drought-stressed areas, are very interested in drought tolerance, only a few are working actively with sea beet (Srivastava et al. 2000).

8.2.2 Salinity Tolerance

The resistance of *Beta maritima* to salt stress is well known and in the early 1980s this trait was used as an indicator of *Beta maritima* gene flow into ruderal beet populations (Evans and Weir 1981). Research has examined betaine accumulation and its relation to salinity comparing sugar beet with *Beta maritima* (Hanson and Wyse 1982). More recent work has compared the effect of salinity on lipid peroxidation and antioxidants in the leaves of sea beet and sugar beet (Bor et al. 2003; Koyro 2000) and evaluated the osmotic adjustment response between the two taxa to try and understand the response to salinity (Bagatta et al. 2008; Koyro and Huchzermeyer 1999).

There is an increasing interest in halophytic crops because the world's supply of freshwater is shrinking and world population growing (Baydara 2008). If more saline water can be used to produce food, it will make available more freshwater for human consumption. There is an interest in using *Beta maritima* as a model system, a potential donor of salt tolerance genes, and even as a potential halophytic cash crop (Koyro et al. 2006; Koyro and Lieth 2008). Sugar beet is not the only crop that could benefit from the salt resistance in the sea beet genome; there is also interest in developing more salt-tolerant fodder beet cultivars (Niazi et al. 2000, 2005; Rozema et al. 1990). This response to saline soils is especially important to areas in the Mid-Eastern and North African areas where both heat and salinity of irrigation water are a problem. Recent work in Egypt looked at gene expression in relation to salt stress (El-Zohairy et al. 2009). Although sugar beet is well adapted to saline areas when compared to other crop plants, at germination it is equally sensitive to saline conditions. Research has looked at gene expression and phenotypic differences in sugar beet and sea beet during this critical time of crop establishment (McGrath et al. 2008; Panella and Lewellen 2007).

8.3 Other Traits

According to Krasochkin (1959) and many other authors, *Beta maritima* collected in the northern sites should be an important resource for increasing the sugar content in sugar beet. Campbell (1989) selected 30 sea beets with very high sugar content in good correlation with the root weight. Dale et al. (1985) ascertained that in sea beet accessions there were plants developing male sterile flowers. These plants produced seed if individually crossed with normal pollen producers of the same accession, thus suggesting the presence of the O-type trait or CMS in *Beta maritima* populations (Sect. 3.10). What is important to remember is that we can never with certainty predict what traits will be of importance in the future. Populations of sea beet existing in situ, undergoing continual coevolution with pests, disease, and the environment, are our insurance policy that we will have the genetic resources to fill future needs (this chapter).

References

Abe H, Tamada T (1986) Association of beet necrotic yellow vein virus with isolates of *Polymyxa betae* Keskin. Ann Phytopathol Soc Jpn 52:235–247

Abegg FA (1936) A genetic factor for the annual habit in beets and linkage relationship. J Agr Res 53:493–511

Achard FC (1803) Anleitung zum Anbau der zur Zuckerfabrication anwendbaren Runkelrüben und zur vortheilhaften Gewinnung des Zuckers aus denselben. Reprinted in: Ostwald's Klassiker der exacten Wissenschaft (1907). Engelmann, Leipzig, Germany

Ahmadinejad A, Okhovat M (1976) Pathogenicity test of some soil-borne fungi on important field crops. Iran J Plant Pathol 12:10–18

Amiri R, Mesbah M, Moghaddam M, Bihamta MR, Mohammadi SA, Norouzi P (2009) A new RAPD marker for beet necrotic yellow vein virus resistance gene in *Beta vulgaris*. Biol Plant 53:112–119

Asher MJC, Hanson LE (2006) Fungal and bacterial diseases. In: Draycott AP (ed) Sugar beet. Blackwell, Oxford, UK, pp 286–315

Asher MJC, Luterbacher MC, Frese L (2001a) Wild *Beta* as a source of resistance to sugar-beet pests and diseases. In: Proceedings of the 64th IIRB congress. Bruges (B), pp 141–152

Asher MJC, Luterbacher MC, Frese L (2001b) Wild *Beta* species as a source of resistance to sugar-beet pests and diseases. In: Proceedings of the 64th congress of the IIRB, 2001. IIRB, Brussels, pp 141–152

Asher MJC, Grimmer MK, Mutasa-Goettgens ES (2009) Selection and characterization of resistance to *Polymyxa betae*, vector of *Beet necrotic yellow vein virus*, derived from wild sea beet. Plant Path 58:250–260

Azorova M, Subikova V (1996) The modified procedure of ELISA for detection of BNYVV. Listy Cukrovarnicke a Reparske 112:297–300

Bagatta M, Pacifico D, Mandolino G (2008) Evaluation of the osmotic adjustment response within the genus *Beta*. J Sugar Beet Res 45:119–133

Barr KJ, Asher MJC, Lewis BG (1995) Resistance to *Polymyxa betae* in wild *Beta* species. Plant Path 44:307

Bartsch D, Clegg J, Ellstrand N (1999) Origin of wild beets and gene flow between *Beta vulgaris* and *B. macrocarpa* in California. Proc Br Crop Prot Counc Symp 72:269–274

Barzen E, Stahl R, Fuchs E, Borchardt DC, Salamini F (1997) Development of coupling-repulsion phase SCAR markers diagnostic for the sugar beet *Rz1* allele conferring resistance to rhizomania. Mol Breed 3:231–238

Baydara EP (2008) Salt stress responsive proteins identification in wild sugar beet (*Beta maritima*) by mass spectrometry. Thesis İzmir Institute of Technology, Izmir, Turkey

Becker JO, Wrona AF, Lewellen RT (1996) Effect of solarization and soil fumigation on sugarbeet cyst nematode population, 1993–1995. Biol Cult Tests Control Plant Dis 11:19

Bennett CW (1960) Sugar beet yellows disease in the United States. US Dep Agr Bull 1218:1–63

Bennett CW (1971) The curly top disease of sugarbeet and other plants. The American Phytopathological Society, St. Paul, MN, USA

Bennett CW, Leach LD (1971) Diseases and their control. In: Johnson RT, Alexander JT, Rush GE, Hawkes GR (eds) Advances in sugarbeet production: principles and practices. The Iowa State University Press, Ames, Iowa (USA), pp 223–285

Bennett CW, Tanrisever A (1958) Curly top disease in Turkey and its relationship to curly top in North America. J ASSBT 10:189–211

Biancardi E, Lewellen RT, de Biaggi M, Erichsen AW, Stevanato P (2002) The origin of rhizomania resistance in sugar beet. Euphytica 127:383–397

Biancardi E, McGrath JM, Panella LW, Lewellen RT, Stevanato P (2010) Sugar beet. In: Bradshaw JE (ed) Root and tuber crops. Springer, New York, NY USA, pp 173–219

Biancardi E, Tamada T (ed) (2016) Rhizomania. Springer Science + Bussiness Media, LLC, New York, NY, pp 281

Bongiovanni GC (1964) La diffusione della rizomania in Italia. Informatore Fitopatologico 10:265

Bongiovanni GC, Gallarate G, Piolanti G (1958) La Barbabietola da Zucchero. Edagricole, Bologna, Italy

Bor M, Özdemir F, Türkan I (2003) The effect of salt stress on lipid peroxidation and antioxidants in leaves of sugar beet *Beta vulgaris* L. and wild beet *Beta maritima* L. Plant Sci 164:77–84

Bosemark NO (1969) Interspecific hybridization in *Beta vulgaris* L.: prospects and value in sugarbeet breeding. J IIRB 4:112–122

Bosemark NO (1971) Use of Mendelian male sterility in recurrent selection and hybrid breeding in beets. Eucarpia Fodder Crops Section, Report of Meeting, Lusigan, France, pp 127–136

Bosemark NO (1979) Genetic Poverty of the sugar beet in Europe. In: Zeven AC (ed) Proceedings of the conference broadening genetic base of crops. Pudoc, Wageningen, The Netherlands, pp 29–35

Bosemark NO (1989) Prospects for beet breeding and use of genetic resources. In: Report on an international workshop on beta genetic resources, Wageningen (the Netherlands), 7–10 Feb 1989. International Crop Network Series. 3. IBPGR, Rome, pp 89–97

Brewbaker HE, Bush HL, Wood RR (1950) A quarter century of progress in sugar beet imporvement by the Great Western Sugar Company. Proc ASSBT 6:202–207

Briddon RW, Stenger DC, Bedford ID, Stanley J, Izadpanah K, Markham PG (1998) Comparison of a beet curly top virus isolate originating from the old world with those from the new world. Eur J Plant Path 104:77–84

Brunt AA, Richards KE (1989) Biology and molecular biology of furoviruses. Adv Virus Res 36:1–32

Buchholtz WF, Meredith CH (1944) Pathogensis of *Aphanomyces cochlioides* on taproots of the sugar beet. Phytopathology 34:485–489

Burlakoti P, Rivera V, Secor GA, Qi A, Del Rio-Mendoza LE, Khan MFR (2012) Comparative pathogenicity and virulence of *Fusarium* species on sugar beet. Plant Dis 96:1291–1296

Burenin VI (2001) Genetic resources of sugarbeet and resistance to diseases (in Russian). Sakharnaya Svekla 7:25–29

Burenin VI, Timoshenko ZV (1985) Blackleg resistance in beet (in Russian). Genet Selek 92:56–62

Büttner G, Pfähler B, Petersen J (2003) 2003: Rhizoctonia root rot in Europe—incidence, economic importance and concept for integrated control. In: Proceedings of the international institute of beet research 1st joint IIRB-ASSBT congress, 26th Feb–1st Mar 2003, San Antonio, USA, pp 897–901

Cai D, Kleine M, Kifle S, Harloff HJ, Sandal NN, Marcker KA, Klein-Lankhorst RM, Salentijn EMJ, Lange W, Stiekema WJ, Wyss U, Grundler FMW, Jung C (1997) Positional cloning of a gene for nematode resistance in sugar beet. Science (Washington) 275:832–834

Campbell LG (1989) *Beta vulgaris* NC-7 collection as a source of high sucrose germplasm. J Sugar Beet Res 26:1–9

Campbell LG (2010) Registration of seven sugarbeet germplasms selected from crosses between cultivated sugarbeet and wild *Beta* species. J Plant Reg 4:149–154

Canova A (1966) Ricerche virologiche nella rizomania della bietola. Rivista Patologia Vegetale 2:3–41

Carsner E (1926) Resistance in sugar beets to curly top. USDA Circ 388:7

Carsner E (1928) The wild beet in California. Facts About Sugar 23:1120–1121

Carsner E (1933) Curly-top resistance in sugar beets and test of the resistant variety U.S. no. 1. USDA Technical. Bulletin 360:68

Christ DS, Varrelmann M (2010) Fusarium in sugarbeet. Zuckerind 136:161–171

Coe GE, Hogaboam GJ (1971) Registration of sugarbeet parental line SP 6322–0. Crop Sci 11:947

Coe GE, Schneider CL (1966) Selecting sugar beet seedlings for resistance to *Aphanomyces cochlioides*. J ASSBT 14:164–167

Coons GH (1936) Improvement of the sugar beet. 1936 Yearbook of Agriculture. USDA, Washington, DC, USA, pp 625–656

Coons GH (1953) Some problems in growing sugar beets. Yearbook of Agriculture, USDA, Washington, DC, USA, pp 509–524

Coons GH (1975) Interspecific hybrids between *Beta vulgaris* L. and the wild species of *Beta*. J ASSBT 18:281–306

Coons GH, Stewart D, Elcock HA (1931) Sugar-beet strains resistant to leaf spot and curly top. Yearbook of Agriculture, pp 493–496

Coons GH, Owen FV, Stewart D (1955) Improvement of the sugar beet in the United States. Adv Agron 7:89–139

Dale MFB, Ford-Lloyd BV, Arnold MH (1985) Variation in some agronomically important characters in a germplasm collection of beet (*Beta Vulgaris* L.). Euphytica 34:449–455

de Biaggi M (2005) Beet yellows. In: Biancardi E, Campbell LG, Skaracis GN, de Biaggi M (eds) Genetics and breeding of sugar beet. Science, Enfield (NH), USA, p 367

de Biaggi M (1987) Mehodes de selection—un cas concret. In: Proceedings of the 50th winter congress of the IIRB. IIRB Brussels, Belgium, pp 157–161

de Biaggi M, Erichsen AW, Lewellen RT, Biancardi E (2003) The discovery of rhizomania resistance traits in sugar beet. In: Proceedings of the international institute of beet research 1st joint IIRB-ASSBT congress, 26th Feb–1st Mar 2003, San Antonio, USA, pp 131–147

de Bock TSM (1986) The genus *Beta*: domestication, taxonomy and interspecific hybridization for plant breeding. Acta Hort 182:335–343

de Lucchi C, Stevanato P, Hanson L, McGrath M, Panella L, de Biaggi M, Broccanello C, Bertaggia M, Sella L, Concheri G (2017) Molecular markers for improving control of soil-borne pathogen *Fusarium oxysporum* in sugar beet. Euphytica 213: 71. https://doi.org/10.1007/s10681-017-1859-7

Doney DL (1993) Broadening the genetic base of sugarbeet. J Sugar Beet Res 30:209–220

Doney DL (1995) USDA-ARS sugarbeet releases. J Sugar Beet Res 32:229–257

Doney DL (1998) *Beta* evaluation and sugar beet enhancement from wild sources. In: Frese L, Panella L, Srivastava HM, Lange W (eds) International *Beta* genetic resources network. A report on the 4th international *Beta* genetic resources workshop and world *Beta* network conference held at the Aegean agricultural Research Institute, Izmir, Turkey, 28 Feb–3 Mar 1996. International crop network series no. 12. International plant genetic resources institute, Rome, pp 62–72

Doney DL, Whitney ED (1969) Screening sugarbeet for resistance to *Heterodera schachtii* Schm. J ASSBT 15:546–552

Doney D, Whitney E (1990) Genetic enhancement in *Beta* for disease resistance using wild relatives: a strong case for the value of genetic conservation. Econ Bot 44:445–451

Doxtator CW, Downie AR (1948) Progess in breeding sugar beets for resistance to Aphanomyces root rot. Proc ASSBT 5:130–136

Doxtator CW, Finkner RE (1954) A summary of results in the breeding for resistance to *Aphanomyces cochlioides* (Drecks) by the American Crystal Sugar Company since 1942. Proc ASSBT 8:94–98

Drechsler C (1929) The beet water mold and several related root parasites. J Agr Res 38:309–361

Duffus JE (1973) The yellowing virus diseases of beet. Adv Virus Res 18:347–386

Duffus JE, Liu H-Y (1991) Unique beet western yellows virus isolates from California and Texas. J Sugar Beet Res 28:68

Duffus JE, Ruppel EG (1993) Diseases. In: Cooke DA, Scott RK (eds) The sugar beet crop. Chapman & Hall, Cambridge, UK, pp 346–427

Duffus JE, Whitney ED, Larsen RC, Liu HY, Lewellen RT (1984) First report in western hemisphere of rhizomania of sugar beet caused by *Beet necrotic yellow vein virus*. Plant Dis 68:251

Dusi AN, Peters D (1999) Beet mosaic virus: vector and host relationships. J Phytopathol 147:293–298

El-Zohairy S, El-Awady A, Eissa HF, El-Khishin DA, Nassar A, McGrath JM (2009) Differential expression of salt stress-related genes in wild *Beta vulgaris*. Egypt J Genet Cytol 38:187–206

Evans A, Weir J (1981) The evolution of weed beet in sugar beet crops. Genet Res Crop Evol 29:301–310

Fischer HE (1989) Origin of the 'Weisse Schlesische Rübe' (white Silesian beet) and resynthesis of sugar beet. Euphytica 41:75–80

Ford-Lloyd BV, Maxted N, Kell S (2009) Prioritization of wild *Beta* species for conservation: the PGR forum experience. In: Frese L, Maggioni L, Lipman E (eds) Report of a working group on *Beta* and the world *Beta* network. Third Joint Meeting, 8–11 Mar 2006, Puerto de la Cruz, Tenerife, Spain. Bioversity International, Rome, Italy

Franc GD, Harveson RM, Kerr ED, Jacobsen BJ (2002) Disease management p. 131–160 in Sugarbeet Production Guide, University of Nebraska, Lincoln, p 210

Francis SA, Luterbacher MC (2003) Identification and exploitation of novel disease resistance genes in sugar beet. Pest Manag Sci 59:225–230

Frese L (2003) Sugar beets and related wild species—from collecting to utilisation. In: Knüpffer H, Ochsmann J (eds) Schriften zu Genetischen Ressourcen, vol 22. Zentralstelle für Agrardokumentation und -information (ZADI). Bonn, Germany pp, pp 170–181

Frese L (2004a) Evaluation and enhancement of *Beta* collections for extensification of agricultural production, In: GENRES CT95 42. Final project report, Reporting period: 1996–2002. Federal centre for breeding research on cultivated plants (BAZ), Braunschweig, Germany, pp 1–114

Frese L (2004b) Rationale for in situ management of wild *Beta* species. Crop Wild Relative 2:4–7

Frese L (2010) Conservation and access to sugar beet germplasm. Sugar Tech 12:207–219

Frese L, Germeier CU (2009) The international database for *Beta* and *in situ* management—potential, role and functions. In: Frese L, Maggioni L, Lipman E (eds) Report of a working group on *Beta* and the world *Beta* network. Third joint meeting, 8–11 Mar 2006, Puerto de la Cruz, Tenerife, Spain. Bioversity International, Rome, Italy, pp 59–74

Frese L, Desprez B, Ziegler D (2001) Potential of genetic resources and breeding strategies for base-broadening in *Beta*. In: Cooper HD, Spillane C, Hodgkin T (eds) Broadening the genetic base of crop production. FAO, IBPRGI jointly with CABI, Rome, Italy, pp 295–309

Friesen TL, Weiland JJ, Aasheim ML, Hunger S, Borchardt DC, Lewellen RT (2006) Identification of a SCAR marker associated with *Bm*, the beet mosaic virus resistance gene, on chromosome 1 of sugar beet. Plant Breed 125:167–172

Fujisawa ST (1976) Transmission of *Beet necrotic yellow vein virus* by *Polymyxa betae*. Ann Phytopathol Soc Jpn 43:583–586

Gao X, Yin B, Borneman J, Becker JO (2008) Assessment of parasitic activity of *Fusarium* strains obtained from a *Heterodera schachtii*-suppressive soil. J Nematol 40:1–6

Gentili P, Poggi G (1986) Ritmo, esperienze italiane contro rizomania e cercospora. Maribo, Bologna, Italy

Geyl L, Garcia Heriz M, Valentin P, Hehn A, Merdinoglu D (1995) Identification and characterization of resistance to rhizomania in an ecotype of *Beta vulgaris* subsp. *maritima*. Plant Path 44:819–828

Gidner S, Lennefors BL, Nilsson NO, Bensefelt J, Johansson E, Gyllenspetz U, Kraft T (2005) QTL mapping of BNYVV resistance from the WB41 source in sugar beet. Genome 48:279–285

Grimmer M, Bean K, Asher M (2007) Mapping of five resistance genes to sugar-beet powdery mildew using AFLP and anchored SNP markers. Theor Appl Genet 115:67–75

Grimmer MK, Bean KMR, Luterbacher MC, Stevens M, Asher MJC (2008a) Beet mild yellowing virus resistance derived from wild and cultivated *Beta* germplasm. Plant Breed 127:315–318

Grimmer MK, Bean KMR, Qi A, Stevens M, Asher MJC (2008b) The action of three *Beet yellows virus* resistance QTLs depends on alleles at a novel genetic locus that controls symptom development. Plant Breed 127:391–397

Grimmer MK, Kraft T, Francis SA, Asher MJC (2008c) QTL mapping of BNYVV resistance from the WB258 source in sugar beet. Plant Breed 127:650–652

Hajjar R, Hodgkin T (2007) The use of wild relatives in crop improvement: a survey of development over the last 20 years. Euphytica 156:1–13

Hanson AD, Wyse R (1982) Biosynthesis, translocation, and accumulation of betaine in sugar beet and its progenitors in relation to salinity. Plant Physiol 70:1191–1198

Hanson LE (2006) First report of fusarium yellows of sugar beet caused by *Fusarium oxysporum* in Michigan. Plant Dis 90:1554

Hanson LE, Hill AL (2004) *Fusarium* species causing Fusarium yellows of sugarbeet. J Sugarbeet Res 41:163–178

Hanson LE, Lewellen RT (2007) Stalk Rot of sugar beet caused by *Fusarium solani* on the Pacific Coast. Plant Dis 91:1204

Hanson LE, Panella L (2002a) *Beta* PIs form the USDA-ARS NPGS evaluated for resistance to *Cercospora beticola*, 2001. Biological cultural tests for control of plant diseases. Am Phytopathol Soc 17:F03. doi:https://doi.org/10.1094/bc17

Hanson LE, Panella L (2002b) Evaluation of *Beta* PIs from the USDA-ARS NPGS for resistance to curly top virus, 2001. Biological and cultural test for control of plant diseases. Am Phytopathol Soc 17:F04. doi:https://doi.org/10.1094/bc17

Hanson LE, Panella L (2002c) Rhizoctonia root rot resistance of *Beta* PIs from the USDA-ARS NPGS, 2001. Biological and cultural test for control of plant diseases. Am Phytopathol Soc 17:F05. doi:https://doi.org/10.1094/bc17

Hanson LE, Panella L (2003a) *Beta* PIs from the USDA-ARS NPGS evaluated for resistance to *Cercospora beticola*, 2002. Biological and cultural test for control of plant diseases. Am Phytopathol Soc 18:F015. doi:https://doi.org/10.1094/bc18

Hanson LE, Panella L (2003b) Evaluation of *Beta* PIs from the USDA-ARS NPGS for resistance to curly top virus, 2002. Biological and cultural test for control of plant diseases. Am Phytopathol Soc 18:F016. doi:https://doi.org/10.1094/bc18

Hanson LE, Panella L (2003c) Rhizoctonia root-rot resistance of *Beta* PIs from the USDA-ARS NPGS, 2002. Biological and cultural test for control of plant diseases. Am Phytopathol Soc 18:F014. doi:https://doi.org/10.1094/bc18

Hanson LE, Panella L (2004a) Evaluation of *Beta* PIs from the USDA-ARS NPGS for resistance to Beet curly top virus, 2003. Biological and cultural test for control of plant diseases. Am Phytopathol Soc 19:FC013. doi:https://doi.org/10.1094/bc19

Hanson LE, Panella L (2004b) Rhizoctonia root-rot resistance of *Beta* PIs from the USDA-ARS NPGS, 2003. Biological and cultural test for control of plant diseases. Am Phytopathol Soc 19:FC012. doi:https://doi.org/10.1094/bc19

Hanson LE, Panella L (2005) Rhizoctonia root rot resistance of *Beta* PIs from the USDA-ARS NPGS, 2004. Biological and cultural test for control of plant diseases. Am Phytopathol Soc 20:FC016. doi:https://doi.org/10.1094/bc20

Hanson LE, Panella L (2006) Rhizoctonia root-rot resistance of *Beta* PIs from the USDA-ARS, NPGS, 2005. Biological and cultural test for control of plant diseases. Am Phytopathol Soc 21:FC011. doi:https://doi.org/10.1094/bc21

Hanson LE, Panella L (2007) Rhizoctonia root rot resistance of *Beta* PIs from the USDA-ARS NPGS, 2006. Plant Dis Manag Rep 1:V023. doi:https://doi.org/10.1094/pdmr01

Hanson L, de Lucchi L, Stevanato P, McGrath M, Panella L, Sella L, de Biaggi M, Concheri G (2018) Root rot symptoms in sugar beet lines caused by *Fusarium oxysporum* f. sp. *betae*. Eur J Plant Pathol 150:589. https://doi.org/10.1007/s10658-017-1302-x

Hanson LE, Duckert T, McGrath JM (2009a) *Beta* PIs from the USDA-ARS NPGS evaluated for resistance to *Cercospora beticola*, 2008. Plant Dis Manag Rep 3:V017. https://doi.org/10.1094/PDMR03

Hanson LE, Duckert T, Goodwill TR, McGrath JM (2010) *Beta* PIs from the USDA-ARS NPGS evaluated for resistance to *Cercospora beticola*, 2009. Plant Dis Manag Rep 4:FC005. doi:https://doi.org/10.1094/pdmr04

Hanson LE, Duckert T, Goodwill TR, McGrath J (2011) *Beta* PIs from the USDA-ARS NPGS evaluated for resistance to *Cercospora beticola*, 2010. Plant Dis Manag Rep 5:FC056. doi:https://doi.org/10.1094/pdmr05

Hanson LA, Hill AL, Jacobsen BJ, Panella LW (2009) Varying response of sugar beet lines to different *Fusarium oxysporum* f.sp. *betae* isolates from the United States. J Sugar Beet Res 46:11–26

Harlan JR, de Wet JMJ (1971) Toward a rational classification of cultivated plants. Taxon 24:509–517

Harveson RM, Rush CM (1997) Genetic variation among *Fusarium oxysporum* isolates from sugar beet as determined by vegetative compatibility. Plant Dis 81:85–88

Harveson RM, Rush CM (2002) The influence of irrigation frequency and cultivar blends on the severity of multiple root diseases in sugar beets. Plant Dis 86:901–908

Hauser S, Stevens M, Mougel C, Smith HG, Fritsch C, Herrbach E, Lemaire O (2000) Biological, serological, and molecular variability suggest three distinct *Polerovirus* species infection beet or rape. Phytopathology 90:460–466

Hautekèete NC, Piquot Y, van Dijk H (2002) Life span in *Beta vulgaris* ssp. *maritima*: the effects of age at first reproduction and disturbance. J Ecol 90:508–516

Hautekèete NC, van Dijk H, Piquot Y, Teriokhin A (2009) Evolutionary optimization of life-history traits in the sea beet *Beta vulgaris* subsp. *maritima*: comparing model to data. Acta Oecol 35:104–116

Hecker RJ, Ruppel EG (1979) Registration of FC 702/4, FC 702/4 (4*X*), FC 705, FC 706, and FC 707 sugarbeet germplasm. Crop Sci 19:935

Hecker RJ, Ruppel EG (1981) Registration of FC 708 and FC 708 CMS sugarbeet germplasm. Crop Sci 21:802

Hecker RJ, Ruppel EG (1991) Registration of Rhizoctonia root rot resistant sugarbeet germplasm FC 710. Crop Sci 31:494

Heijbroek W (1977) Partial resistance of sugarbeet to beet cyst eelworm (*Heterodera schachtii* Schm.). Euphytica 26:257–262

Heijbroek W, McFarlane JS, Doney DL (1977) Breeding for tolerance to beet-cyst eelworm *Heterodera schachtii* Schm. in sugarbeet. Euphytica 26:557–564

Herr LJ (1996) Sugar beet diseases incited by *Rhizoctonia* spp. In: Sneh B (ed) *Rhizoctonia* species: taxonomy, molecular biology, ecology, pathology and disease control. Kluwer, Dordrecht, pp 341–350

Hill AL, Reeves PA, Larson RL, Fenwick AL, Hanson LE and Panella L (2011) Genetic variability among isolates of *Fusarium oxysporum* from sugar beet. Plant Pathol 60(3): 496–505. doi:https://doi.org/10.1111/j.1365-3059.2010.02394.x

Hijner JA (1951) De gevoeligheid van wilde bieten voor het bietecysteaaltje (*Heterodera schachtii*). Meded Inst Rat Suikerprod 21:1–13

Hjerdin A, Säll T, Nilsson NO, Bornmann CH, Halldén C (1994) Genetic Variation among wild and cultivated beets of the section *Beta* as revealed by RFLP analysis. J Sugar Beet Res 31:59–67

Holtschulte B (2000) *Cercsopora beticola*—worldwide distribution and incidence. In: Asher MJC, Holtsculhte B, Richard-Molard M, Rosso G, Steinrücken G, Beckers R (eds) Advances in sugar beet research, vol 2. *Cercospora beticola* Sacc. biology, agronomic influence and control measures in sugar beet. IIRB, Brussels, Belgium, pp 5–16

Jacobsen BJ, Franc GD (2009) Cercospora leaf spot. In: Harveson RM, Hanson LE, Hein GL (eds) Compendium of beet diseases and pests. American Phytopathological Society Press, St. Paul, MN, pp 7–10

Janssen GKW, Nihlgard M, Kraft T (2003) Mapping of resistance genes to powdery mildew (*Erysiphe betae*) in sugarbeet. In: Proceedings of the 1st joint IIRB-ASSBT congress held in San Antonio TX, 26th Feb–1st Mar, 2003. IIRB, Brussels, Belgium, 2-26-2003, pp 175–180

Jones PD, Lister DH, Jaggard KW, Pidgeon JD (2003) Future climate impact on the productivity of sugar beet (*Beta vulgaris* L.) in Europe. Clim Change 58:93–108

Jung C, Pillen K, Frese L, Fähr S, Melchinger AE (1993) Phylogenetic relationships between cultivated and wild species of the genus *Beta* revealed by DNA "fingerprinting". Theor Appl Genet 86:449–457

Jung C, Herrmann RG, Eibl C, Kleine M (1994) Molecular analysis of a translocation in sugar beet carrying a gene for nematode resistance from *Beta procumbens*. J Sugar Beet Res 31:27–42

Kaufmann A, Koenig R, Lessemann DE (1992) Tissue printing-immunoblotting reveals an uneven distribution of beet necrotic yellow vein and beet soil-borne viruses in sugarbeets. Arch Virol 126:329–335

Kazantseva LA (1975) Evaluation of a collection of beetroot and mangel for resistance to black leg (in Russian). Byul Vses Ord Len Ord Druzhby Nar Inst Rast Imeni N I Vavilova 50:69–71

Keskin B (1964) *Polymyxa betae* n. sp., a parasite in the roots of *Beta vulgaris* torunefort, particularly during the early growth of the sugar beet. Arch Mikrobiol 19:374

Kingsnorth CS, Kingsnorth AJ, Lyons PA, Chwarszczynska DM, Asher MJC (2003) Real-time analysis of *Polymyxa betae* GST expression in infected sugar beet. Mol Plant Pathol 4:171–176

Koyro HW (2000) Effect of high NaCl-salinity on plant growth, leaf morphology, and ion composition in leaf tissues of *Beta vulgaris* ssp. *maritima*. J Appl Bot 74:67–73

Koyro HW, Huchzermeyer B (1999) Influence of high NaCl salinity on growth, water and osmotic relations of the halophyte *Beta vulgaris* ssp. *maritima*. Development of the quick check. In: Liethe H, Moschenko M, Lohmann M, Koyro HW, Hamdy A (eds) Progress in biometerology halophyte uses in different climates I. Ecological and ecophysiological research. Bakchuys, Leinden, The Netherlands, pp 43–64

Koyro HW, Lieth H (2008) Global water crisis: the potential of cash crop halophytes to reduce the dilemma. In: Lieth H, García Sucre M, Herzog, B (eds) Mangroves and halophytes: restoration and utilisation. Tasks for vegetation science vol 34. Springer Science + Business Media, B.V., The Netherlands, pp 7–19. doi:https://doi.org/10.1007/978-1-4020-6720-4_2

Koyro HW, Daoud S, Harrouni C, Huchzermeyer B (2006) Strategies of a potential cash crop halophyte (*Beta vulgaris* ssp. *maritima*) to avoid salt injury. Trop Ecol 47:191–200

Krasochkin VT (1959) Review of the species of the genus *Beta*. Trudy Po Prikladnoi Botanike. Genetik i Selektsii 32:3–35

Landa BB, Navas-Cortés JA, Hervás A, Jiménez-Díaz RM (2001) Influence of temperature and inoculum density of *Fusarium oxysporum* f. sp. *ciceris* on suppression of Fusarium wilt of chickpea by rhizosphere bacteria. Phytopathology 91:807–816. https://doi.org/10.1094/PHYTO.2001.91.8.807

Lange W, de Bock T (1994) Pre-breeding for nematode resistance in beet. J Sugar Beet Res 31:13–26

Lange W, Brandenburg WA, de Bock TSM (1999) Taxonomy and cultonomy of beet (*Beta vulgaris* L.). Bot J Linn Soc 130:81–96

Lehmann W, Karl E, Schliephake E (1983) Vergleich von Methoden zur Prufung der Resistenz von Kulturpflanzen gegen Aphiden. Tag-Ber Akad Land-Wiss Berlin 216:667–677

Lewellen RT (1973) Inheritance of beet mosaic virus resistance in sugarbeet. Phytopathology 63:877–881

Lewellen RT (1988) Selection for resistance to rhizomania in sugar beet. In: Proceedings of the 5th international congress plant pathology. Kyoto, Japan, p 455

Lewellen RT (1992) Use of plant introductions to improve populations and hybrids of sugarbeet. Use of plant introductions in cultivar development, part 2. Crop Science Society of America, Madison, WI (USA), pp 117–135

Lewellen RT (1994) Registration of C790-6, C790-15, and C790-54 parental lines of sugarbeet. Crop Sci 34:319–320

Lewellen RT (1995a) Performance of near-isolines of sugarbeet with resistance to rhizomania from different sources. In: Proceeding of the 58th IIRB congress. IIRB. IIRB, Brussels, Belgium, pp 83–92

Lewellen RT (1995b) Registration of sugarbeet germplasm lines with multiple disease resistance: C39, C39R, C39R-6, C47, C47R, C93, and C94. Crop Sci 35:596–597

Lewellen RT (1997a) Registration of 11 sugarbeet germplasm C79 lines with resistance to rhizomania. Crop Sci 37:1026

Lewellen RT (1997b) Registration of sugarbeet germplasm lines C78, C80, and C82. Crop Sci 37:1037

Lewellen RT (1998) Registration of C76-89-5 parental line of sugarbeet. Crop Sci 38:905

Lewellen RT (2000a) Registration of powdery mildew resistant sugarbeet germplasms CP01 and CP02. Crop Sci 40:1515

Lewellen RT (2000b) Registration of rhizomania resistant sugarbeet × Beta vulgaris subsp. maritima germplasms C26, C27, and C51. Crop Sci 40:1513–1515

Lewellen RT (2004a) Registration of CP03, CP04, CP05, and CP06 sugarbeet germplasms with resistance to powdery mildew, rhizomania, and other diseases. Crop Sci 44:1886–1887

Lewellen RT (2004b) Registration of CP07 and CP08 sugarbeet germplasms with resistance to powdery mildew, rhizomania, and other diseases. Crop Sci 44:2276–2277

Lewellen RT (2004c) Registration of sugar beet germplasm lines C67/2, C69/2, C78/3, and C80/2 with resistance to virus yellows and rhizomania. Crop Sci 44:358–359

Lewellen RT (2004d) Registration of sugarbeet germplasm lines C927-4, C929-62, C930-19, and C930-35 with resistance to rhizomania, virus yellows and bolting. Crop Sci 44:359–361

Lewellen RT (2006a) Registration of C931, C941, CR168, and CZ25/2 self-fertile, genetic-male-sterile facilitated, random-mated, sugarbeet germplasm populations. Crop Sci 46:1412–1413

Lewellen RT (2006b) Registration of CN12 and CN72 sugarbeet germplasm populations with resistance to cyst nematode. Crop Sci 46:1414–1415

Lewellen RT (2007) Registration of CN927-202, CN926-11-3-22, and CN921-306 sugarbeet cyst nematode resistant sugarbeet lines. J Plant Reg 1:167–169

Lewellen RT, Biancardi E (1990) Breeding and performance of rhizomania resistant sugar beet. In: 53rd congress of the international institute for sugar beet research, 14–15 Feb 1990, Brussels (Belgium), Institut International de Recherches Betteravieres, Brussels, Belgium, pp 69–87

Lewellen RT, Biancardi E (2005) Beet mosaic. In: Biancardi E, Campbell LG, Skaracis GN, de Biaggi M (eds) Genetics and breeding of sugar beet. Science, Enfield, NH, USA, pp 79–80

Lewellen RT, Pakish LM (2005) Performance of sugarbeet cyst nematode resistant cultivars and a search for sources of resistance. J Sugar Beet Res 42(1 & 2):48

Lewellen RT, Schrandt JK (2001) Inheritance of powdery mildew resistance in sugar beet derived from Beta vulgaris ssp. maritima. Plant Dis 85:627–631

Lewellen RT, Skoyen IO (1984) Beet western yellows can cause heavy losses in sugarbeet. Calif Agric 38:4–5

Lewellen RT, Whitney ED (1976) Inheritance of resistance to race C2 of Cercospora beticola in sugarbeet. Crop Sci 16:558–561

Lewellen RT, Whitney ED (1993) Registration of germplasm lines developed from composite crosses of sugar beet × Beta maritima. Crop Sci 33:882–883

Lewellen RT, Wrona AF (1997) Solarization and host-plant resistance as alternatives to soil fumigation to control rhizomania in sugarbeet. In: Proceedings of the 60th IIRB congress. IIRB, Brussels, Belgium, pp 189–201

Lewellen RT, Skoyen IO, Whitney ED, McFarlane JS (1982) Registration of 11 sugarbeet germ plasm lines with resistance to virus yellows. Crop Sci 22:900–901

Lewellen RT, Whitney ED, Skoyen IO (1985) Registration of C37 sugarbeet parental line. Crop Sci 25:375

Lewellen RT, Skoyen IO, Erichsen AW (1987) Breeding sugarbeet for resistance to rhizomania: evaluation of host-plant reactions and selections for and inheritance of resistance. In: Proceedings of the 50th winter congress of the IIRB. IIRB, Brussels, Belgium, pp 139–156

Lewellen RT, Wisler GC, Liu H-Y, Kaffka SR, Sears JL, Duffus JE (1999) Reaction of sugarbeet breeding lines and hybrids to beet chlorosis luteovirus. J Sugar Beet Res 36:76

Liu H-Y, Lewellen RT (2007) Distribution and molecular characterization of resistance-breaking isolates of *Beet necrotic yellow vein virus* in the United States. Plant Dis 91:847–851

Liu H-Y, Lewellen RT (2008) Suppression of resistance-breaking *Beet necrotic yellow vein virus* isolates by beet oak-leaf virus in sugar beet. Plant Dis 92:1043–1047

Liu H-Y, Wisler GC, Sears JL, Duffus JE (1999) Beet chlorosis virus—a new luteovirus affecting sugar beet. J Sugar Beet Res 36:69

Liu H-Y, Sears JL, Lewellen RT (2005) Occurrence of resistance-breaking *Beet necrotic yellow vein virus* of sugar beet. Plant Dis 89:464–468

Lowe HJB, Russell GE (1969) Inherited resistance of sugar beet to aphid colonization. Ann Appl Biol 63:337–344

Luterbacher MC, Smith JM, Asher MJC, Frese L (2000) Disease resistance in collections of *Beta* species. J Sugar Beet Res 37:39–47

Luterbacher MC, Asher MJC, DeAmbrogio E, Biancardi E, Stevanato P, Frese L (2004) Sources of resistance to diseases of sugar beet in related *Beta* germplasm: I. Foliar diseases. Euphytica 139:105–121

Luterbacher MC, Asher MJC, Beyer W, Mandolino G, Scholten OE, Frese L, Biancardi E, Stevanato P, Mechelke W, Slyvchenko O (2005) Sources of resistance to diseases of sugar beet in related *Beta* germplasm: II. Soil-borne diseases. Euphytica 141:49–63

Margara J, Touvin H (1955) Sur le possibilité d'obtention de types de betteraves tolerants au virus de la jaunisse. Comptes Rendus de l'Academie des Sciences de France (Paris) 41:650–655

Maxted N, Ford-Lloyd BV, Jury SL, Kell SP, Scholten MA (2006) Towards a definition of a crop wild relative. Biodivers Conserv 15:2673–2685

McFarlane JS (1971) Variety development. In: Johnson RT (ed) Advances in sugarbeet production: principles and practices. The Iowa State University Press, Ames, IA, pp 402–435

McFarlane JS (1975) Naturally occurring hybrids between sugarbeet and *Beta macrocarpa* in the Imperial Valley of California. J ASSBT 18:245–251

McFarlane JS, Bennett CW (1963) Occurrence of yellows resistance in the sugar beet with an appraisal of the opportunities for developing resistant varieties. J ASSBT 12:403–514

McFarlane JS, Price C (1952) A new non-bolting, curly-top-resistant sugar beet variety, U.S. 75. Proc ASSBT 7:384–386

McFarlane JS, Skoyen IO, Lewellen RT (1971) Registration of sugarbeet parental lines. Crop Sci 11:946–947

McFarlane JS, Savitsky H, Steele A (1982) Breeding for resistance to the sugarbeet nematode. J ASSBT 21:311–323

McGrath JM (2006) Registration of EL53 sugarbeet germplasm with smooth-root and moderate resistance to rhizoctonia crown and root rot. Crop Sci 46:2334–2335

McGrath JM, Derrico CA, Yu Y (1999) Genetic diversity in selected, historical US sugarbeet germplasm and *Beta vulgaris* ssp. *maritima*. Theor Appl Genet 98:968–976

McGrath JM, Elawady A, El-Khishin D, Naegele RP, Carr KM, de los Reyes B (2008) Sugar beet germination: phenotypic selection and molecular profiling to identify genes involved in abiotic stress response. Acta Hort 782:35–49

Mesken M, Lekkerkerker B (1988) Selectie op partiele resistentie tegen het bietecysteaaltje in kruisingen van suiker-en voederbieten met *B. maritima*. Prophyta, Bijlage Januari, pp 68–71

Mita G, Dani M, Casciari P, Pasquali A, Selva E, Minganti C, Piccardi P (1991) Assessment of the degree of genetic variation in beet based on RFLP analysis and the taxonomy of *Beta*. Euphytica 55:1–6

Munerati O (1932) Sull' incrocio della barbabietola coltivata con la beta selvaggia della costa adriatica. L'Industria Saccarifera Italiana 25:303–304

Munerati O, Zapparoli TV (1915) di alcune anomalie della *Beta vulgaris* L. Atti Regia Accademia dei Lincei 25:1239

Munerati O, Mezzadroli C, Zapparoli TV (1913a) Osservazioni sulla *Beta maritima* L. nel triennio 1910–1912. Stanz Sper Ag Ital 46:415–445

Munerati O, Mezzadroli G, Zapparoli TV (1913b) Osservazioni sulla *Beta maritima* L., nel triennio 1910–1912. Sta Sperimentali Agr Ital 46:415–445

Murphy AM (1946) Sugar beet and curly top history in Southern Idaho 1912–1945. Proc ASSBT 4:408–412

Mutasa ES, Ward E, Adams MJ, Collier CR, Chwarszczynska DM, Asher MJC (1993) A sensitive DNA probe for the detection of *Polymyxa betae* in sugar beet roots. Physiol Mol Plant Pathol 43:379–390

Mutasa ES, Chwarszczynska DM, Adams MJ, Ward E, Asher MJC (1995) Development of PCR for the detection of *Polymyxa betae* in sugar beet roots and its application in field studies. Physiol Mol Plant Pathol 47:303–313

Mutasa ES, Chwarszczynska DM, Asher MJC (1996) Single tube, nested PCR for the diagnosis of *Polymyxa betae* infection in sugar beet roots and colorimetric analysis of amplified products. Phytopathology 86:493–497

Niazi BH, Rozema J, Broekmann RA, Saline M (2000) Dynamics of growth and water relation of fodder and sea beet in response to salinity. J Agron Crop Sci 184:101–110

Niazi BH, Athar M, Saum M, Rozema J (2005) Growth and ionic relations of fodderbeet and seabeet under saline environments. Int J Environ Sci Technol 2:113–120

Nouhi A, Amiri R, Haghnazari A, Saba J, Mesbah M (2008) Tagging of resistance gene(s) to rhizomania disease in sugar beet (*Beta vulgaris* L.). Afr J Biotechnol 7:430–433

Ober ES, Luterbacher MC (2002) Genotypic variation for drought tolerance in *Beta vulgaris*. Ann Bot 89:917–924

Ober E, Rajabi A (2010) Abiotic stress in sugar beet. Sugar Tech 12:294–298

Ober ES, Guarise M, Smith CHG, Luterbacher MC (2004a) Evaluation of drought tolerance in *Beta* germplasm. In: Frese L, Germeier CU, Lipman E, Maggioni L (eds) Report of a working group on *Beta* and world *Beta* network. Second joint meeting, 23–26 Oct 2002, Bologna, Italy. IIRB, Rome, Italy, pp 112–113

Ober ES, Clark CJA, Le Bloa M, Royal JKW, Pidgeon JD (2004b) Assessing the genetic resources to improve drought tolerance in sugar beet: agronomic traits of diverse genotypes under droughted and irrigated conditions. Field Crops Res 90:213–234

Ober ES, Le Bloa M, Royal A, Jaggard KW, Pidgeon JD (2005) Evaluation of physiological traits as indirect selection criteria for drought tolerance in sugar beet. Field Crops Res 91:231–249

Ogata N, Taguchi K, Kuranouchi T, Tanaka M (2000) Breeding for root rot resistant varieties. 3. Resitance of seed parents bred in NARCH. Proc Jap Soc Sugar Beet Technol 42:13–19

Oltmann W, Burba M, Bolz G (1984) Die Qualität der Zuckerrübe, Bedeutung. Beurteilungskriterien und Züchterische Massnahmen zu ihre Verbesserung, Berlin, Germany

Owen FV (1942) Inheritance of cross- and self-sterility in *Beta vulgaris* L. J Agric Res 64:679–698

Owen FV (1945) Cytoplasmically inherited male-sterility in sugar beets. J Agric Res 71:423–440

Owen FV, Abegg FA, Murphy AM, Tolman B, Price C, Larmer FG, Carsner E (1939) Curly-top-resistant sugar-beet varieties in 1938. United States Department of Agriculture, Washington, D.C., Circular No. 513, p 10

Owen FV, Murphy AM, Ryser GK (1946) Inbred lines from curly-top-resistant varieties of sugar beet. Proc ASSBT 4:246–252

Panella L (1998a) Screening and utilizing *Beta* genetic resources with resistance to *Rhizoctonia* root rot and *Cercospora* leaf spot in a sugar beet breeding program. In: Frese L, Panella L, Srivastava HM, Lange W (eds) International *Beta* genetic resources network. A report on the 4th international *Beta* genetic resources workshop and world *Beta* network conference held at the Aegean agricultural research institute, Izmir, Turkey, 28 Feb–3 Mar 1996. International Crop Network Series No. 12. International Plant Genetic Resources Institute, Rome, pp 62–72

Panella L (1998b) Screening of *Beta* PIs from the USDA-ARS national plant germplasm system (NPGS) for resistance to curly top virus, 1997. Biological and cultural test for control of plant diseases. Am Phytopathol Soc 13:149

Panella L (1999a) Evaluation of *Beta* PIs from the USDA-ARS national plant germplasm system for resistance to curly top virus, 1998. Biological and cultural test for control of plant diseases. Am Phytopathol Soc 14:152

Panella L (1999b) Evaluation of *Beta* PIs from the USDA-ARS national plant germplasm system for resistance to Rhizoctonia root rot, 1998. Biological and cultural test for control of plant diseases. Am Phytopathol Soc 14:154

Panella L (1999c) Evaluation of *Beta* PIs from the USDA-ARS NPGS for resistance to Cercospora leaf spot, 1998. Biological and cultural test for control of plant diseases. Am Phytopathol Soc 14:153

Panella L (2000a) Evaluation of *Beta* PIs from the USDA-ARS NPGS for resistance to Cercospora leaf spot, 1999. Biological and cultural test for control of plant diseases. Am Phytopathol Soc 15:34

Panella L (2000b) Evaluation of *Beta* PIs from the USDA-ARS NPGS for resistance to curly top virus, 1999. Biological and cultural test for control of plant diseases. Am Phytopathol Soc 15:33

Panella L (2000c) Evaluation of *Beta* PIs from the USDA-ARS NPGS for resistance to Rhizoctonia root rot, 1999. Biological and cultural test for control of plant diseases. Am Phytopathol Soc 15:35

Panella L (2005a) Black root. In: Biancardi E, Campbell LG, Skaracis GN, de Biaggi M (eds) Genetics and breeding of sugar beet. Science, Enfield, NH, pp 101–102

Panella L (2005b) Curly top. In: Biancardi E, Campbell LG, Skaracis GN, de Biaggi M (eds) Genetics and breeding of sugar beet. Science, Enfield, NH, pp 74–76

Panella L (2005c) Root rots. In: Biancardi E, Campbell LG, Skaracis GN, de Biaggi M (eds) Genetics and breeding of sugar beet. Science, Enfield, NH, pp 95–99

Panella L, Biancardi E (2016) Genetic resistances. In: Biancardi E, Tamada T (eds) Rhizomania. Springer Science + Bussiness Media, LLC, New York, NY, pp 195–220

Panella L, Frese L (2000) Cercospora resistance in *Beta* species and the development of resistant sugar beet lines. In: Asher MJC, Holtschulte B, Richard-Molard M, Rosso G, Steinrücken G, Beckers R (eds) Advances in sugar beet research, vol 2. *Cercospora beticola* Sacc. biology, agronomic influence and control measures in sugar beet. IIRB, Brussels, Belgium, pp 163–176

Panella L, Frese L (2003) *Beta* germplasm evaluation data in two databases, GRIN & IDBB. In: Proceedings from the 1st joint IIRB ASSBT congress, 26th Feb–1st Mar 2003, pp 233–241

Panella L, Hanson LE (2001a) *Beta* PIs from the USDA-ARS NPGS evaluated for resistance to Cercospora leaf spot, 2000. Biological and cultural tests for control of plant diseases. Am Phytopathol Soc 16:F10. doi:https://doi.org/10.1094/bc16

Panella L, Hanson LE (2001b) Evaluation of *Beta* PIs from the USDA-ARS NPGS for resistance to curly top virus, 2000. Biological and cultural test for control of plant diseases. Am Phytopathol Soc 16:F9. doi:https://doi.org/10.1094/bc16

Panella L, Hanson LE (2001c) Rhizoctonia root rot resistance of *Beta* PIs from the USDA-ARS NPGS, 2000. Biological and cultural tests for control of plant diseases. Am Phytopathol Soc 16:F11. doi:https://doi.org/10.1094/bc16

Panella L, Lewellen RT (2005) Registration of FC301, monogerm, O-type sugarbeet population with multiple disease resistance. Crop Sci 45:2666–2667

Panella L, Lewellen RT (2007) Broadening the genetic base of sugar beet: introgression from wild relatives. Euphytica 154:382–400

Panella L, McGrath JM (2010) The history of public breeding for resistance to Cercospora leaf spot in North America. In: Lartey RT, Weiland JJ, Panella L, Crous PW, Windels CE (eds) Cercospora leaf spot of sugar beet and related species. APS, St. Paul, MN, pp 141–156

Panella L, Ruppel EG (1998) Screening of *Beta* PIs from the USDA-ARS national plant germplasm system for resistance to Rhizoctonia root rot, 1997. Biological and cultural test for control of plant diseases. Am Phytopathol Soc 13:151

Panella L, Strausbaugh CA (2011a) Beet curly top resistance of USDA-ARS national plant germplasm system Plant introductions, 2009. Plant Dis Manag Rep 5:FC065. doi:https://doi.org/10.1094/pdmr05

Panella L, Strausbaugh CA (2011b) Beet curly top resistance of USDA-ARS national plant germplasm system plant introductions, 2010. Plant Dis Manag Rep 5:FC066. doi:https://doi.org/10.1094/pdmr05

Panella L, Strausbaugh CA (2013) Beet curly top resistance in USDA-ARS plant introductions, 2012. Plant disease management reports 7:FC121. Online https://doi.org/10.1094/pdmr07

Panella L, Fenwick AL, Hill AL, McClintock M, Vagher T (2008) Rhizoctonia root rot resistance of *Beta* PIs from the USDA-ARS NPGS, 2007. Plant Dis Manag Rep 2:V057. doi:https://doi.org/10.1094/PDMR02

Panella L, Fenwick AL, Hill AL, Vagher T, Webb KM (2010) Rhizoctonia crown and root rot resistance of *Beta* PI from the USDA-ARS NPGS, 2009. Plant Dis Manag Rep 4:FC004. doi:https://doi.org/10.1094/pdmr04

Panella L, Fenwick AL, Stevanato P, Eujayl I, Strausbaugh CA, Richardson KL, Wintermantel WM, Lewellen RT (2018) Registration of FC1740 and FC1741 multigerm, rhizomania-resistant sugar beet germplasm with resistance to multiple diseases. J Plant Reg 12:257–263. doi:https://doi.org/10.3198/jpr2017.07.0042crg

Panella L, Hanson L, McGrath JM, Fenwick AL, Stevanato P, Frese L, and Lewellen RT (2015b) Registration of FC305 multigerm sugarbeet germplasm selected from a cross to a crop wild relative. J Plant Reg. 9:115–120. doi:https://doi.org/10.3198/jpr2014.08.0052crg

Panella L, Lewellen RT, Webb KM (2011a) Registration of FC1018, FC1019, FC1020, and FC1022 multigerm sugarbeet pollinator germplasms with disease resistance. J Plant Reg 5:233–240

Panella L, Ruppel EG, Lioviæ I, Kristek A (1998) Screening of *Beta* PIs from the NPGS for resistance to Cercospora leaf spot at multiple locations, 1997. Biological and cultural test for control of plant diseases. Am Phytopathol Soc 13:150

Panella L, Vagher T, Fenwick AL, Webb KM (2011b) Rhizoctonia crown and root rot resistance of *Beta* PIs from the USDA-ARS national plant germplasm system, 2010. Plant Dis Manag Rep 5:FC067. doi:https://doi.org/10.1094/pdmr05

Panella L, Vagher T and Fenwick AL (2012) Rhizoctonia crown and root rot resistance of *Beta* PIs from the USDA-ARS, national plant germplasm system, 2011. Plant Dis Manag Rep 6:FC083. doi:https://doi.org/10.1094/pdmr06

Panella L, Vagher T, Fenwick AL (2013) Evaluation of Beta PIs from the USDA-ARS, NPGS for Rhizoctonia crown and root rot resistance, 2012. Plant Dis Manag Rep 7:FC117. doi:https://doi.org/10.1094/pdmr07

Panella L, Vagher T, Fenwick AL (2014) Evaluation of Beta PIs from the USDA-ARS, NPGS for Rhizoctonia crown and root rot resistance, 2013. Plant Dis Manag Rep 8:FC178. doi:https://doi.org/10.1094/pdmr08:fc178

Panella L, Vagher T, Fenwick AL (2015a) Rhizoctonia crown and root rot resistance evaluation of *Beta* PIs in Fort Collins, CO, 2014. Plant Dis Manag Rep 9:FC137. doi:https://doi.org/10.1094/pdmr09:fc137

Panella L, Vagher T, Fenwick AL (2016) Rhizoctonia crown and root rot resistance evaluation of *Beta* PIs in Fort Collins, CO, 2015. Plant Dis Manag Rep 10:FC167. doi:https://doi.org/10.1094/pdmr10

Paul H, Henken B, Bock TSM, Lange W (1992) Resistance tp *Polymyxa betae* in *Beta* species of the section *Procumbentes*, in hybrids with *B. vulgaris* and in monosomic chromosome additions of *B. procumbens* in *B. vulgaris*. Plant Breed 109:265–273

Paul H, Henken B, Scholten OE, de Bock T, Lange W (1994) Resistance to *Polymoxa betae* and *Beet necrotic yellow vein virus* in *Beta* species of the Section *Corollinae*. J Sugar Beet Res 31:1–6

Pavli OI, Stevanato P, Biancardi E, Skaracis GN (2011) Achievements and prospects in breeding for rhizomania resistance in sugar beet. Field Crops Res 122:165–172

Pegadaraju V, Nipper R, Hulke B, Qi L, Schultz Q (2013) De novo sequencing of sunflower genome for SNP discovery using RAD (Restriction site Associated DNA) approach. BMC Genom 14:556

Pidgeon JD, Werker AR, Jaggard KW, Richter GM, Lister DH, Jones PD (2001) Climatic impact on the productivity of sugar beet in Europe, 1961–1995. Agric Forest Meteorol 109:27–37

Pidgeon JD, Jaggard KW, Lister DH, Richter GM, Jones PD (2004) Climate impact on the productivity of sugarbeet in Europe. Zuckerind 129:20–26

Pritchard FJ (1916) Some recent investigations in sugar-beet breeding. Bot Gazette 62:425–465

Rasmussen J (1932) Näjra undersokiningen over *Beta maritima*. Bot Notiser 33–36

Richardson KL, Mackey B, Hellier B (2019) Resistance in *Beta vulgaris* L. subsp. *maritima* (L.) Thell. to the *Rz1*-breaking strain of rhizomania. Genet Resour Crop Evol 66:929–939

Rietberg H, Hijner JA (1956) Die Bekampfung der Vergilbungskrankheit der Ruben in den Niederlanden. Zucker 9:483–485

Rozema J, Zaheer SH, Niazi BH, Linders H, Broekman R (1990) Salt tolerance of *Beta vulgaris* L.: a comparison of the growth of seabeet and fodderbeet in response to salinity. In: Lieth H (ed) Towards the rational use of high salinity tolerant plants. Kluwer, Dordrecht, pp 193–197

Ruppel EG (1991) Pathogenicity of *Fusarium* spp. from diseased sugar beets and variation among sugar beet isolates of *F. oxysporum*. Plant Dis 75:486–489

Rush CM, Liu H-Y, Lewellen RT, Acosta-Leal R (2006) The continuing saga of rhizomania of sugar beets in the United States. Plant Dis 90:4–15

Savitsky VF (1952) Methods and results of breeding work with monogerm beets. Proc ASSBT 7:344–350

Savitsky H (1975) Hybridization between *Beta vulgaris* and *B. procumbens* and transmission of nematode (*Heterodera schachtii*) resistance to sugar beet. Can J Genet Cytol 17:197–209

Savitsky H (1978) Nematode (*Heterodera schachtii*) resistance and meiosis in diploid plants from interspecific *Beta vulgaris* × *B. procumbens* hybrids. Can J Genet Cytol 20:177–186

Schlösser LA (1957) Cercoploy—ein Fortschritt in de Cercospora-Reistenzzüchtung. Zucker 10(489–492):539

Schneider F (1937) Sur un croisement de la betterave a sucre avec *Beta procumbens*. Inst Belge Amelior Betterave 5:544–545

Schneider CL (1954) Methods of inoculating sugar beets with *Aphanomyces cochlioides* Drechs. Proc ASSBT 8:247–251

Schneider CL, Gaskill JO (1962) Tests of foreign introductions of *Beta vulgaris* L. for resistance to *Aphanomyces cochliodes* Drechs. and *Rhizoctonia solani* Kühn. J ASSBT 11:656–660

Scholten OE, Lange W (2000) Breeding for resistance to rhizomania in sugar beet: a review. Euphytica 112:219–231

Scholten OE, Jansen RC, Keiser LCP, de Bock TSM, Lange W (1996) Major genes for resistance to beet necrotic yellow vein virus (BNYVV) in *Beta vulgaris*. Euphytica 91:331–339

Scholten OE, de Bock TSM, Klein-Lankhorst R, Lange W (1999) Inheritance of resistance to beet necrotic yellow vein virus in *Beta vulgaris* conferred by a second gene for resistance. Theor Appl Genet 99:740–746

Shane WW, Teng PS (1992) Impact of Cercospora leaf spot on root weight, sugar yield, and purity of *Beta vulgaris*. Plant Dis 76:812–820

Shaw B, Thomas TH, Cooke DT (2002) Response of sugar beet (*Beta vulgaris* L.) to drought and nutrient deficiency stress. Plant Growth Regul 37:77–83

Shepherd TJ, Hills FJ, Hall DH (1964) Losses caused by beet mosaic virus in California grown sugar beets. J ASSBT 13:244–251

Skaracis GN, Biancardi E (2000) Breeding for cercospora resistance in sugar beet. In: Asher MJC, Holtsculhte B, Richard-Molard M, Rosso G, Steinrücken G, Beckers R (eds) Advances in sugar beet research, vol 2. *Cercospora beticola* Sacc. biology, agronomic influence and control measures in sugar beet. IIRB, Brussels, Belgium, pp 177–195

Skuderna AW (1925) Sugar beet breeding in the Arkansas Valley of Colorado. J Am Soc Agron 17:631–634

Smith HG, Hallsworth PB (1990) The effects of yellowing viruses on yield of sugar beet in field trials, 1985 and 1987. Ann Appl Biol 116:503–511

Smith GA, Ruppel EG (1973) Association of Cercospora leaf spot, gross sucrose, percentage sucrose, and root weight in sugarbeet. Can J Plant Sci 53:695–696

Srivastava HM, Shahi HN, Kumar R, Bhatnagar S (2000) Genetic diversity in *Beta vulgaris* ssp. *maritima* under subtropical climate of north India. J Sugar Beet Res 37:79–87

Stevanato P, Biancardi E, Saccomani M, de Biaggi M, Mandolino G (2004) The sea beet of the Po delta. In: Frese L, Germeier CU, Lipman E, Maggioni L (eds) 2004. Report of a working group on *Beta* and world *Beta* network. Second joint meeting, 23–26 Oct 2002, Bologna, Italy. IPGRI, Rome, Italy, pp 104–107

Stevanato P, Broccanello C, Biscarini F, Del Corvo M, Sablok G, Panella L, Stella A, Concheri G (2014a) High-throughput RAD-SNP genotyping for characterization of sugar beet genotypes. Plant Mol Biol Rep 32:691–696

Stevanato P, de Biaggi M, Broccanello C, Biancardi E, Saccomani M (2015) Molecular genotyping of "Rizor" and "Holly" rhizomania resistances in sugar beet. Euphytica 204:1–5

Stevanato P, Trebbi D, Panella L, Richardson K, Broccanello C, Pakish L, Fenwick A, Saccomani M (2014b) Identification and validation of a SNP marker linked to the gene *HsBvm-1* for nematode resistance in sugar beet. Plant Mol Biol Rep 33:474–479

Stevanato P, Trebbi D, Norouzi P, Broccanello C, Saccomani M (2012) Identification of SNP markers linked to the *Rz1* gene in sugar beet. Int Sugar J 114:715–718

Stevens M, Hallsworth PB (2003) The effects of Beet chlorosis virus (BChV) on the yield of sugar beet. In: 1st Joint IIRB-ASSBT congress, 26 Feb–1 Mar 2003, San Antonio, TX, USA, pp 805–808

Stevens M, Hallsworth PB, Smith HG (2004) The effects of Beet mild yellowing virus and Beet chlorosis virus on the yield of UK field-grown sugar beet in 1997, 1999, and 2000. Ann Appl Biol 144:113–119

Stevens M, Patron NJ, Dolby CA, Weekes R, Hallsworth PB, Lemaire O, Smith HG (2005) Distribtion and properties of geographically distinct isolates of sugar beet yellowing viruses. Plant Pathol 54:100–107

Stevens M, Liu H-Y, Lemaire O (2006) Virus diseases. In: Draycott AP (ed) Sugar beet. Blackwell, Oxford, UK, pp 256–285

Stewart D (1931) Sugar-beet yellows caused by *Fusarium conglutinans* var. *betae*. Phytopathology 21:59–70

Strausbaugh CA, Wintermantel WM, Gillen AM, Eujayl IA (2008) Curly top survey in the western United States. Phytopathology 98:1212–1217

Strausbaugh CA, Panella L (2014) Beet curly top resistance in USDA-ARS plant introduction lines, 2013. Plant Dis Manag Rep 8:FC171. doi:https://doi.org/10.1094/pdmr08:fc171

Strausbaugh CA, Panella L (2015) Beet curly top resistance in USDA-ARS plant introduction lines, 2014. Plant Dis Manag Rep 9:FC091. doi:https://doi.org/10.1094/pdmr09

Strausbaugh CA, Panella L (2016) Beet curly top resistance in USDA-ARS plant introduction lines, 2015. Plant Dis Manag Rep 10:FC056. doi:https://doi.org/10.1094/pdmr10

Strausbaugh CA, Panella L (2017) Beet curly top resistance in USDA-ARS plant introductions lines, 2016. Plant Dis Manag Rep 11:V082. (ARIS Log# 336269; Submitted Nov 30, 2016; Accepted Jan 23, 2017; Published March 14, 2017)

Stump WL, Franc GD, Harveson RM, Wilson RG (2004) Strobilurin fungicide timing for Rhizoctonia root and crown rot suppression in sugarbeet. J Sugar Beet Res 41:17–38

Tamada T, Asher M (2016a) Ecology and Epidemiology. In: Biancardi E, Tamada T (eds) Rhizomania. Springer Science + Bussiness Media, LLC, New York, NY, pp 155–174

Tamada T, Asher M (2016b) The plasmodiophorid protist *Polymyxa betae*. In: Biancardi E, Tamada T (eds) Rhizomania. Springer Science + Bussiness Media, LLC, New York, NY, pp 135–154

Tamada T, Baba T (1973) *Beet necrotic yellow vein virus* from rizomania-affected sugar beet in Japan. Ann Phytopathol Soc Jpn 39:325–332

Thomas TH, Asher MJC, Smith HG, Clarke NA, Mutasa ES, Stevens M, Thompson JR (1993) The development of diagnostics for evaluating *Beta* germplasm. J Sugar Beet Res 30:261–266

USDA-ARS (2011a) National genetic resources program. Germplasm resources information network (GRIN). [Online Database] National Germplasm Resources Laboratory, Beltsville, Maryland. http://www.ars-grin.gov/cgi-bin/npgs/html/desc.pl?49075. Accessed 28 Apr 2011

USDA-ARS (2011b) National genetic resources program. Germplasm resources information network (GRIN). [Online Database] National Germplasm Resources Laboratory, Beltsville, Maryland. http://www.ars-grin.gov/cgi-bin/npgs/html/desc_find.pl. Accessed 06 May 2011

USDA-ARS (2011c) National genetic resources program. Germplasm resources information network (GRIN). [Online Database] National Germplasm Resources Laboratory, Beltsville, Maryland. http://www.ars-grin.gov/cgi-bin/npgs/html/desc_find.pl. Accessed 02 May 2011

van Geyt JPC, Lange W, Oleo M, de Bock TSM (1990) Natural variation within the genus *Beta* and its possible use for breeding sugar beet: a review. Euphytica 49:57–76

Wagmann K, Hautekèete NC, Piquot Y, van Dijk H (2010) Potential for evolutionary change in the seasonal timing of germination in sea beet (*Beta vulgaris* ssp. *maritima*) mediated by seed dormancy. Genetica 138:763–773

Watson MA (1940) Studies on the transmission of sugar-beet yellows virus by the aphid, *Myzus persicae* (Sulz.). Proc Royal Soc London B Biol Sci 128:525–552

Weiland JJ, Lewellen RT (1999) Generation of molecular genetic markers associated with resistance to powdery mildew (*Erysiphe polygoni* DC) in sugarbeet (*Beta vulgaris*). In: Proceedings of the congress of the international society of plant-microbe interaction, 9 July 1999, p 215

Weiland JJ, Yu MH (2003) A cleaved amplified polymorphic sequence (CAPS) marker associated with root-knot nematode resistance in sugarbeet. Crop Sci 43:1814–1818

Whitney ED (1989a) *Beta maritima* as a source of powdery mildew resistance in sugarbeet. Plant Dis 73:487–489

Whitney ED (1989b) Identification, distribution, and testing for resistance to rhizomania in *Beta maritima*. Plant Dis 73:287–290

Whitney ED, Lewellen RT, Skoyen IO (1983) Reactions of sugar beet to powdery mildew: genetic variation, association among testing procedures, and results of resistance breeding. Phytopathology 73:182–185

Windels CE, Harveson RM (2009) Aphanomyces root rot. In: Harveson RM, Hanson LE, Hein GL (eds) Compendium of beet disease and pests. The American Phytopatholgical Society Press, St. Paul, MN, pp 24–27

Windels CE, Jacobsen BJ, Harveson RM (2009) Rhizoctonia root and crown rot. In: Harveson RM, Hanson LE, Hein GL (eds) Compendium of beet diseases and pests. APS, St. Paul, MN, pp 33–36

Wisler GC, Liu H-Y, Duffus JE (1994) *Beet necrotic yellow vein virus* and its relationship to eight sugar beet furo-like viruses from the U.S.A. Plant Dis 78:995–1001

Wisler GC, Lewellen RT, Sears JL, Wasson JW, Liu H, Wintermantel WM (2003) Interactions between *Beet necrotic yellow vein virus* and *Beet soilborne mosaic virus* in sugar beet. Plant Dis 87:1170–1175

Yu MH (1995) Identification of a *Beta maritima* source of resistance to root-knot nematode for sugarbeet. Crop Sci 35:1288–1290

Yu MH (1996) Registration of root-knot nematode resistant beet germplasm M66. Crop Sci 36:469

Yu MH (1997) Registration of Mi-1 root-knot nematode resistant beet germplasm line. Crop Sci 37:295

Yu MH (2001) Registration of M6-1 root-knot nematode resistant sugarbeet germplasm. Crop Sci 41:278–279

Yu MH (2002a) Registration of M1-2 beet germplasm resistant to root-knot nematode. Crop Sci 43:317–318

Yu MH (2002b) Registration of sugarbeet germplasm M1-3 resistant to root-knot nematode. Crop Sci 42:1756–1757

Yu Y (2004) Genetics of Aphanomyces disease resistance in sugarbeet (*Beta vulgaris*), AFLP mapping and QTL analyses. PhD Dissertation, Michigan State University

Yu MH (2005) Cyst nematode. In: Biancardi E, Campbell LG, Skaracis GN, de Biaggi M (eds) Genetics and breeding of sugar beet. Science, Enfield, NH, pp 103–109

Yu MH, Lewellen RT (2004) Registration of root-knot nematode-resistant sugarbeet germplasm M6-2. Crop Sci 44:1502–1503

Yu MH, Roberts PA (2002) Selection of root-knot nematode resistant sugar beet from field plantings. Nematology 2:240

Yu MH, Heijbroek W, Pakish LM (1999) The sea beet source of resistance to multiple species of root-knot nematode. Euphytica 108:151–155

Yu MH, Pakish LM, Zhou H (2001) An isozyme marker for resistance to root-knot nematode in sugarbeet. Crop Sci 41:1051–1053

Zossimovich VP (1939) New hybrids between wild and sugar beets that are resistant to *Cercospora*. Selektsiya i Semenovodstvo, USSR 1939:1–16

Chapter 9
Cultivated Offspring

Enrico Biancardi

Abstract Sea beet was first harvested wild for leaves to be eaten as a vegetable and potherb. Once domestication had begun, root and hypocotyl were slowly enlarged through selection and used after cooking. Fodder and sugar beet appeared, respectively, around 1000 and 200 years ago in Central Europe. Sugar beet had become one of the more important crops and, consequently, was more studied and selected than the rest of the beet types. Some of the progresses obtained in breeding sugar beet (monogermy, male sterility, and some resistances) has been utilized for other types. Whether the interest shown today for the green fuels will develop it into another crop is yet to be determined.

Keywords Garden beet · Red beet · Leaf beet · Swiss chard · Sugar beet · Fodder beet · Energy beet

The first effect of mass selection initiated with the domestication was the delay of the reproductive phase in the cultivated beet. This resulted in the leaves and later, the roots remaining edible for a longer time, while the plant increased its potential to accumulate carbohydrates in the storage tissues (Fig. 9.1). The development of an enlarged root and hypocotyl is considered a consequence of early selection toward biennial types (Simmonds 1976). The wide variety and employment of cultivated beet demonstrates the genetic plasticity of the plant and the remarkable morphological and physiological changes resulting from selection (Fig. 9.2). In all types of cultivated beet (leaf, garden, fodder, and sugar beet), domestication often diminished several traits of the ancestor and produced a crop with an increased sensitivity to disease and environmental stresses. Moreover, it reduced the genetic variation available in cultivated types (Bosemark 1979; Lewellen 1992; McGrath et al. 1999; Richards et al. 2004). Pathogens do not discriminate among the beet types. Therefore, the genetic disease resistances selected into sugar beet were transferred to other types. The same happened for other important traits, such as monogermy, CMS, tetraploidy.

E. Biancardi (✉)
Stazione Sperimentale di Bieticoltura, Viale Amendola 82, 45100 Rovigo, Italy
e-mail: enrico.biancardi@alice.it

E. Biancardi et al. (eds.), *Beta maritima*,
https://doi.org/10.1007/978-3-030-28748-1_9

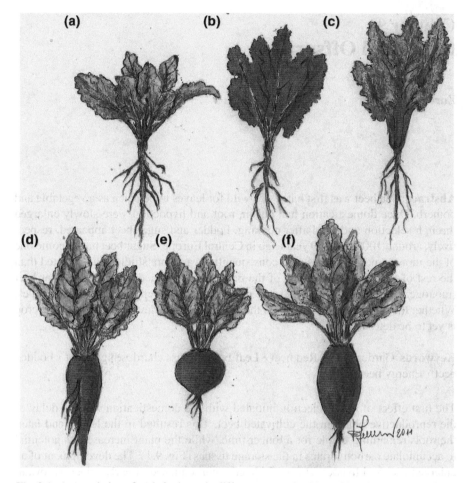

Fig. 9.1 Accumulation of carbohydrates in different types of cultivated beets compared with the wild ancestor **a** = *Beta maritima*; **b** = leaf beet; **c** = leaf (silver) beet; **d** = sugar beet; **e** = garden beet; **f** = fodder beet (from Heinisch 1960 modified)

The different parts of the beet crop used for food and feed are purely vegetative and do not depend on flowering for their development.

9.1 Leaf Beet

The first use of beets harvested in the wild was for the leaves. Leaf beet (also called chard, Swiss chard, beet greens, spinach beet, etc.) played an important role both as food and a medicinal herb. Currently, numerous varieties are available and are consumed in many styles around the world (Goldman and Navazio 2008). Like other

Fig. 9.2 *Beta vulgaris* selected for different uses. Above from left: red or garden beets and fodder beets. Below from left: leaf beets, sugar beets, and ethanol beets

minor vegetable crops, the cultivated area is highly variable from year to year and there is not much information available on the world acreage. The leaf beet has been selected for leaves to be used as a potherb and medicinal herb. However, there has been no selection on the roots (Fig. 9.3). Therefore, the roots have maintained the shape of sea beet, whereas the leaves show a large variety of form, structure, and color (red, yellow, pink, etc.). The variety of colors (Fig. 9.4) have earned them the name "novelty beet" by some (Schrader and Mayberry 2003). If harvested early, the leaves can be eaten raw as in a salad. Once they are more developed, they can be cooked and

Fig. 9.3 Leaf beets

Fig. 9.4 Leaf beets with colored petiols

used the same as spinach. Some varieties of Swiss chard develop brightly colored expanded petioles and midribs, which can be prepared in many styles together or without the leaf laminas. Those varieties called also silver chards are thought to be derived from the Adriatic Sea beet (Krasochkin 1960). Chard and particularly the colored chards are now extensively used in commercial prepackaged salad mixes to add color, flavor, and bulk.

9.2   Garden Beet

According to ENVEG (2003), the area of garden beet cultivation in Western Europe was around 8,000 ha, mainly located in France and the UK. Acreage for canning rarely exceeds 6,000 ha/year in the USA (McGrath et al. 2007b). More current data are unavailable. Evidence of the cultivation of garden beet (also called table beet, beetroot, or red beet) dates back to Roman times, if not before. In seventeenth century, the most used form of the root were: (i) round or globe shaped; (ii) flattened; (iii) cylindrical (Goldman and Navazio 2008).

The swollen and dark red-colored hypocotyl (Figs. 9.5 and 9.6) is used as a cooked vegetable, and, more recently, is canned or vacuum packaged after cooking. Pink, yellow, and ringed varieties (e.g., "heritage varieties") also can be found but are not as widespread as the red rooted table beet. The yellow color varies from a very pale yellow to dark orange and is mainly due to betaxanthin pigments (Goldman and Austin 2000).

Fig. 9.5 Red beets

Fig. 9.6 Red beet with sections showing the differently colored rings

The red ranges from pale to very dark (Baranski et al. 2001; Eagen and Goldman 1996). This color, present also on petioles and leaves, is due to the varying concentrations of anthocyanins (betalain) and betaxanthin pigments. When the former is absent, the yellow appears (Goldman and Austin 2000). Alternating bright and dark rings in the section are visible, especially in red varieties (Baranski et al. 2001; Goldman and Austin 2000; Goldman and Navazio 2008). The roots have the possibility to be stored for months, and this aptitude was very useful in northern Europe, as source of fresh vegetable during the winter.

The red pigment is employed as a natural dye in the food industry with increasing applications due to its antioxidant properties (Eagen and Goldman 1996; McGrath et al. 2007b). The betalain concentration in the root was rapidly improved by selection (Eagen and Goldman 1996; Wolyn and Gabelmann 1990). Wild beets with deep red roots are frequent in Transcaucasia especially at high altitudes (Krasochkin 1960). Both leaf and garden beet reach market size in about 6–10 weeks (Schrader and Mayberry 2003).

9.3 Fodder Beet

The crop, also called "mangel" or "mangold" or "forage beet", lost much of its importance in the course in the last century (Figs. 9.7 and 9.8). Indeed, the cultivated area decreased rapidly after the WWII due to the intensive manpower requirement for cultivation and the spread of other forage crops (corn, soybean, alfalfa, etc.) that have higher yield, greater feed value, and long-lasting storable (Henry 2010). Cultivation currently is confined to some European countries including France, Denmark, Belgium, and Ireland, all encompassing about 60,000 ha (Henry 2010). Both leaves and roots are used for feeding cattle and other animals. Roots have the advantage that they can be stored and used as fresh feed over the winter.

Fodder beets are characterized by broad roots and are classified according to their shape and dry matter content. Depending on the proportion of the hypocotyl, the taproot is more or less buried and may different shapes and colours. It is important to distinguish the color of the skin from that of the flesh. In the buried part, the skin can be white to grayish, yellow, orange, pink, red, and all the intermediate colours. In the above-ground part, the addition of chlorophyll to the other pigments produces a green coloring or intermediate greenish, yellowish, and reddish (Frandsen 1958). The color has the same traits described for garden beets. Fodder beet cultivation today is fully mechanized and takes full advantage of some of the breeding progress made in sugar beet (monogerm seed, male sterility, polyploidy, hybrid cultivars, disease resistances, etc.). Many modern varieties are half sugar beet and half traditional low dry matter fodder beet. In other words, the seed is harvested on monogerm CMS pollinated by traditional fodder beet varieties. In the past decades, the storage of the taproots became not convenient for the manpower required and also for the low dry matter content (about 25%) if compared with other feeds, like soybean and corn (about 80%).

Fig. 9.7 Drawing of fodder beet (Reichenbach and Reichenbach 1909)

9.4 Sugar Beet

Among the many destinations of the beet crop, sugar extraction is the latest, beginning in full with the first beet sugar factory, which went into operation in 1804 at Cunern, Germany. The crop was literally invented by Achard (Achard 1907; Biancardi 2005; Coons 1936; Francis 2006; von Lippmann 1925; Zossimovitch 1934) who, among other things, gave the first written description of mass selection applied to crops (Becker-Dillingen 1928). Fischer (1989) hypothesized that the first sugar beet varieties were derived from crosses of white fodder beet with leaf beets. Jung et al. (1993) disagreed with this hypothesis after examining the diversity of the crops revealed through DNA "fingerprinting". Given the relatively high sugar content of the North Atlantic sea beets, it is possible that these plants were initially used for increasing the trait in sugar beets, at the time they were cultivated for fodder (Krasochkin 1936). Unintended hybridization between cultivated crops and sea beet also was possible.

Fig. 9.8 Drawing of fodder
beet

From the beginning, there were commercial and political problems between sugar
beet and sugarcane, which at the time was intensively cultivated, especially in the
Americas. The competition has continued to present. According to Le Couter and
Burreson (2003), sugar was one of the commodities that changed world history as a
result of world trade, embargoes, and blockades, social upheaval as a consequence
of slavery, habitat destruction, etc. Even today, despite the fact that global sugar
consumption is increasing along with its international price, sugar beet suffers the
consequences of complex political and economic factors, which currently seem to
favor the cane. The world sugar production is currently around 167 Mt, of which
around 20% is produced by sugar beet (www.fao.org), down from 43% in 1971
(Goodshall 2011). According to Betteravier Française, the European Union is the
largest beet sugar producer (13.7 Mt, which is more than 50% of the world's beet
sugar), and also the leader in yield and processing quality (Jaggard et al. 2010).
Over the past 5 years, the European Union has reduced its acreage by about 15%.
The same is true to varying degrees for other producing countries in Europe. The

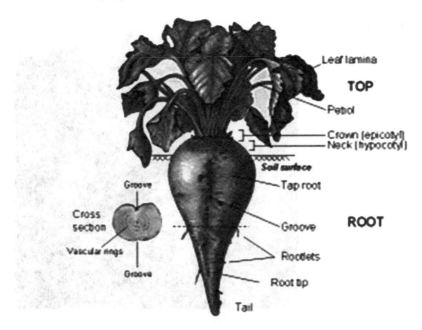

Fig. 9.9 Morphology of sugar beet. **a** = leaves + petiols (Biancardi et al. 2010)

falling prices of beets will force farmers to increase production per area and reduce the input costs to control biotic and abiotic limiting factors. The search for new and more effective genetic resistance will be, as has been said before, the most important goal for the sugar beet breeder. Survival of the crop (at least in its current dimensions) will depend on this progress. A very concrete approach to reduce grower's costs is the search for useful traits in wild species of genus *Beta*. Among these species, as has been mentioned above, *Beta maritima* is the most tested, cross compatible, and promising.

The sugar beetroot has been deeply modified through breeding (Figs. 9.9 and 9.10). The root is cone-shaped and uniform in shape and color (Artschwager 1926). The idiotype across current sugar beet varieties is similar, therefore, it is quite difficult, if not impossible, to distinguish them based on the morphology of roots and leaves alone (Bandlow 1955). The skin and the flesh of the roots are always ivory white because pigments interference with commercial sugar extraction and crystallized sucrose should be absolutely white. The true taproot extends downward from just below the lowest leaf scars to the tail. The circumference of the root is slightly flattened on two opposing sides. Each side contains a groove of varying depth that descends almost vertically or slightly turned (to the right in the northern hemisphere) the length of the taproot (Theurer 1993). Numerous small, thread-like lateral fibrous roots develop from the grooves (Artschwager 1926), and frequently can reach the depth of 2.5 m or more (Stevanato et al. 2010; Ulbrich 1934). The taproot section displays 8–13 concentric rings of vascular and parenchyma tissue surrounding a

Fig. 9.10 Bolting winter
beets at Rovigo, Italy

central, star-shaped core. The rings contain bundles of vascular tissue separated by
rays of parenchyma tissue. The vacuoles in the parenchyma cells are adapted for
sucrose storage (Wyse 1984). Selection reduced the size of the crown and neck
(hypocotyl), which diminished losses caused by scalping during mechanical harvest.
Also to reduce harvest losses, the length of the taproot, the depth of the lateral
grooves, and the protrusion of the top of the root (crown) from the soil were reduced
or made more homogenous. By lowering the depth of the grooves, the amount of
soil adhering to the beet at harvest (i.e., the soil tare) was reduced. Uniformity
in the shape and weight of the roots also helped make mechanical harvest more
efficient. This regularity mainly depends on a complete stand. Emphasis has been
put on developing high germination potential of the seed to obtain uniform field
emergence. Therefore, germination quality of the seed has been greatly improved,
especially through processing technology and by means of better protection of the
seedlings against the fungal diseases in the early stages of development. The form
of the roots is correlated with the soil traits: as a consequence, the variety features
become quite completely masked.

Current hybrid sugar beet cultivars are produced on 2n (diploid), monogerm, CMS
seed-bearing parents. The genetic monogerm seed trait was identified and selected in
the USA in 1948. This development allowed the elimination of manual singling of the
seedlings, which had been necessary to obtain the stands of about 80,000–100,000
plants per hectare (Coe and Stewart 1977; Savitsky 1950, 1952a, b, 1954). Pollinators
of hybrid cultivars can be 2n or 4n (tetraploid), producing 2n or 3n (triploid) hybrids,
respectively. One or both parents may carry enhanced morphological and agronomic
traits and resistances to diseases and pests. For quantitatively inherited (additive

genetic variance) resistance and morphological traits, the best expression is if both carry equal levels of resistance. If resistance occurs in only one parent, then there is usually an advantage for having that resistance in the 4n pollinator, e.g., for resistance to cercospora leaf spot, curly top virus. For qualitatively inherited (dominant) resistance and traits, the effective factor can be on either side of the hybrid. Choice of parental components of commercial hybrids depends upon the most favorable combination of all production traits and may rely upon general and specific additive, dominant, and epistatic (heterotic) gene combinations. The most common disease resistances are against rhizomania, CLS, and recently cyst nematode. Increasingly important, especially in the USA, are transgenic derived resistances against herbicides, like Roundup®, Liberty®, etc., carried by GM varieties (McGinnis et al. 2010). The use of transgenes is soon expected to extend to disease and pest resistance and specific metabolic products.

In the Mediterranean areas, sugar beet is planted at the end of February, then later in the more northern and colder growing areas. Sowing in autumn is possible where winter temperatures are relatively mild so the beet crop survives and is not "vernalized," causing flowering in the winter crop. This extended growing period (9 vs. 6–7 months) normally gives higher sugar yields. Autumn sowing has been proposed or hypothesized farther north by several authors beginning with Achard himself (Jaggard and Werker 1999; Pohl-Orf et al. 1999), but many attempts have been unsuccessful, mainly due to breeders inability to improve the frost resistance of varieties (Rosso et al. 2000). Winter beets need to have a very regular stand at the beginning of the winter with a full canopy to help protect against cold. Especially in the case of an inadequate stand, isolated beets are more subject to freezing and sugar yield reduction. The potential for freezing then is a conundrum because the percentage of beets killed by the winter frost can't be predicted at the time of sowing, and the final stand may be either too high or too low. In areas with winter temperatures that drop below freezing, induction of bolting and flowering (Fig. 9.10) occurs creating a problem which remains to be solved (Biancardi 1999). The percentage sugar (sucrose) in the root at harvest (also called sugar content, polarization, and polarimeter degree) is about 16–18% by fresh weight and is inversely correlated to root yield (Simmonds 1994). Both traits have been improved impressively through plant breeding (Artschwager 1930; McFarlane 1971; McGrath et al. 2007a).

Normally, only 90% or a little more of the sugar present in the root, can be extracted by the factory. This extraction value depends on the root's processing quality. Beet quality or juice purity are approximately proportional to the ratio between sugar content and impurity factors such as alpha-amino nitrate, sodium, and potassium ions (Goodshall 2011). Beet quality is influenced by environment, variety, and cultural practices. Of course, the value depends on the efficiency of the factory as well. Unextracted sugar remains incorporated in the molasses, a co-product with important uses in the food, ethanol, and feed industries.

9.5 Energy Beet

Sugar beet has a great potential for use as feedstock for the production of biogas, ethanol, ETBE, etc. (Panella 2011; Panella and Kaffka 2010). Jøersbo (2007) estimated a production of 5.7 m^3 ha^{-1} ethanol (ethyl alcohol) from sugar beet, compared with 2.6 m3 ha^{-1} produced using wheat. But the energetic and economic balance of producing biofuel from crop plants is still controversial (Cassman and Liska 2007; Young 2009). Breeding genotypes more adapted as a biofuel feedstock may have different objectives, depending on whether the whole beet, the pulp (also called marc), the molasses, or some combination is to be used. Currently, the sugar beet varieties used for energy are selected for maximum production of crystallized sugar. If the goal is only to produce biofuel through fermentation and distillation, the components of extractable sugar yield (root yield × sucrose content × purity) can be modified. Instead of industrial extraction of sugar, where high sucrose content and juice purity are critical, the ethanol varieties can be slanted toward higher sugar yield with purity not relevant (Biancardi and Pavarin 1993; von Felde 2008). Without the necessity of maintaining the parameters involved in processing quality, the genotypes selected for biofuel production might be fairly different from sugar beet. In this case, beets selected as biofuel feedstock would become more and more unsuitable for traditional processing in a sugar factory. This freedom from the traditional requirements for processing quality could allow plant breeders to better increase sugar yield.

Most likely, it would be necessary to differentiate energy beets so as not to confuse them with sugar beet, due to the obvious lower price of the former. Perhaps, the "red skin" trait from the garden beet will be used. Initially, the cross sugar × red beets causes about a 25% reduction in sugar content and yield. But 4–5 back-crosses would quickly recover the loss (Biancardi and Pavarin 1993). Theurer et al. (1987) reported that the real sugar yield potential is lower in fodder beet and in their hybrids with sugar beet, if compared to normal sugar beet varieties. However, the cited trials were run with the best-adapted sugar beet hybrids competing against not adapted fodder and half-fodder beet varieties. Biofuel production through sucrose fermentation could provide a good opportunity to expand the area both of sugar- and energy beet cultivation.

Not withstanding the increasing diffusion of "green economy", the employment of renewable energy sources such as ethanol extracted from beets, sugarcane, wheat, corn is still marginal due to the quite high production costs and the relatively low prices of crude oil and gas. Moreover, the major handicap for beets are the energy requirements for the extraction, about 50% of the ethanol energy obtained (Chavanne and Sadones). The energy balance of ethanol produced from beets is particularly negative if compared with sugarcane, where the processing costs are lower since the energy for factories is totally supplied by the bagasse, the by-product remaining after crushing of the stalks. The material is dry enough to be burned. If compared to corn and wheat, the main weakness of beet is the poor storability of the roots due to their high water content (around 75%).

For historical and policy issues, sugar beet and sugarcane cultivation areas have always been strictly separated and the border can be well traced on the world map. With the expected climatic changes, beet could move quite easily to the South and the cane more easily to the North building an overlapping band of conspicuous acreage, where the coexistence of both crops become possible. In this belt, both crops could be cultivated putting together, among other things, the main part of the extraction facilities and utilizing the energy surplus achieved by the bagasse. In this perspective, the economic gap among renewable and nonrenewable sources could be reduced greatly. The possible choice of the more convenient alcohol crops could take into account a number of variables, such as long term social, environmental, agronomical, and pathological problems, in other words the long-term sustainability of the crops. These evaluations may invalidate any economic analyses (Bowen et al. 2010).

9.6 Ornamental Beet

de Vilmorin (1923) described a type of beet developing very colored leaves and petioles (all shades of red, yellow, green) and grown in gardens for ornamental use. He referred to the plant that was introduced in England around 1840, and, which was likely a variety derived from red beet, in spite of the name "*Beta chilensis*" which means "beet coming from Chile," unlikely refers to its place of origin. A similar variety, bred for the same ornamental purpose, was briefly described by Gerard and Poggi (1636), Bauhin (1561), and de Lobel (1576), but it was ignored by Linnè and by subsequent authors as well. Chilean chard as ornamental plant is cited also by Henriette's website (Henreitte's 2011).

The ornamental beet appeared frequently at the end of 1800 on the American gardener's catalogues and popular encyclopedias, as in the "Gardner's dictionary" (Miller 1768). Sturtevant (1919) listed the name of ornamental varieties available at the time: Crimson-veined Brazilian, Golden-veined Brazilian, Scarlet-veined Brazilian, Scarlet-ribbed Chilean, Yellow-ribbed Chilean, and Red-stalked Chilean. Similar varieties are sold currently on several websites. The variety "Bulls Blood" with dark violet leaves, petioles, and roots, is the most widespread and easy to grow also in tops by not professional gardeners. Since entirely edible, the variety is considered an excellent substitute of radicchio leaves in row salad "ready to eat".

9.7 Weed Beets

Weed beet is a general term comprising all unplanted, spontaneous beets growing within crops in arable fields. Therefore, it could be considered an unwanted by-product of beet crops. Before the 60s, in south Europe they were "inland sea" beets, and in north Europe they were progeny of sugar beet bolters whose seeds has been vernalized by low temperature during ripening or storage in the soil seed bank.

After the 60s, the need to avoid seed vernalization forced breeders to move the sugar beet seed multiplication areas toward Mediterranean regions where crosses with inland wild beets as pollen donors occurred in nurseries, therefore producing hybrids subsequently sown in northern regions (Boudry et al. 1993). In that last case, they display a dominant bolting allele inherited from the wild parent, so that they are typically annual. Weed beets are poor competitors so that they can grow and reproduce mainly in sugar beet and crops like peas (Sester et al. 2004), but they produce abundant progeny that shed by gravity and form patches of related seedlings in subsequent years (Arnaud et al. 2011). Since they share the same genetic equipment as sugar beet, there is no selective herbicide available to destroy them, so that more and more seeds enriched the soil seed bank. At the same time, monogerm varieties were released, which made unnecessary the laborious hand-thinning performed to homogenize the crop density on the row, but which also resulted in no more control of the weed beet seedlings. Therefore, weed beets became a threat in sugar beets as they caused high yield losses: up to one million plants per hectare were often recorded, which imposed the use of several, repeated weed control practices, and sometimes caused the abandon of the crop (Desplanque et al. 2002). Nowadays, there are still teams of farm workers dedicated to uproot weed beets as soon as possible in the season. As a consequence, biotechnologically engineered herbicide-tolerant varieties are highly desirable, but care might be taken to avoid any tolerant hybrid in certified seeds and to dug out any bolter, otherwise gene flow and raising herbicide-tolerant seed bank would be unavoidable (Darmency et al. 2007). Indeed, weediness is dominant in hybrids and their progeny, so that any favorable gene is readily selected without fitness cost (Vigouroux and Darmency 2017). In addition, gene flow and weed beet control management is a regional matter as pollen flows over long distances (Darmency et al. 2009). Another solution would be to better manage weed beet control through gaining more knowledge on the plant biology and its interaction with the farming tools. Experiments were set up to study the population biology of weed beets, especially the soil seed bank dynamics (Sester et al. 2004, 2007; Landová et al. 2010). The data were used to parametrize models to simulate and predict the effect of various farming systems on the weed beet populations (Colbach et al. 2011) and the management of herbicide-resistant sugar beet varieties (Tricault et al. 2009). Such models go through the entire life cycle by daily calculation of transition equations from one step to the next according to biological parameters and environmental variables. The effects of the farming practices (e.g., sowing, cultivation and harvest dates, cultivation modes, fertilizers and pesticide use, crop rotation) are included in interaction with the previous equations, which finally provides estimates of the weed beet adult plant populations and seed bank amount and quality. An important result is that optimal timing of weed management operations is often more important than their exact efficacy (Colbach et al. 2011).

References

Achard FC (1907) Anleitung zum Anbau der zur Zuckerfabrication anwendbaren Runkelrüben und zur vortheilhaften Gewinnung des Zuckers aus denselben. Reprinted in: Ostwald's Klassiker der exacten Wissenschaft. Engelmann, Lipsia, Germany

Arnaud J-F, Cuguen J, Fénart S (2011) Metapopulation structure and fine-scaled genetic structuring in crop-wild hybrid weed beets. Heredity 107:395–404

Artschwager E (1926) Anatomy of the vegetative organs of sugar beet. J Agr Res 33:143–176

Artschwager E (1930) A study of the structure of sugar beets in relation to sugar content and type. J Agr Res 40:867–915

Bandlow G (1955) Die Genetik der *Beta vulgaris*-Rüben. Züchter 25:104–122

Baranski R, Grzebelus D, Frese L (2001) Estimation of genetic diversity in a collection of the Garden Beet Group. Euphytica 122:19–29

Bauhin J (1561) *Historia plantarum* universalis nova et absolutissima. Ebroduni

Becker-Dillingen J (1928) Die Wurzelfrüchte (Rüben). In: Parey P (ed) Handbuch des Hackfruchtbaues und Handelspflanzbaues. Berlin, Germany

Biancardi E, McGrath JM, Panella LW, Lewellen RT, Stevanato P (2010) Sugar beet. In: Bradshaw JE (ed) Root and tuber crops. Springer Science + Business Media, LLC, New York, NY, USA, pp 173–219

Biancardi E (1999) Miglioramento genetico. In: Casarini B, Biancardi E, Ranalli P (eds) La Barbabietola da Zucchero in Ambiente Mediterraneo. Edagricole, Bologna, Italy, pp 45–57

Biancardi E (2005) Brief history of sugar beet cultivation. In: Biancardi E, Campbell LG, Skaracis GN, de Biaggi M (eds) Genetics and breeding of sugar beet. Science Publishers Inc, Enfield (NH), USA, pp 3–9

Biancardi E, Pavarin G (1993) La barbabietola per impieghi energetici. L'Informatore Agrario 5:87–89

Bosemark NO (1979) Genetic poverty of the sugar beet in Europe. In: Zeven AC (ed) Proceeding of the conference on broadening genet. Base of Crops. Pudoc, Wageningen, The Netherlands, pp 29–35

Boudry P, Mörchen M, Saumitou-Laprade P, Vernet P, van Dijk H (1993) The origin and evolution of weed beets: consequences for the breeding and release of herbicide resistant transgenic sugar beets. Theor Appl Genet 87:471–478

Bowen E, Kennedy CK, Miranda K (2010) Ethanol from sugar beet. A process and economic analysis. Report Worchester Polytechnic Institute, NE USA

Cassman KG, Liska AJ (2007) Food and fuel for all: realistic or foolish? Biofuels Bioprod Bioref 1:18–23

Coe GE, Stewart D (1977) Cytoplasmic male sterility, self fertility, and monogermness in *Beta maritima* L. J ASSBT 19:257–261

Colbach N, Chauvel B, Darmency H, Tricault Y (2011) Sensitivity of weed emergence and dynamics to life-traits of annual spring-emerging weeds in contrasting cropping systems, using weed beet (*Beta vulgaris* ssp. *vulgaris*) as an example. J Agric Sci 149:679–700

Coons GH (1936) Improvement of the sugar beet. 1936 yearbook of agriculture. USDA, Washington, DC, pp 625–656

Darmency H, Vigouroux Y, Gestat de Garambé T, Richard- Molard M, Muchembled C (2007) Transgene escape in sugar beet production fields: data from six years farm scale monitoring. Environ Biosafety Res 6:197–206

Darmency H, Klein EK, Gestat de Garambé T, Gouyon P-H, Molard MR, Muchembled C (2009) Pollen dispersal in sugar beet production fields. Theor Appl Genet 118:1083–1092

Desplanque B, Hautekeete N, van Dijk H (2002) Transgenic weed beets: possible, probable, avoidable? J Appl Ecol 39:561–571

de Lobel M (1576) *Plantarum* seu stirpium historia… etc. Anterwep, Belgium de Vilmorin JL (1923) L' Hérédité de la betterave cultivée. Gauthier-Villars, Paris, France

Eagen KA, Goldman IL (1996) Assessment of RAPD marker frequencies over cycles of recurrent selection for pigment concentration and percent solids in red beet (*Beta vulgaris* L.). Mol Breed 2:107–115

ENVEG (2003) Horticulture Research International http://www.hri.ac.uk/ENVEG, 25 Aug 2010. Wellesbourne, Warwick, UK

Fischer HE (1989) Origin of the 'Weisse Schlesische Rübe' (white Silesian beet) and resynthesis of sugar beet. Euphytica 41:75–80

Francis SA (2006) Development of sugar beet. In: Draycott AP (ed) Sugar beet. Blackwell Publishing Ltd, Oxford, UK, pp 9–29

Frandsen KJ (1958) Breeding of forage beet. In: Romer R, Rudorf W (eds) Handbuch der Pflanzenzüchtung. Paul Parey, Berlin, Germany, pp 284–311

Gerard P, Poggi G (1636) The herbal, or general history of plants. London, UK

Goldman IL, Austin D (2000) Linkage among the R, Y and Bl loci in table beet. Theor Appl Genet 100:337–343

Goldman IL, Navazio JP (2008) Table Beet. In: Prohens J, Nuez F (eds) Vegetables I Asteraceae, Brassicaceae, Chenopodicaceae, and Cucurbitaceae. Springer, New York, pp 219–238

Goodshall MA (2011) Sugar and other sweeteners. In: Kent JA (ed) Kent and Riegel's handbook of industrial chemistry and biotechnology, Part 2. Springer, pp 1657–1653

Heinisch O (1960) Zur Terminologie in der Zuckerrübenzüchtung. Tagungsberichte, Deutsche Akademie der Landwirtschaftswissenschaften zu Berlin 48:71–74

Henreitte's (2011) Henriette's herbal homepage. http://www.henriettesherbal.com/. Accessed 25 Sept 2011

Henry C (2010) Fodder beet. In: Bradshow JE (ed) Root and tuber crops. Springer, NY, USA, pp 221–243

Jaggard KW, Qi A, Ober ES (2010) Possible change to arable crop yields by 2050. Phil Trans R Soc B 365:2835–2851

Jaggard KW, Werker AR (1999) An evaluation of the potential benefits and costs of autumn-sown sugar beets in NW Europe. J Agric Sci 132:91–102

Jøersbo M (2007) Sugar beet. In: Pua EC, Davey MR (eds) Transgenic crops, vol 4. Springer, Berlin, Heidelberg, Germany, pp 355–379

Jung C, Pillen K, Frese L, Fähr S, Melchinger AE (1993) Phylogenetic relationships between cultivated and wild species of the genus *Beta* revealed by DNA "fingerprinting". Theor Appl Genet 86:449–457

Krasochkin VT (1936) New facts in beet-root breeding. Bull Appl Bot (Leningrad) 19:27

Krasochkin VT (1960) Beet. Gos. Izdat. S.H. Lit, Moskva-Leningrad

Landová M, Hamouzová K, Soukup J, Jursík M, Holec J, Squire GR (2010) Population density and soil seed bank of weed beet as influenced by crop sequence and soil tillage. Plant Soil Environ 56:541–549

Le Couter P, Burreson J (2003) Napolean's button. Micron Geological Ltd., Vancouver, Canada

Lewellen RT (1992) Use of plant introductions to improve populations and hybrids of sugarbeet. Use of plant introductions in cultivar development, Part 2. Crop Science Society of America, Madison, WI (USA), pp 117–135

McFarlane JS (1971) Variety development. In: Johnson RT (ed) Advances in sugarbeet production: principles and practices. The Iowa State University Press, Ames, IA, pp 402–435

McGinnis EE, Meyer MH, Smith AG (2010) Sweet and sour: a scientific and legal look at herbicide-tolerant sugar beet. Plant Celltpc

McGrath JM, Derrico CA, Yu Y (1999) Genetic diversity in selected, historical US sugarbeet germplasm and *Beta vulgaris* ssp. *maritima*. Theor Appl Genet 98:968–976

McGrath JM, Saccomani M, Stevanato P, Biancardi E (2007a) Beet. In: Kole C (ed) Vegetables. Springer, Berlin Heidelberg, pp 191–207

McGrath JM, Trebbi D, Fenwick A, Panella L, Schulz B, Laurent V, Barnes S, Murray SC (2007b) An open-source first-generation molecular genetic map from a sugarbeet × table beet cross and its extension to physical mapping. Crop Sci 47:27–44

Miller P (1768) Gardener's dictionary. Printed by Francis Rivington et al, London, UK

Panella L (2011) Sugar beet as an energy crop. Sugar Tech 12:288–293

Panella L, Kaffka SR (2010) Sugar beet (*Beta vulgaris* L) as a biofuel feedstock in the United States. Sustainability of the sugar and sugar? Ethanol Industries. American Chemical Society, pp 163–175

Pohl-Orf M, Brand U, Driessen S, Hesse PR, Lehnen M, Morak C, Mucher T, Saeglitz C, von Soosten C, Bartsch D (1999) Overwintering of genetically modified sugar beet, *Beta vulgaris* L. subsp. *vulgaris*, as a source for dispersal of transgenic pollen. Euphytica 108:181–186

Reichenbach L, Reichenbach HG (1909) Icones florae Germanicae et Helveticae. Sumptibus Federici de Zezschwitz, Lipisia, Germany

Richards CM, Brownson M, Mitchell SE, Kresovich S, Panella L (2004) Polymorphic microsatellite markers for inferring diversity in wild and domesticated sugar beet (*Beta vulgaris*). Mol Ecol Notes 4:243–245

Rosso F, Meriggi P, Amaducci MT, Venturi G (2000) Winter sugar beet above the 42nd degree of latitude North. Sementi Elette 46:23–24

Savitsky H (1950) A method of determining self-fertility of self sterility in sugar beet based upon the stage of ovule development shortly after flowering. Proc ASSBT 6:198–201

Savitsky H (1954) Self-sterility and self- fertility in monogerm sugar beets. Proc ASSBT 8:29–33

Savitsky VF (1952a) Methods and results of breeding work with monogerm beets. Proc Am Soc Sugar Beet Technol 7:344–350

Savitsky VF (1952b) Monogerm sugar beets in the United States. Proc Am Soc Sugar Beet Technol 7:156–159

Schrader WL, Mayberry KS (2003) Beet and Swiss chard production in California. University of California, division agriculture and natural resources. Publication 8096, pp 1–10

Sester M, Delanoy M, Colbach N, Darmency H (2004) Crop and density effects on weed beet growth and reproduction. Weed Res 44:50–59

Sester M, Dürr C, Darmency H, Colbach N (2007) Modelling the effects of cropping systems on the seed bank dynamics and the emergence of weed beet. Ecol Model 204:47–58

Simmonds NW (1976) Evolution of crop plants. Longman, London, UK

Simmonds NW (1994) Yield and sugar content in sugar beet. Int Sugar J 96:413–416

Stevanato P, Zavalloni C, Marchetti R, Bertaggia M, Saccomani M, McGrath JM, Panella LW, Biancardi E (2010) Relationship between subsoil nitrogen availability and sugarbeet processing quality. Agron J 102:17–22

Sturtevant J (1919) Notes on edible plants. JB Lyon and Co., Albany, New York, USA

Theurer JC (1993) Pre-breeding to change sugarbeet root architecture. J Sugar Beet Res 30:221–240

Theurer JC, Doney DL, Smith GA, Lewellen RT, Hogaboam GJ, Bugbee WM, Gallian JJ (1987) Potential ethanol production from sugar beet and fodder beet. Crop Sci 27:1034–1040

Tricault Y, Darmency H, Colbach N (2009) Identifying key components of weed beet management using sensitivity analyses of the GeneSys-Beet model in GM sugar beet. Weed Res 49:581–591

Ulbrich E (1934) Chenopodiaceae. In: Engler A, Harms H (eds) Die Natürlichen Pflanzenfamilien. Wilhelm Engelmann, Leipzig, pp 375–584

Vigouroux Y, Darmency H (2017) Assessing fitness parameters of hybrids between weed beets and transgenic sugar beets. Plant Breed 136:969–976

von Felde A (2008) Trends and developments in energy plant breeding—special features of sugarbeet. Zuckerind 133:342–345

von Lippmann EO (1925) Geschichte der Rübe (*Beta*) als Kulturpflanze. Verlag Julius Springer, Berlin, Germany

Wolyn DJ, Gabelmann WH (1990) Selection for betalain pigment concentration and total dissolved solids in red table beets. J Am Soc Hort Sci 115:165–169

Wyse RE (1984) The sugar beet and chemistry: the sugar beet and sucrose formation. In: McGinnis RA (ed) Beet sugar technology. Beet Sugar Development Foundation, Fort Collins, CO, USA, pp 17–24

Young A (2009) Finding the balance between food and biofuels. Environ Sci Pollut Res 16:117–119

Zossimovitch V (1934) Wild species of beets in Transcaucasia. VNIS2-3

Chapter 10
Application of Biotechnology

J. Mitchell McGrath and Piergiorgio Stevanato

Abstract Modern genetic analyses are evolving quickly and have unprecedented ability to provide clarity and context to the genetic control of traits important for survival of crop wild relatives and sustainability of cropping systems. The ability to examine whole genomes at single nucleotide resolution complements the ongoing genetic marker approaches, allowing easily surveyed and inexpensive genetic marker surveys of germplasm and populations to be correlated with specific genes, and where known, to anticipate differences in regulation between wild and crop genes. Genome sequences allow this integration, and for sugar beet at least two high-quality reference genome assemblies are currently available. Such genome assemblies form the foundation of survey activities designed to identify genes controlling traits, examine the extent and distribution of genetic variation in the species, and assess completeness of germplasm collections. Genome archeology can reveal genetic responses to selection and perhaps predict germplasm accessions with higher probability of contributing traits useful in sustainable beet agriculture. Such genome-enabled investigations are only newly available and thus their potential will only be limited by availability of nucleotide sequence data coupled with geographic and phenotypic characterization of germplasm held in gene banks and breeding programs. Directed engineering for improvement of traits will require knowledge of their genetic control and assessment of their diversity.

Keywords *Beta maritima* · Genome evolution · DNA marker technology · Genome editing

J. M. McGrath (✉)
USDA-ARS Sugarbeet and Bean Unit, Plant Soil and Microbial Sciences, Michigan State University, 1066 Bogue Street 360 PSSB, East Lansing, MI 48824-1325, USA
e-mail: mitchmcg@msu.edu

P. Stevanato
DAFNAE, University of Padua, Padova, Italy

E. Biancardi et al. (eds.), *Beta maritima*,
https://doi.org/10.1007/978-3-030-28748-1_10

237

10.1 Evolution of DNA Marker Technology (Piergiorgio Stevanato)

Numerous genetic maps based on morphological, isozyme, and all flavors of molecular markers have been published, along with genotypic and phenotypic associations that provide context and clarity to populations (reviewed in McGrath et al. 2007; McGrath and Panella 2019). Their utility has been somewhat limited since map resolution and integrating mapped markers with reference genomic intervals is just only the beginning. Beet's nine linkage groups were named with assistance from Schondelmaier and Jung's (1997) molecular linkage groups assigned to the Butterfass (1964) trisomic series, and portability was improved with publicly available SSRs (Laurent et al. 2007; McGrath et al. 2007). Numerous sugar beet sequence collections have become available and many have been applied to molecular marker development, particularly microsatellites (SSRs) and single nucleotide polymorphisms (SNPs) (Fugate et al. 2014; Holtgrawe et al. 2014).

Molecular markers are not used very widely for sugar beet traits other than routine use of markers for the rhizomania resistance genes *Rz1* and *Rz2* (Norouzi et al. 2015; Panella et al. 2018). The *Rz2* gene was recently identified and sequenced (Capistrano-Gossmann et al. 2017). Markers developed for resistance to both Aphanomyces root rot and Cercospora leaf spot are being used routinely in Japanese sugar beet breeding programs (Taguchi et al. 2010, 2011). These researchers also developed molecular markers involved in CMS seed parent breeding (e.g., *Rf1* and *Rf2*) (Hagihara et al. 2005; Matsuhira et al. 2012; Honma et al. 2014). Public germplasm is being screened for resistance to the sugar beet cyst nematode with a marker developed by researchers in Italy and the US (Stevanato et al. 2014), as is screening of sugar beet germplasm for resistance to Fusarium diseases with recently reported markers (de Lucchi et al. 2017).

10.2 Genomes and Genome Editing (J. Mitchell McGrath)

Beta vulgaris has nine chromosomes in the haploid state. Chromosomes are morphologically similar at mitotic metaphase (Paesold et al. 2012). The DNA content of *Beta vulgaris* ranges from 714 to 758 Mb per haploid genome (Arumuganathan and Earle 1991), although few accessions have been tested and there could be a larger genome size range. Highly repetitive DNA sequences comprise a majority of the beet genome (Dohm et al. 2014), consisting of ribosomal DNA repeats, numerous families of shorter nucleosome-size repeat units present at 10^5-10^6 copies, various classes of transposable elements, and centromeric heterochromatic (Schmidt and Heslop-Harrison 1998; Heitkam et al. 2014; Kowar et al. 2016; Schwichtenberg et al. 2016; Zakrzewski et al. 2017). Each chromosome has a characteristic pattern of repeat-sequence distribution, suggesting sugar beet is fully diploidized with little or no duplication of the primary chromosome set (Halldén et al. 1998; Dohm et al. 2014).

A reference-quality genome assembly of sugar beet (RefBeet) is proving to be highly informative regarding gene content located at over 40,000 scaffolds (Dohm et al. 2014). A reference-quality genome sequence of USDA-ARS germplasm release 'EL10' is less well annotated, however, is less fragmented with >95% of the genome assembled into nine chromosome-sized contigs (McGrath et al. 2013, 2016; Funk et al. 2018). EL10 scaffolds show high concordance with the RefBeet genome sequence. Of note, the entire first linkage group described in beet (Keller 1936) is present in the EL10 genome assembly, and that of the *R-Y-B* group on Chromosome 2. Each of these genes has been recently identified and cloned [*R*, for the red alkaloid betalains, a novel cytochrome P450, (Hatlestad et al. 2012), *Y*, a Myb transcription factor (Hatlestad et al. 2014), and *B* for the annual bolting gene; Pin et al. 2012]. Genome assemblies are likely to become of increasing interest and value in describing and exploiting genetic variation in the wild and cultivated beet types. RefBeet predicts 26,923 protein-coding genes with transcript support (Minoche et al. 2015) and EL10 predicts 24,255 protein-coding genes (McGrath et al. 2016).

Molecular marker evidence suggests greater diversity is present in *Beta maritima* than the cultivated crop types. Typically, wild beets reside in highly diverse populations, although their heterozygosity is not necessarily greater that those found in 'traditional' open-pollinated cultivars or hybrids (McGrath et al. 1999; Andrello et al. 2017). Genetic differentiation between *Beta maritima* populations varies considerably, most often correlated with location or geography. Marker analyses suggested little or no separation of cultivated and wild *Beta vulgaris* spp. *maritima* forms, but did suggest *Beta vulgaris* spp. *maritima* accessions can be placed in groups centered on either Mediterranean or Atlantic regions (Andrello et al. 2016). Cultivated beets may contain only a third of the allelic diversity present in sea beet (Jung et al. 1993; Hansen et al. 1999; Fenárt et al. 2008; Saccomani et al. 2009). While diversity is reduced in sugar beet relative to sea beets, evidence is consistent with all crop types having been selected from the wild sea beet germplasm pool. Wild species diversity is viewed as a potential source of novel agronomic alleles (Frese et al. 2001), and it is clear that much genetic and allelic diversity remains to be explored for crop improvement. Diversity per se is not precisely relevant for crop improvement, since as yet, specific agronomic loci in the cultivated crops have not been identified. Identifying variants at these loci in wild populations may identify alleles contributing to increased agronomic performance, or at least may help in determining gene functions, once identified. In part, the AKER Project has succeeded in substituting short marker-defined fragments containing wild and unadapted alleles across elite germplasm linkage groups and is in the process of testing these in hybrid performance trials (Henry et al. 2019). Population genetic differentiation measures may be informative as to which loci have been under selection, and thus likely to be involved in agronomic performance (Galewski and McGrath 2019).

10.3 Development and Perspectives (J. Mitchell McGrath and Piergiorgio Stevanato)

The essence of plant breeding is to facilitate recombination and select the most desirable offspring from crosses between adapted germplasm or between adapted and unadapted germplasm. Crosses using *Beta maritima* as a parent with any of the crop types fall into the latter category. Here, the target is often a novel disease resistance locus, or perhaps a few loci, but beyond this small number of targets, it is very difficult to grow or screen sufficient population sizes to recover a trait whose genetic control is uncertain at the start. As we begin the genomics era, we can see the promise but are uncertain to the outcome. One might easily imagine, for instance, that given the anticipated rapid progress in deducing the genetics of agronomic and disease resistances in the crop types, one can genomically survey *Beta maritima* populations for novel agronomic and disease resistance alleles and systematically replace these in the crop types. Further, novel alleles that remain to be discovered through genome surveys may be present in the wild germplasm that could be usefully deployed in the cultivated material.

The genome sequence space of both wild and cultivated *Beta* species remains to be surveyed in a global fashion. The nature of genetic differentiation within and between populations of *Beta* germplasm needs to be understood. Preliminary analyses from linkage disequilibrium and genome wide association studies (Würschum and Kraft 2015; Würschum et al. 2011; Mangin et al. 2015), which survey past recombinations, suggest recombination in sugar beet may be restricted by chromosomal location, or more accurately, that selection has preserved some chromosome regions in a more-or-less unchanged state. Adetunji et al. (2014) concluded that linkage decay, or the amount of recombination between closely linked loci, depends on the population interrogated, and clues as to the nature of this conservation of recombination and selection are just beginning to emerge. For instance, Adetunji et al. (2014) suggested four chromosomes show persistent linkage disequilibrium in sugar beet, suggesting genes or adaptive complexes important for breeding progress and agronomic performance are present on these chromosomes. Mangin et al. (2015) also detected persistent effects of selection in sugar beet populations, and the effects were detected on Chromosomes 3 and 9 in common with Adetunji et al. (2014). It is likely that such population structure is present in *Beta maritima*, at least at the level of groups of populations that have diversified differentially relative to the species as a whole. It may be more difficult to detect such relics of population structure due to the high heterozygosity in most wild populations; however, they could be informative for detecting adaptive genes and alleles, particularly from accessions collected under extreme environments.

10.4 Climate Change (J. Mitchell McGrath and Piergiorgio Stevanato)

Climate change is likely to have a dramatic impact on the distribution and evolution of wild and cultivated types of *Beta*. The ultimate impact depends on the extent and scale of change, as well as its rapidity. As covered elsewhere (this volume), the genus *Beta* has survived at least one major glaciation event. During this glaciation, refugia toward the southern tip of Iberia and perhaps a larger distribution through northern Africa have since spread through the Mediterranean. Refugia in Anatolia may also have harbored isolated populations, and these may have been spread more toward east and south than extant populations are found today.

Beta vulgaris is a cool season species. Although most or all Mediterranean types are phenotypically annual, this was perhaps originally manifest as a drought avoidance response to limited moisture during the hot, dry summer season. Such a strategy has been very effective in the spread of *Beta* taxa, and the fact that annuality is largely controlled by a single gene makes the scenario somewhat plausible. Timing of this spread, according to ancient texts (Chapters in this volume), suggests that refugia radiations from east to west might have met in the Venice lagoon in the historical era, or spread there from one refugia arena or the other. In current times, the spread of *Beta maritima* northward along the Atlantic coast into Scandinavia is still occurring. Here, one might postulate that the lack of annuality is an advantageous trait, thus this particular reproductive strategy may thus be quite labile.

Populations within the *Beta vulgaris* species complex are capable of large morphological variation, evidenced by the selection of crop types from the wild species and the diversity seen among the wild types. In any future climate change scenario, this plasticity cannot be overlooked. It is impossible to predict with certainty what responses will be evolutionarily advantageous, only that the species has the capacity to change dramatically and responds well to selection. And this plasticity likely operates and continues each generation, thus the pace of climate change is unlikely to overcome the adaptive responses by the species, albeit local extant populations may be drastically affected or extinguished by higher temperatures, higher mean sea level, altered rainfall patterns, and/or severe extreme weather events. Thus, the prediction is that these taxa will survive, although not always in the habitats they currently enjoy, and even evolve into forms we may not recognize as *Beta maritima* today. Indeed, species within *Beta* we recognize today are derived from ancient forms, and are recognizable as such; however, the profusion of species names and breeding systems suggests that some of what is recognized as different species today (largely based on geography and morphology) may be incipient and the result of genetic drift, perhaps following an island geography evolutionary model (Levin 2001, 2004) since many *Beta maritima* populations today are somewhat isolated.

One expectation is that natural populations would move northwards and to higher elevations in a warming climate. This also applies to cultivated types, where one expectation may be that crop types will have to be more resilient to both higher temperatures and altered moisture patterns if the crop is to be sustainable in current growing areas. To some extent, some regions, especially rain-fed areas, have experienced altered climates already. Searching and screening wild and unadapted germplasm for tolerance to such environments has been ongoing for some years. Relatively few *Beta maritima* types have shown tolerance, and some have been used in commercial hybrid applications. The good news is that genetic variability exists for these and related traits (e.g., enhanced salt tolerance, improved nutrient acquisition). The bad news is that the genetic control for such future adaptive traits is currently unknown. Screening germplasm is expensive, but even one wild accession's beneficial introgression will return huge benefits for growers and processors. The challenge then is how best to screen myriad germplasm for abiotic and biotic stress tolerance traits in an efficient manner. Fortunately, molecular and genomic analyses will help in categorizing populations, and genetic variants within populations, and serve to focus breeding efforts on the more promising materials. Coupled with an improved understanding of the geographical distribution of genetic diversity, accessing wild populations in 'diversity-defined' areas (Andrello et al. 2016, 2017; Manel et al. 2018) subject to the types of stressors desired in cultivated materials may be one way to more efficiently harvest beneficial genes and alleles for beet improvement.

10.5 Germplasm Conservation and Utilization (J. Mitchell McGrath and Piergiorgio Stevanato)

Much has been written regarding conservation of *Beta* germplasm and much has been put into practice. Of course, it is a more difficult and long-term endeavor to utilize this germplasm and that is the focus here. With over 2500 accessions of various types of germplasm held in the public trust (e.g., crop wild relatives, landraces, and improved germplasm), it is unlikely that any one breeding program can screen, incorporate, and select improved germplasm in a systematic fashion, although USDA has an ongoing program to screen 30 *Beta* accessions from their GRIN collection for seven important diseases and pests of sugar beet each year. Success using this approach is more or less serendipitous, but where successful, can contribute essential traits to meeting long-standing issues for growers. There have not been many alternatives than screening germplasm for serendipitous discoveries. But what if there were? What might this look like?

There are a number of considerations in using crop wild relative germplasm for beet improvement. Beyond the obvious things such as crossability, fertility, and fecundity of constructed populations, the question might be posed as to whether it would be more productive to substitute variants for genes of known benefit (e.g., identified agronomic and disease resistances), or to introgress novel genes in regions of low genetic diversity in existing cultivars. Of course, both could be useful and should be done, but this question presupposes that agronomic and disease resistance targets are known, and this is largely not the case at the moment. Thus, the first step in finding alternative utilization strategies must be to identify regions of cultivar genomes that either contribute or do not contribute to agronomic performance. Fortunately, the requisite step of defining a largely complete reference genome (e.g., nucleotide sequence and assembly) has been completed to the first approximation in at least two instances (Dohm et al. 2014; Funk et al. 2018).

Two very different approaches are being taken today that could help identify such genomic regions. First, short chromosomal segments (~1–5 cM) from wild and/or unadapted germplasm are being systematically introgressed across the genome of an elite sugar beet line (Henry et al. 2019). Screening the available germplasm holdings identified geographic regions harboring similar genetic diversity, and thus narrowed the scope of wide hybrid introgressions that represent unique genetic diversity to as few as 30 germplasms. These germplasms introgressed systematically in small chunks, substituting each agronomic allele with an unadapted allele, and can illuminate regions of the elite genome that contribute to agronomic performance. Agronomic performance of these introgression lines, in hybrids with elite seed parents, is currently being evaluated in a 2-year trial over multiple environments.

A second approach to determining the number and position of agronomic genes is also underway (Galewski and McGrath 2019). This approach relies on sequencing populations of germplasm with defined phenotypes, beginning with crop types (e.g., sugar, fodder, table, and leaf). Determining allele frequencies in these populations reveals genomic regions that have undergone selection (or mutation, migration, and/or drift), such that regions of high kinship between populations suggest regions of their genomes that have been transmitted through generations relatively intact and thus presumably are important for some biochemical, metabolic, or developmental process(es). Allele frequency differences also show regions of the genome that have low heterozygosity, and when these differ between crop types, suggests that genes in these regions are important for that particular crop's or population's performance.

Both approaches identify genome regions that have undergone selection. To date, the number of these regions is comparatively small relative to the total number of genes in beet, perhaps on the order of 1%. Substituting these 1% of agronomic genes seems a less daunting task than substituting all ca. 25,000 genes of the beet genome, and may be a first target of future allelic-substitution introgression activities. It would be a relatively small effort to sequence candidate introgression populations and determine allelic status using whole genome sequencing, and from lines showing promising variants at the sequence level, to then cross these into elite germplasm and determine their effects. One caveat includes recognition that all genes in some way or another contribute to performance, thus the approach of allele substitution (or allele editing in the case that route may appear more promising in the future) implies that alleles in current elite lines are suboptimal. This must be determined on a case-by-case basis. It also assumes that genes act individually, and it may be true that most genes act in concert to contribute to performance. In this case, substituting one allele may reduce performance but substituting alleles from a number of genes affecting a biochemical pathway, or process, may have a more dramatic effect on performance. At the moment, there is no convenient, high-throughput way to establish interacting genes. However, whole genome sequences can determine 'same or different' in a convenient fashion, and thus has a role in characterization of the wild germplasm.

Other gene and allele discovery approaches are sure to be developed. All will have to account for the basic biology of the species, that is, most populations are out-crossing, geographically restricted to some extent, and more or less heterozygous. Should a program of germplasm screening at the nucleotide sequencing level be deployed, the minimum allele frequency detected issue needs to be addressed. Currently, either one gamete was sampled from each population in the case of the allele-substitution introgression lines described above, or 50 gametes from each population in the population sequencing approach described above does not fully survey the depth of allelic diversity available in wild populations.

It may be argued that storing and regenerating seeds from collected wild germplasm may be supplanted by comprehensive surveys of their gene content, since these are presumably what is important for germplasm enhancement (via gene editing). Perhaps for the most vulnerable wild populations in danger of extinction, whole genome sequencing may be the only way to preserve their genetic information. It is likely very unwise to suggest that sequence databases replace germplasm collections, although this may be unavoidable in the future. A case in point is the survival of *Beta nana*, which is the only alpine species of *Beta* having a narrow altitudinal distribution range above 1800 m, where the effect of global warming could be disastrous for the survival of this species (Frese et al. 2009).

10.6   Transgenes Spread (J. Mitchell McGrath and Piergiorgio Stevanato)

Most of the species in section *Beta* belong to the primary genepool of cultivated beet, while those of sections *Corollinae* and *Nanae*, and the genus *Patellifolia* belong to secondary and tertiary genepools (Ford-Lloyd 2005; Kadereit et al. 2006; McGrath et al. 2011; McGrath and Jung 2016). Risk of transgene spread and incidental genepool mixing is likely only to be an issue within section *Beta*. Weed beet is generally fully interfertile with cultivated beets as well as wild beet (de Bock 1986; Ford-Lloyd 1986; Ford-Lloyd and Hawkes 1986). This has been a long-recognized risk in Europe (Hornsey and Arnold 1979; Longden 1976), especially in southern seed production areas where sea beet populations and domestic beet seed production both occur (Boudry et al. 1993; Desplanque et al. 1999). *Beta maritima* is not native to the United States (de Bock 1986), thus weed beets generally have not been a persistent problem in the United States (except for California). Carsner (1938) mentioned annual beets along the California coast and in the San Joaquin Valley that were similar to 'hybrids between cultivated sugar beets and wild forms of foreign origin' but noted wild beets in the Imperial Valley were different from these sea beet hybrids (likely *Beta macrocarpa*) (Bartsch and Ellstrand 1999; Dahlberg and Brewbaker 1948; McFarlane 1975). Thus, it is possible that transgenes might spread throughout these populations if an opportunity exists (e.g., synchronous flowering of wild and crop types). Since the majority of beet seed in the United States is produced hundreds of miles away in Oregon, an overlap in flowering or crop and these California introductions seems unlikely.

Transgene spread via germplasm exchange is a risk that the USDA Agricultural Research Service (ARS) has recently addressed. USDA ARS germplasm curators, plant breeders, and researchers are responsible for implementing best management practices that ensure commercially deregulated transgenes, such as the glyphosate resistance in sugar beet, do not contaminate accessions housed in the National Plant Germplasm System, and if such transgenes are detected, procedures to mitigate and correct these risks. The risks of transgene spread in USDA ARS breeding stocks is small but not negligible, and is primarily associated with breeding stocks acquired from external sources because no USDA ARS breeding program currently incorporates deregulated genetically engineered traits. The vast majority of commercial seed sold in North America incorporates such traits. Consequently, open-pollinated seed production fields within pollinating distance of commercial sugar crop production fields with 'bolting beets' constitutes one possible risk. Monitoring for transgenes could be a financial burden for public breeding programs and germplasm curators, therefore monitoring of transgenes is indicated at critical control points that include assurance that steps have been taken to prevent transgene contamination, specific testing for transgene presence in germplasm releases at the time of deposit into the germplasm repository, and mandatory testing of specific germplasm accessions before distribution when documentation is lacking that best management practices have been followed, or when a breach of such practices is documented. Procedures and recommended practices will be reviewed regularly and updated as needed.

10.7 Identification of New Traits (J. Mitchell McGrath and Piergiorgio Stevanato)

Disease management is crucial for beets where the same pathogens affect all crop types (Harveson et al. 2009). Adequate levels of genetic resistance or tolerance to many biotic and abiotic stressors are urgently needed (Biancardi et al. 2005, 2010; Panella and Lewellen 2007; Biancardi and Tamada 2016). Most public breeding in the United States is geared toward improving disease and stress tolerance which is released to industry and the general public as improved germplasm. Field methods for disease resistance selection are summarized in Panella et al. (2008).

Genomics-enabled inquiry to detail aspects of the plant immune system should allow application to improve resistance breeding (Shigenaga et al. 2017; Mauch-Mani et al. 2017). Resistance genes (R-genes) play a role in recognition and response to pathogen attack, and in plants such R-genes tend to be clustered at a few loci containing variable numbers of R-genes per cluster, as is the case in sugar beet (Hunger et al. 2003; Lein et al. 2007). Analysis of the *Rz* gene region on Chromosome

3, where $Rz1$ and $Rz2$ rhizomania resistance genes reside (Scholten et al. 1996, 1999; Grimmer et al. 2008; Capistrano-Gossmann et al. 2017), revealed 25 NB-ARC type R-genes in a 10 Mb region in the genome sequence of EL10 (Funk et al. 2018). A comprehensive list of R-genes in *Beta* might help in characterizing and deploying novel resistance genes and alleles, and efficiently determining copy number variation at R-gene loci in wild species will certainly help in deciding which wild germplasm accessions may be more promising for crop improvement.

Salinity tolerance is necessary in areas such as Iran, north Africa, parts of Spain, and the western United States. However, it should be noted that biotic stress is more important than abiotic stress in all situations (Norouzi et al. 2017). Germinating seeds are sensitive to saline soils (Khayamim et al. 2014), and here salt tolerant germination is required for good emergence and stand establishment. Germplasm release 'EL56' (PI 663211) was selected for germination in 150 mM sodium chloride (McGrath 2011) and may be a source of genes contributing to enhanced stand establishment in saline soils. Adult beets are more tolerant to salt. However, breeding may have reduced the level of salinity tolerance in crop versus wild populations, thus additional gains may be possible through introgressive hybridization with sea beet (Rozema et al. 2015). It is also possible that epigenetic marks affect differences in salinity responses between wild and crop beets (Yolcu et al. 2016). In regions where drought is the dominant abiotic stress, germplasm may be available to meet these needs of intermittently droughty climates in temperate and non-irrigated regions, such as England, Poland, Serbia, and the eastern United States (Ober and Luterbacher 2002; Ober et al. 2004; Rajabi et al. 2009). Nutrient use efficiency is a desired trait, although difficult to select. Screening wild germplasm in nutrient depletion experiments showed root structure and growth patterns were different among *Beta maritima* populations (Saccomani et al. 2009).

10.8 Conclusion (J. Mitchell McGrath and Piergiorgio Stevanato)

Identification of beet genes may be identified directly through analyses of sequenced DNA (Ries et al. 2016; Capistrano-Gossmann et al. 2017). High-quality genome assemblies revolutionize our understanding of beet genetics because the complete genome represents the entire sequence of nucleotides and allow identification of genetic variation and traits with high precision. Today, the cost of whole genome sequencing (Fig. 10.1) and mapping them to a high-quality genome assembly is so low as to be a preferred method over creation of physical and genetic maps. Large amounts of data are generated, which requires new skill sets to analyze and interpret. However, examination of a new set of nucleotide sequences with a well-annotated reference genome puts all variation in genome-scale context, and highlights similarities and differences at the single nucleotide level. Since the availability of beet genome sequences is relatively new, current reference genome assemblies will likely

Fig. 10.1 NGS instrument adopted by leading institutions for whole genome sequencing project

undergo continuous improvement. And, a single assembly will likely not capture all genetic diversity within a species, so adding new genomes to the growing list of reference genome assemblies is important for our understanding of beet genome evolution.

References

Adetunji I, Willems G, Tschoep H, Bürkholz A, Barnes S, Boer M, Malosetti M, Horemans S, Eeuwijk F (2014) Genetic diversity and linkage disequilibrium analysis in elite sugar beet breeding lines and wild beet accessions. Theor Appl Genet 127:559–571

Andrello M, Henry K, Devaux P, Desprez B, Manel S (2016) Taxonomic, spatial and adaptive genetic variation of *Beta* section *Beta*. Theor Appl Genet 129:257–271

Andrello M, Henry K, Devaux P, Verdelet D, Desprez B, Manel S (2017) Insights into the genetic relationships among plants of *Beta* section *Beta* using SNP markers. Theor Appl Genet 130:1857–1866

Arumuganathan K, Earle ED (1991) Nuclear DNA content of some important plant species. Plant Mol Biol Rep 9:208–218

Bartsch D, Ellstrand N (1999) Genetic evidence for the origin of Californian wild beets (genus *Beta*). Theor Appl Genet 99:1120–1130

Biancardi E, Campbell LG, Skaracis GN, de Biaggi M (2005) Genetics and breeding of sugar beet. Science Publishers, Enfield (NH), USA

Biancardi E, Tamada T (2016) Rhizomania. Springer, Heidelberg

Biancardi E, McGrath JM, Panella LW, Lewellen RT, Stevanato P (2010) Sugar beet. In: Bradshaw JE (ed) Tuber and root crops, Handbook of Plant Breeding, vol 7. Springer, New York, pp 173–220

Boudry P, Mörchen M, Saumitou-Laprade P, Vernet P, Dijk H (1993) The origin and evolution of weed beets: consequences for the breeding and release of herbicide-resistant transgenic sugar beets. Theor Appl Genet 87:471–478

Butterfass T (1964) Die chloroplastenzahlen in verschiedenartigen zellen trisomer zuckerruben (*Beta vulgaris* L.). Z Bot 52:46–77

Capistrano-Gossmann GG, Ries D, Holtgrawe D, Minoche A, Kraft T, Frerichmann SLM, Rosleff Soerensen T, Dohm JC, Gonzalez I, Schilhabel M, Varrelmann M, Tschoep H, Uphoff H, Schutze K, Borchardt D, Toerjek O, Mechelke W, Lein JC, Schechert AW, Frese L, Himmelbauer H, Weisshaar B, Kopisch-Obuch FJ (2017) Crop wild relative populations of *Beta vulgaris* allow direct mapping of agronomically important genes. Nat Commun 8:15708. https://doi.org/10.1038/ncomms15708

Carsner E (1938) Wild beets in California. Proc Am Soc Sugar Beet Technol 1:79

Dahlberg HW, Brewbaker HE (1948) A promising sugar beet hybrid of the Milpitas wild type x commercial. Proc Am Soc Sugar Beet Technol 5:175–178

de Bock TSM (1986) The genus *Beta*: Domestication, taxonomy and interspecific hybridization for plant breeding. Acta Hort 182:335–343

de Lucchi C, Stevanato P, Hanson LE, McGrath JM, Panella L, de Biaggi M, Broccanello C, Bertaggia M, Sella L, Concheri G (2017) Molecular markers for improving control of soil-borne pathogen *Fusarium oxysporum* in sugar beet. Euphytica 213:71. doi:https://doi.org/10.1007/s10681-017-1859-7

Desplanque B, Boudry P, Broomberg K, Saumitou-Laprade P, Cuguen J, van Dijk H (1999) Genetic diversity and gene flow between wild, cultivated and weedy forms of *Beta vulgaris* L. (Chenopodiaceae), assessed by RFLP and microsatellite markers. Theor Appl Genet 98:1194–1201

Dohm JC, Minoche AE, Holtgräwe D, Capella-Gutiérrez S, Zakrzewski F, Tafer H, Rupp O, Sörensen TR, Stracke R, Reinhardt R, Goesmann A, Schulz B, Stadler PF, Schmidt T, Gabaldón T, Lehrach H, Weisshaar B, Himmelbauer H (2014) The genome of the recently domesticated crop plant sugar beet (*Beta vulgaris*). Nature 505:546–549

Fenárt S, Arnaud J, de Cauwer I, Cuguen J (2008) Nuclear and cytoplasmic genetic diversity in weed beet and sugar beet accessions compared to wild relatives: new insights into the genetic relationships within the *Beta vulgaris* complex species. Theor Appl Genet 116:1063–1077

Ford-Lloyd BV (1986) Infraspecific variation in wild and cultivated beets and its effect upon infraspecific classification. In: Styles BT (ed) Infraspecific classification of wild and cultivated plants, vol 29, Syst Assoc Spec, pp 331–334

Ford-Lloyd BV, Hawkes JG (1986) Weed beets, their origin and classification. Acta Hort 82:399–404

Ford-Lloyd BV (2005) Sources of genetic variation, genus *Beta*. In: Biancardi E, Campbell L, Skaracis GN, de Biaggi M (eds) Genetics and breeding of sugar beet. Science Publishers, Enfield, New Hampshire, USA, pp 25–33

Frese L, Desprez B, Ziegler D (2001) Potential of genetic resources and breeding strategies for base-broadening in *Beta*. In: Cooper HD, Spillane C, Hodgkin T (eds) Broadening the genetic base of crop production. FAO, IBPRGI jointly with CABI Publishing, Italy, pp 295–309

Frese L, Hannan R, Hellier B, Samaras S, Panella L (2009) Survey of *Beta nana* in Greece. In: Frese L, Germeier CU, Lipman E, Maggioni L (eds) Report of the ECP/GR *Beta* working group and world *beta* network. Third joint meeting 8–10 March 2006, Tenerife, Spain. Bioversity International, Rome, Italy

Fugate KK, Fajardo D, Schlautman B, Ferrareze JP, Bolton MD, Campell LG, Wiesman E, Zalapa J (2014) Generation and characterization of a sugar beet transcriptome and transcript-based SSR markers. Plant Genome 7. doi:https://doi.org/10.3835/plantgenome2013.11.0038

Funk A, Galewski P, McGrath JM (2018) Nucleotide-binding resistance gene signatures in sugar beet, insights from a new reference genome. Plant J 95:659–671

Galewski P, McGrath M (2019) Quantifying genetic diversity in cultivated beet (*Beta vulgaris*) using a pooled population sequencing strategy. Plant and Animal Genome Meeting, Genomics of

Gewnebanks. 12 Jan 2019. https://plan.core-apps.com/pag_2019/abstract/012b143a-73d1-437e-9f20-8e4821c2b5f7. Accessed 31 Mar 2019

Grimmer MK, Kraft T, Francis SA, Asher MJC (2008) QTL mapping of BNYVV resistance from the WB258 source in sugar beet. Plant Breed 127:650–652

Hagihara E, Itchoda N, Habu Y, Iida S, Mikami T, Kubo T (2005) Molecular mapping of a fertility restorer gene for Owen cytoplasmic male sterility in sugar beet. Theor Appl Genet 111:250–255

Halldén C, Ahrén D, Hjerdin A, Säll T, Nilsson NO (1998) No conserved homoeologous regions found in the sugar beet genome. J Sugar Beet Res 35:1–13

Hansen M, Kraft T, Christiansson M, Nilsson N-O (1999) Evaluation of AFLP in Beta. Theor Appl Genet 98:845–852

Harveson RM, Hanson LE, Hein GL (2009) Compendium of beet diseases and pests, 2nd edn. APS Press, St. Paul, Minnesota, USA

Hatlestad GJ, Akhavan NA, Sunnadeniya RM, Elam L, Cargyle S, Hembd A, Gonzalez A, McGrath JM, Lloyd AM (2014) The beet *Y* locus is a co-opted anthocyanin MYB that regulates betalain pathway structural genes. Nat Genet 47:92–96

Hatlestad GJ, Sunnadeniya RM, Akhavan NA, Gonzalez A, Goldman IL, McGrath JM, Lloyd AM (2012) The beet *R* locus encodes a new cytochrome P450 required for red betalain production. Nat Genet 44:816–820

Heitkam T, Holtgrawe D, Dohm JC, Minoche AE, Himmelbauer H, Weisshaar B, Schmidt T (2014) Profiling of extensively diversified plant LINEs reveals distinct plant-specific subclades. Plant J 79:385–397

Henry KB, Desprez B, Devaux B, Devaux PJ, Goudemand E, Guillaume O, Henry N, Mangin B, Pegot-Espagnet P, Willems G (2019) Discovery of genetic diversity of interest by comparing a progeny from (sugar beet elite x exotic) crosses with a sugar beet elite panel. Plant and Animal Genome Meeting, Sugar Beet Workshop, 12 Jan 2019. https://static.coreapps.net/pag-2019/handouts/3e2dbaad-f4c6-47fd-a9a4-78fcddc2ca21_1.pdf. Accessed 31 Mar 2019

Holtgrawe D, Sorensen TR, Viehover P, Schneider J, Schulz B, Borchardt D, Kraft T, Himmelbauer H, Weisshaar B (2014) Reliable *in silico* identification of sequence polymorphisms and their application for extending the genetic map of sugar beet (*Beta vulgaris*). PLoS ONE 9:e110113. https://doi.org/10.1371/journal.pone.0110113

Honma Y, Taguchi K, Hiyama H, Yui-Kurino R, Mikami T, Kubo T (2014) Molecular mapping of *restorer-of-fertility 2* gene identified from a sugar beet (*Beta vulgaris* L. ssp. *vulgaris*) homozygous for the non-restoring *restorer-of-fertility 1* allele. Theor Appl Genet 127:2567–2574

Hornsey KG, Arnold MH (1979) The origins of wild beet. Ann Appl Biol 92:279–285

Hunger S, Gaspero G, Möhring S (2003) Isolation and linkage analysis of expressed disease-resistance gene analogues of sugar beet (*Beta vulgaris* L.). Genome 82:70–82

Jung C, Pillen K, Frese L, Fahr S, Melchinger A (1993) Phylogenetic-relationships between cultivated and wild species of the genus *Beta* revealed by DNA fingerprinting. Theor Appl Genet 86:449–457

Kadereit G, Hohmann S, Kadereit JW (2006) A synopsis of *Chenopodiaceae* subfam. *Betoideae* and notes on the taxonomy of *Beta*. Willdenowia 36:9–19

Keller W (1936) Inheritance of some major color types in beets. J Agric Res 52:27–38

Khayamim S, Tavakkol Afshari R, Sadeghian SY, Poustini K, Rouzbeh F, Abbasi Z (2014) Seed germination, plant establishment, and yield of sugar beet genotypes under salinity stress. J Agr Sci Tech 16:779–790

Kowar T, Zakrzewski F, Macas J, Kobližková A, Viehoever P, Weisshaar B, Schmidt T (2016) Repeat composition of CenH3-chromatin and H3K9me2-marked heterochromatin in sugar beet (*Beta vulgaris*). BMC Plant Biol 16:120

Laurent V, Devaux P, Thiel T, Viard F, Mielordt S, Touzet P, Quillet M (2007) Comparative effectiveness of sugar beet microsatellite markers isolated from genomic libraries and GenBank ESTs to map the sugar beet genome. Theor Appl Genet 115:793–805

Lein JC, Asbach K, Tian Y, Schulte D, Li C, Koch G, Jung C, Cai D (2007) Resistance gene analogues are clustered on Chromosome 3 of sugar beet and cosegregate with QTL for rhizomania resistance. Genome 50:61–71

Levin DA (2001) 50 years of plant speciation. Taxon 50:69–91

Levin DA (2004) Ecological speciation: crossing the divide. Syst Bot 29:807–816

Longden PC (1976) Annual beet: problems and prospects. Pestic Sci 7:422–425

Manel S, Andrello M, Henry K, Verdelet D, Darracq A, Guerin P-E, Desprez B, Devaux P (2018) Predicting genotype environmental range from genome– environment associations. Mol Ecol 27:2823–2833

Mangin B, Sandron F, Henry K, Devaux B, Willems G, Devaux P, Goudemand E (2015) Breeding patterns and cultivated beets origins by genetic diversity and linkage disequilibrium analyses. Theor Appl Genet 128:2255–2271

Matsuhira H, Kagami H, Kurata M, Kitazaki K, Matsunaga M, Hamaguchi Y, Hagihara E, Ueda M, Harada M, Muramatsu A, Yui-Kurino R, Taguchi K, Tamagake H, Mikami T, Kubo T (2012) Unusual and typical features of a novel restorer-of-fertility gene of sugar beet (*Beta vulgaris* L.). Genetics 192:1347–1358

Mauch-Mani B, Baccelli I, Luna E, Flors V (2017) Defense priming: an adaptive part of induced resistance. Ann Rev Plant Biol 68:485–512

McFarlane JS (1975) Naturally occurring hybrids between sugarbeet and *Beta macrocarpa* in the imperial valley of California. J Am Soc Sugar Beet Technol 18:245–251

McGrath J, Derrico C, Yu Y (1999) Genetic diversity in selected, historical US sugarbeet germplasm and *Beta vulgaris* ssp *maritima*. Theor Appl Genet 98:968–976

McGrath JM, Panella L (2019) Sugar beet breeding. In: Goldman I (ed) Plant breeding reviews, vol 42. pp 167–218

McGrath JM (2011) Notice of release of EL56 sugar beet germplasm with high levels of tolerance to salinity during germination. USDA-ARS Germplasm Release

McGrath JM, Jung C (2016) Chapter 16: use of polyploids, interspecific, and intergeneric wide hybrids in sugar beet improvement. In: Mason AS (ed) Polyploidy and interspecific hybridization in crop improvement. Science Publishers (CRC Press), Boca Raton, FL, pp 408–420

McGrath JM, Townsend B, Bogden B, Wittendorp J, Davenport K, Daligault H, Johnson S, Hastie A, Bocklandt S, Darracq A, Willems G, Barnes S, Galewski P, Funk A, Pulman J, Lui T, Childs K (2016) Towards a gold reference assembly for sugar beet 'C869'. Annual beet sugar development foundation research report. Beet Sugar Development Foundation (Available from the authors), Denver, Colorado

McGrath JM, Panella LW, Frese L (2011) *Beta*. In: Kole C (ed) Wild crop relatives: genomic and breeding resources, industrial crops. Springer, Heidelberg, pp 1–28

McGrath JM, Saccomani M, Stevanato P, Biancardi E (2007) Beet. In: Kole C (ed) Genome mapping and molecular breeding in plants, vol 5, Vegetables. Springer, New York, pp 191–207

McGrath JM, Drou N, Waite D, Swarbreck D, Mutasa-Gottgens E, Barnes S, Townsend B (2013) The 'C869' sugar beet genome: a draft assembly. International Plant & Animal Genome XXI. W735

Minoche AE, Dohm DC, Schneider J, Holtgräwe D, Viehöver P, Montfort M, Sörensen TR, Weisshaar B, Himmelbauer H (2015) Exploiting single-molecule transcript sequencing for eukaryotic gene prediction. Genome Biol 16:184

Norouzi P, Sabzehzari M, Stevanato P (2015) Efficiency of some molecular markers linked to rhizomania resistance gene (*Rz1*) for marker assisted selection in sugar beet. J Crop Sci Biotech 18:319–323

Norouzi P, Stevanato P, Mahmoudi SB, Fasahat P, Biancardi E (2017) Molecular progress in sugar beet breeding for resistance to biotic stresses in sub-arid conditions-current status and perspectives. J Crop Sci Biotech 20:99–105

Ober ES, Luterbacher MC (2002) Genotypic variation for drought tolerance in *Beta vulgaris*. Ann Bot 89:917–924

Ober ES, Clark CJA, Le Bloa M, Royal A, Jaggard KW, Pidgeon JD (2004) Assessing the genetic resources to improve drought tolerance in sugar beet: agronomic traits of diverse genotypes under droughted and irrigated conditions. Field Crops Res 90:213–234

Paesold S, Borchart D, Schmidt T, Dechyeva D (2012) A sugar beet (Beta vulgaris L.) reference FISH karyotype for chromosome and chromosome-arm identification, integration of genetic linkage groups and analysis of major repeat family distribution. Plant J 72:600–611

Panella L, Fenwick AL, Stevanato P, Eujayl I, Strausbaugh CA, Richardson KL, Wintermantel WM, Lewellen RT (2018) Registration of FC1740 and FC1741 multigerm, rhizomania-resistant sugar beet germplasm with resistance to multiple diseases. J Plant Regist 12:257–263

Panella L, Lewellen R (2007) Broadening the genetic base of sugar beet: Introgression from wild relatives. Euphytica 154: 383–400

Panella L, Lewellen RT, Hanson LE (2008) Breeding for multiple disease resistance in sugar beet: registration of FC220 and FC221. J Plant Regist 2:146–155

Pin PA, Zhang W, Vogt SH, Dally N, Büttner B, Schulze-Buxloh G, Jelly NS, Chia TYP, Mutasa-Göttgens ES, Dohm JC, Himmelbauer H, Weisshaar B, Kraus J, Gielen JJL, Lommel M, Weyens G, Wahl B, Schechert A, Nilsson O, Jung C, Kraft T, Müller AE (2012) The role of a pseudo-response regulator gene in life cycle adaptation and domestication of beet. Curr Biol 22:1095–1101

Rajabi A, Ober ES, Griffiths H (2009) Genotypic variation for water use efficiency, carbon isotope discrimination, and potential surrogate measures in sugar beet. Field Crops Res 112:172–181

Ries D, Holtgräwe D, Viehöver P, Weisshaar B (2016) Rapid gene identification in sugar beet using deep sequencing of DNA from phenotypic pools selected from breeding panels. BMC Genom 17:236. https://doi.org/10.1186/s12864-016-2566-9

Rozema J, Cornelisse D, Zhang Y, Li H, Bruning B, Katschnig D, Broekman R, Ji B, van Bodegom P (2015) Comparing salt tolerance of beet cultivars and their halophytic ancestor: consequences of domestication and breeding programmes. AoB PLANTS 7:plu083. doi:https://doi.org/10.1093/aobpla/plu083

Saccomani M, Stevanato P, Trebbi D, McGrath JM, Biancardi E (2009) Molecular and morphophysiological characterization of sea, ruderal and cultivated beets. Euphytica 169:19–29

Schmidt T, Heslop-Harrison JS (1998) Genomes, genes and junk: the large-scale organization of plant chromosomes. Trends Plant Sci 3:195–199

Scholten O, de Bock T, Klein-Lankhorst RM, Lange W (1999) Inheritance of resistance to beet necrotic yellow vein virus in Beta vulgaris conferred by a second gene for resistance. Theor Appl Genet 99:740–746

Scholten O, Jansen RC, Keizer L, de Bock T, Lange W (1996) Major genes for resistance to beet necrotic yellow vein virus (BNYVV) in Beta vulgaris. Euphytica 91:331–339

Schondelmaier J, Jung C (1997) Chromosomal assignment of the nine linkage groups of sugar beet (Beta vulgaris L.) using primary trisomics. Theor Appl Genet 95:590–596

Schwichtenberg K, Wenke T, Zakrzewski F, Seibt KM, Minoche A, Dohm JC, Weisshaar B, Himmelbauer H, Schmidt T (2016) Diversification, evolution and methylation of short interspersed nuclear element families in sugar beet and related Amaranthaceae species. Plant J 85:229–244

Shigenaga AM, Berens ML, Tsuda K, Argueso CT (2017) Towards engineering of hormonal crosstalk in plant immunity. Curr Opin Plant Biol 38:164–172

Stevanato P, Trebbi D, Panella L, Richardson K, Broccanello C, Pakish L, Fenwick AL, Saccomani M (2014) Identification and validation of a SNP marker linked to the gene HsBvm-1 for nematode resistance in sugar beet. Plant Mol Biol Rep 33:474–479

Taguchi K, Okazaki K, Takahashi H, Kubo T, Mikami T (2010) Molecular mapping of a gene conferring resistance to Aphanomyces root rot (black root) in sugar beet (Beta vulgaris L.). Euphytica 173:409–418

Taguchi K, Kubo T, Takahashi H, Abe H, Paterson AH (2011) Identification and precise mapping of resistant QTLs of Cercospora leaf spot resistance in sugar beet (Beta vulgaris L.). G3: Genes Genomes Genet 1:283–291

Würschum T, Kraft T (2015) Evaluation of multi-locus models for genome-wide association studies: a case study in sugar beet. Heredity 114:281–290

Würschum T, Maurer HP, Kraft T, Janssen G, Nilsson C, Reif JC (2011) Genome-wide association mapping of agronomic traits in sugar beet. Theor Appl Genet 123:1121–1131

Yolcu S, Ozdemir F, Güler A, Bor M (2016) Histone acetylation influences the transcriptional activation of *POX* in *Beta vulgaris* L. and *Beta maritima* L. under salt stress. Plant Physiol Biochem 100:37–46

Zakrzewski F, Schmidt M, van Lijsebettens M, Schmidt T (2017) DNA methylation of retrotransposons, DNA transposons and genes in sugar beet (*Beta vulgaris* L.). Plant J 90:1156–1175

Appendix A
Scientists and Researchers Involved in *Beta maritima*

The research and breeding activities involving sea beet began in Europe (Germany, Italy, Austria-Hungary, etc.) at the beginning of the 1900s and reached the USA 25 around years later. Since the end of the First World War, a large part of the research in USA has been centered at the USDA-ARS stations in collaboration with seed companies and university scientists. Much of these early researches was published in the Proceedings of ASSBT, Journal of the ASSBT, and Journal of Sugar Beet Research. Breeding developments by the USDA-ARS were often officially released worldwide and documented in Crop Science and Journal of Plant Registrations (Doney 1995). It can be said that the collaborations on research in sugar beet in general and *Beta maritima* in particular have been quite rare in Europe, mainly due to the prevalence of private seed companies. There are only a few exceptions (SIGMEA, AKER, etc.).

In the last two decades, several researches at the University of Lille, France have initiated major studies on the population genetics of *Beta maritima*. Other European researchers also have worked on the plant (University of Birmingham, Wageningen, Rovigo, Braunschweig, Brooms Barn, Kiel, etc.) often in collaboration with their American colleagues. Some seed companies located at Massa Lombarda, Einbeck, Rilland, and Landskrona, have sporadically collaborated as well. Sea beet localization and seed conservation activities are carried out by international organizations including the ECPGR *Beta* Working Group and the World Beta Network. Current activities are sponsored by Biodiversity International, the USDA-ARS National Plant Germplasm System, private seed companies, and sugar beet industries. Basic books and chapters have been edited with the collaboration of ISCI-CRA, Italy; USDA-ARS Stations; Okayama University, Japan; Heilongjiang University, China.

Wilhelm Rimpau obtained hybrids between sea beet and differently colored sugar beets using systems of individual isolation. He classified *Beta maritima* as an annual plant and interpreted the early flowering (bolting) in the first year of cultivated varieties as a return to the ancestral behavior Rimpau (1891). After observation of several hybrid generations between the two species, he believed that "*Beta maritima* is rather similar to *Beta vulgaris.*"

E. Biancardi et al. (eds.), *Beta maritima*,
https://doi.org/10.1007/978-3-030-28748-1

253

Franz Schindler began his research in 1890 by planting in pots and field plots the seed of *Beta maritima* collected at Montpellier, France (von Proskowetz 1892). In both cases, most plants flowered about 2 months after sowing and were crossed with cultivated varieties. Differences in the diameter of pollen and other features of the root (sugar content, fibrousness, etc.) were found between sea beet and the cultivated varieties. At the end of the experiment, Schindler (1891) emphasized the ability of the species belonging to the family Chenopodiaceae to vary the time and the physiology of flowering depending on environmental conditions. According to Rimpau, Schindler expressed the opinion that there are not enough differences between the cultivated beet (*Beta vulgaris*) and *Beta maritima* to consider them as different species.

Emanuel von Proskowetz continued experiments on *Beta maritima* with a small amount of seed received from Schindler. This work lasted two decades and should be considered the first authoritative report on morphology, physiology, and genetics of *Beta maritima*. Seed was sown under normal field conditions and the roots were harvested and analyzed over the following years. Morphological and chemical differences among sea beets grown under wild conditions and cultivated sea beet in two succeeding years were shown. The differences induced by the two environments were notable. von Proskowetz observed that the color of the roots was not uniform and ranged from deep red (30% of individuals) to white (4%). All plants flowered and produced seed the first year (von Proskowetz 1894). In the second generation, all plants bolted except for 19 plants that demonstrated biennial behavior. A small percentage of lines with yellow roots also was detected. The seed of annual plants continued to produce both annual and some biennial beets. Biennial lines retained that characteristic and produced roots were more and more similar (in shape, size, sugar content, etc.) to cultivated varieties. The author wrote that he was convinced that the *Beta maritima* and *Beta vulgaris* were actually a single species, even though there was great morphological variability mainly due to the environment, their natural tendency to variation, and artificial selection. von Proskowetz (1895) also noted that sea beet was an excellent example of the theory of mutation. The equivalence between the two species was criticized by Coons (1975), who described the evident morphological and physiological differences. According to von Proskowetz (1895) and Waldstein and Kitaibel (1864), *Beta trigyna* was a cultivated beet returned to the wild.

Ottavio Munerati established in 1913 the "Regia Stazione Sperimentale di Bieticoltura" at Rovigo, Italy. He initiated experiments on *Beta maritima* with seed collected in 1909 at the mouth of the Po di Levante (Fig. A.1), about 20 km distance from sugar beet fields (Munerati et al. 1913). He increased several collections of sea beet under isolation and began making crosses with commercial varieties. In order to eliminate the undesirable qualities of sea beet, the sugar beet x *Beta maritima* hybrids were backcrossed several times to sugar beet. Selected backcross lines tended to flower later, possessed higher sugar content, and displayed a more regular shape to the roots. More importantly, they were endowed with a high degree of resistance to CLS, to drought, and to root rot. After more than 20 years of recombination and

Fig. A.1 *Beta maritima* at the mouth of Po di Levante River, in the same site where Munerati sampled the seed for the first selections (picture made by Donà dalle Rose, July 1951)

selection, the roots had become almost identical to their cultivated parents in shape, weight, and sugar content. In 1935, some improved lines, including RO581, were sent to the United States where, according to Coons (1954, 1975), they were instrumental in the substantial progress made in sugar yield under severe CLS conditions. Munerati, probably, did not realize entirely just how important his discoveries and developments would be (Munerati 1946). Even today, the Munerati sources account for most of the known resistance to CLS. He investigated annualism, bolting, and carried out a number of experiments on the life cycle and other life history traits of *Beta maritima*. Translations to English of his work brought to attention the value of *Beta maritima* as a useful genetic and plant breeding resource (Coons 1975) (Fig. A.2).

The *Beta maritima* of the Po Delta from which Munerati et al. (1913) isolated the resistance to CLS deserves to be briefly mentioned. When Beguinot (1910) explored this area, *Beta maritima* was localized close to the salty lagoons separating the mainland from the sea. In the terminal branches of the rivers, the lower parts of the banks are normally submerged by the tide, which may be very high during winter storms with wind blowing from South. In particular, sea beet was localized at the south bank of the most northern branch of the Po River, called "Po di Levante". Here Munerati et al. (1913) gathered the seeds of *Beta maritima* growing close to the mouth of the river. During further explorations, sea beet was found neither on the northern banks, nor on beaches, nor on the sandy islands newly formed inside the lagoon (Biancardi and de Biaggi 1979). Some plants are present today on the terminal part of the south bank, although it is more common on the southern bank up to about 300 m from the mouth. Biancardi and de Biaggi (1979) confirmed the observations made by Beguinot (1910) and Munerati et al. (1913). Sea beet never grows directly on the sand, preferring instead sites near the salty water, but among the

Fig. A.2 Stalk of sugar beet
bearing male-sterile and
monogerm flowers (Savitsky
1949)

stones or concrete placed for protection from erosion by the waves. The preference
for soils almost in contact with salt water is probably due to the sensitivity of *Beta
maritima* to competition from weeds. But this advantage is costly. Developing under
extremely difficult conditions, the life of the plants depends on the frequency of rains.
In the case of long-lasting drought, the number of plants in the populations decreases
rapidly (Bartsch et al. 2003; Marchesetti 1897; von Proskowetz 1910) (Fig. A.3).

Jacques de Vilmorin recalled that at the Kew and Montpellier Herbaria he had seen
specimens of *Beta maritima* coming from Malacca, Mexico, Uruguay, and from the
Lido of Venice (see Aldrovandi and Zanichelli). At the Herbarium of Edinburgh,
there were samples coming from China. This book can be considered the first organic
description of genus *Beta* including wild and cultivated species.

Dudok van Heel published some early observation on the inheritance in sugar beet
and on its probable origin. A cross of *Beta maritima* by sugar beet was recorded, in
which biennial forms of *Beta maritima* were chosen and the F_2 generation selected
to eliminate bolters, and then grouped into thick- and thin-leaved forms. The formers
were more like *Beta maritima* in their major traits and the latter quite similar to sugar
beet. Sugar content was then determined and the best beets were used to establish a
series of individual strains in each of the two groups. The thin-leaved group showed
a much higher sugar content than the thick-leaved one, and produced seed superior in
germination capacity, but included strains with more bolters. Strains BM1 and BM9

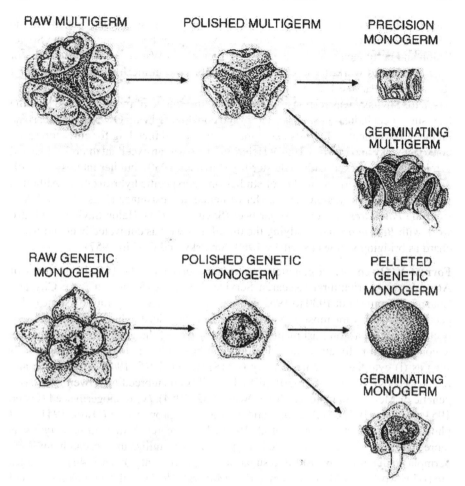

RAW MULTIGERM   POLISHED MULTIGERM   PRECISION
MONOGERM

GERMINATING
MULTIGERM

RAW GENETIC   POLISHED GENETIC   PELLETED
MONOGERM   MONOGERM   GENETIC
MONOGERM

GERMINATING
MONOGERM

Fig. A.3 Multigerm (above) and monogerm seeds of *Beta vulgaris* (BIancardi et al. 2006)

from the thick- and thin-leaved forms, respectively, were carried on. BM9 had not only a lower sugar content but also a lower root weight and the roots showed much more branching. In the F_3 (in which the number of strains was unfortunately greatly reduced), the strain BM9b had leaves resembling the *maritima* type much more than *vulgaris* and was inferior to BM1a in regard to branching of the root. By the time, BMlal and BM9b4 reached the F_4 (though the latter's defects were still evident), the shape had been enormously improved and a reduction in the number of bolters was also evident. The pronounced reddish coloration typical of *Beta maritima* also persisted.

Victor and Helen Savitsky Both were employed by the sugar beet industry in the USDA-ARS between 1947 and 1986. They were sugar beet scientists in the Soviet Union and emigrated to the United States after World War II. A short biography of their arrival and work in the U.S. has been published in the Journal of Sugar Beet Research (McFarlane 1993a, b).

Victor Savitsky was responsible for finding the source of monogerm seed for the U.S. sugar beet industry (Savitsky 1952) and collaborated with Owen and McFarlane on the development of hybrids using male sterility and breeding for curly top resistance (Savitsky and Murphy 1954). Helen Savitsky was an excellent microscopist and cytologist. She worked with male sterility (Savitsky 1950), but her interest in cytology led her to perform a number of studies on interspecific hybridization. Although her principal focus was on the transfer of nematode resistance (Figs. A.4 and A.5) from the *Patellares* section to sugar beet (Savitsky 1975), Helen Savitsky did also work with *Beta maritima*, studying the use of it as well as cultivated beet and Swiss chard as bridging species (Savitsky 1960; Savitsky and Gaskill 1957).

Forrest V. Owen was a geneticist and plant breeder for the U.S. Department of Agriculture's-Agricultural Research Service (USDA-ARS) at Salt Lake City and Logan UT, from about 1930 to 1962. He was considered a true genius by some of his peers and probably the most important American geneticist to have worked on sugar beet. Among his accomplishments are the discovery of the genetics and techniques to convert open-pollinated and synthetic sugar beet cultivars to commercial hybrids such as (i) cytoplasmic male sterility (CMS) (Owen 1945, 1948), (ii) O-type and restorer genes (Owen 1948, 1950), (iii) self-fertility and inbreeding (Owen 1942), (iv) genetic or Mendelian male sterility (Owen 1952, 1954), (v) monogerm seed (Owen 1954; Savitsky 1952), (vi) modifications of asexual propagation (Owen 1941), and photothermal induction (Owen et al. 1940). His research with *Beta maritima* was more subtitled, but he never missed an opportunity to utilize and research any *Beta* germplasm that might be useful to sugar beet improvement (Owen 1944). Examples are (i) his work with Munerati's material (Munerati 1932) on the annual gene *B* and the use of annualism to reduce generation and breeding time (Owen and McFarlane 1958) and to produce a rapid means to index for O-type (Owen 1948, 1950), (ii) curly top resistance that may have been derived from outcrosses of sugar beet to wild beet (Owen et al. 1939, 1946), (iii) self-fertility (*Sf*) from wild beet (Owen 1942), and (iv) *Cercospora* (CLS) resistance through accessions from Italy that are known to have *Beta maritima* sources (Munerati 1946). Owen's annual male-sterile tester likely retained some of the *Beta maritima* traits obtained from Munerati's material. Jones and Davis (1944) are given credit for the discovery and use in onions of CMS to produce hybrid crop varieties. This credit could have gone to Owen and sugar beet. In the early 1940s, Owen had completely worked out the use of CMS (Fig. A.4) to produce hybrid sugar beet (Owen 1945, 1948).

George H. Coons was pathologist for the USDA-ARS in Beltsville, MD from about 1925 to 1955. He investigated the diseases of sugar beet and host-plant resistance and was involved with the development of breeding lines, parental lines, and cultivars with resistance to cercospora leaf spot, *Aphanomyces*, virus yellows, and curly top

Fig. A.4 Progeny line (below) showing resistance under severe cyst nematode attack, Brawley CA, USA) (Lewellen 2007)

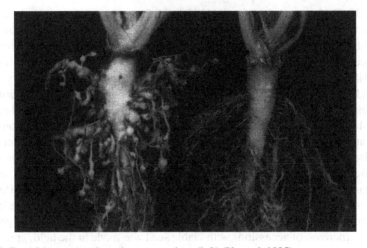

Fig. A.5 Root-knot-nematode-resistant sugar beet (left) (Yu et al. 1995)

virus (Coons 1936, 1953a, b; Coons et al. 1950, 1955) in collaboration with Abegg, Bennett, Bilgen, Bockstahler, Brewbaker, Carsner, Coe, Dahlberg, Deming, Gaskill, Hogaboam, Owen, Stewart, and others. He summarized sugar beet breeding for disease resistance in the USA. (Coons 1936, 1953a; Coons et al. 1955). After retirement, he continued as a collaborator in the USDA-ARS at Beltsville MD. In 1925 (on loan to the USDA from Michigan State University) and 1935, the USDA sent him to Europe to collect *Beta maritima*. Fifty years after his first trip, he returned to Europe and the Middle East to collect the wild species of *Beta* (Coons 1975). Coons also collected or made arrangements for collections of *Beta* subsp. to be shipped to the USA in trips made in 1951 and 1971. The taxonomy of *Beta* subsp. was of continuing interest to Coons. During his 1925 trip, he studied the collections of the genus *Beta* in the herbaria at Kew, UK, and Museum of Natural History, Paris. Collections were made primarily of *Beta maritima* along the coasts of western and southern France, the southeastern coast of England, and the coast of Italy near the Po River

Delta with emphasis on resistance to CLS. In Italy, in 1935, he met Munerati and made arrangement for CLS-resistant germplasm line R481 from Rovigo to be sent to America along with similar germplasm from the seed companies at Cesena and Mezzano, Italy. Earlier Dahlberg, a breeder for Great Western Sugar Company, Longmont, CO, had the 1913 paper by Munerati (1913) translated into English, which made possible the use of *Beta maritima* as a source of resistance genes to cercospora became known (Coons 1975; Dahlberg 1938). In 1935, Coons visited many European countries and their herbaria, including Turkey and Russia, collecting all *Beta* spp. or making arrangements for collections to be sent to USA. In 1951, 1971, and 1975, many of the same locations were revisited and new collections made where possible. These accessions were stored at the Beltsville, MD greenhouse headhouses in various states of disarray and loss and the materials were gathered and organized by McFarlane and Coe and shipped to Salinas in 1969.

John S. McFarlane worked for the USDA-ARS as a sugar beet research geneticist at Salinas CA from 1947 to 1982. He was assigned the responsibility of developing parental lines and commercial hybrids with adaptation to California, specifically for resistance to curly top virus, downy mildew, and bolting (McFarlane 1969; McFarlane et al. 1948). He worked collaborative with Owen, Victor and Helen Savitsky, Coons, Carsner, Murphy, Price, Coe, Lewellen et al. He had an interest in all *Beta* germplasm resources and their preservation and after 1970 this became his principal research focus (McFarlane 1975). Beginning in 1925, Coons had made four collection trips to Europe and brought back seeds of most of the wild *Beta* species. Arrangements were made also with researchers in Europe to make additional collections and have these sent to Beltsville, Europe, and Japan. Efforts were made by Coons, Stewart, and Coe to increase these materials. Good seed storage facilities were lacking at Beltsville and accessions were being lost. The Beltsville collection was sent to Salinas in 1969. These accessions plus the material already at Salinas was the basis for his preservation work. Increase of accessions with viable seed was made in the field, greenhouse, or isolation chambers. All accessions were assigned a Wild Beet (WB) number. For many years, these assigned numbers identified the materials until the National Germplasm System (GRIN) could get PI numbers assigned. For example, for *Beta maritima* about 65 accessions were successfully rescued, increased, partially characterized, and placed in storage with WB numbers ranging from WB29 to WB319. This collection became the material for the subsequent disease-resistant research of USDA-ARS researchers at Salinas and their collaborators, and has yielded resistance genes for powdery mildew, rhizomania, cyst nematode, and root-knot nematode (Doney et al. 1990).

Devon Doney while serving as Chairman of the US Crop Advisory Committee for *Beta*, became involved with a number of *Beta maritima* localizations: 1985—Southern Italy, Sardinia, and Corsica; 1987—England, Wales, and Ireland; and 1989—France, Denmark, and Egypt.

All samples collected were shared with the host country. A grant from the USDA provided funds to evaluate systematically the *Beta maritima* collection for morphological characteristics by Doney (Fargo, North Dakota), rhizomania resistance by

Lewellen (Salinas, California), resistance to Erwinia root rot by Whitney (Salinas, California), and CLS resistance by Ruppel (Fort Collins, CO). Disease resistance has been found and used to enhancement resistance for the above diseases in commercial sugar beets. An enhancement effort to cross *Beta maritima* to commercial sugar beet begun by Doney (Fargo, North Dakota) produced several new lines of promising germplasm released by Larry Campbell (USDA-ARS, Fargo ND).

Efforts to introduce exotic germplasm into cultivated sugar beet expanded after the transfer of Doney from the sugar beet research unit at Logan, Utah to Fargo in 1982 (Doney 1993). Four lines that Doney selected from sugar beet x *Beta maritima* source populations were released in 1994 (Doney 1995). These lines have since been backcrossed to a cultivated sugar beet (Theurer 1978) to increase the sucrose concentration to a more useful level. Doney continued to develop cultivated x wild *Beta* subsp. populations until his retirement in 1996. Early-generation selection in the populations that became F1017 to F1023 was initiated by Doney and released by Campbell (2010). Approximately, 30 populations derived from cultivated sugar beet x *Beta maritima* crosses are currently in the Fargo breeding program. The *Beta maritima* parents in these crosses include accessions from the USDA *Beta* collection originally collected in Belgium, Denmark, France, Guernsey and Jersey Islands, and the United Kingdom. Typically, these populations undergo five–seven cycles of mass selection to reduce the frequency of bolters and plants with multiple crowns and to obtain a single dominant taproot. This usually is followed by at least two cycles of mass selection for sucrose concentration, based upon analysis of individual roots. During his tenure at Fargo, Doney leads wild *Beta* collecting expeditions to Egypt and throughout Europe. Collections made on these trips have increased the diversity within the USDA-*Beta* collection substantially (Doney 1993).

Robert T. Lewellen was a research geneticist for the ARS-USDA at Salinas CA from 1966 to 2008. His research was on the genetics and improvement (enhancement) of sugar beet. Initially, he worked only within developed sugar beet breeding material. After 1980, research and development of parental lines and commercial cultivars were reduced. With Whitney, the Wild Beet (WB) accessions numbered by McFarlane from WB29 to WB319 were screened for reaction to diseases, particularly rhizomania caused by BNYVV, virus yellows caused by *Beet yellows virus*, *Beet western yellows virus*, and *Beet chlorosis virus*, powdery mildew caused by *Erysiphe polygoni*, cyst nematode (*Heterodera schachtii*), and other pests and traits. The sea beet lines increased by McFarlane at Salinas were individually crossed to sugar beet in the greenhouse. The sugar beet x *Beta maritima* F$_2$s were grown in a field under rhizomania conditions and mass selected for resistance to rhizomania and increased in bulk to form line R22, released as germplasm line C50 (Lewellen and Whitney 1993). Over five cycles of selection, R22 was improved for non-bolting, resistance to rhizomania, virus yellow, powdery mildew, root and crown conformation, and root and sugar yield. The improved population was released as C51 (Lewellen 2000). Individual and specific sets of *Beta maritima* accessions also were crossed to sugar beet. From these C48, C58, C79-2 to C79-11, etc. were developed (Lewellen 1997; Lewellen and Whitney 1993). C48 involved specifically WB41 and

WB42 derived from Denmark in the 1950s. Resistance genes $Rz2$ and $Rz3$ were found in these lines and populations R22 (C51) was backcrossed into sugar beet and the populations reselected for favorable traits. From R22 and backcrosses to sugar beet, partial resistance to sugar beet cyst nematode (SBCN) was identified. WB242 was thought to have contributed the resistance to SBCN and specific progenies were selected and advanced with nematode resistance. From WB97 and WB242, gene(s) (*Pm*) conditioning high resistance to *Erysiphe polygoni* syn *betae* were identified and transferred to sugar beet (Lewellen and Schrandt 2001). From the collections of Doney and others, some accessions of *Beta maritima* were evaluated at Salinas in replicated yield and disease evaluation trials. Those that specifically showed resistance to rhizomania were selected and increased in bulk to form broadly based *Beta maritima* populations, coded R23, C26, and C27 (Lewellen 2000). Populations R23 was deliberately left broadly based and composed only of *Beta maritima* germplasm with mild selection pressure exerted only for disease resistance, agronomic type, and non-bolting. These populations should facilitate initial screening of a wide sea beet germplasm base from western Europe in more agronomically acceptable idiotypes. He edited the book "*Beta maritima*: the origin of beets" together with Biancardi and Panella.

Enrico Biancardi began working at the ISCI—Rovigo (ex Stazione Sperimentale di Bieticoltura). Since 1980, he became responsible for the breeding research of the Station and released several male-sterile lines and genetic monogerm varieties, which found commercial development. Using traditional selection systems, applied mainly on crosses with *Beta maritima*, he obtained improvements in resistance to rhizomania and to cercospora leaf spot. Together with de Biaggi and Stevanato, he localized new sites/populations of *Beta maritima* along the Italian coasts. He holds a number of reports at international conferences and authored books, book chapters, and about 150 publications. He coordinated the section "Sugar beet" for the fourth volume of the "Handbook of plant breeding" printed by Springer. In collaboration with Panella and Lewellen, he edited the book "*Beta maritima*: the origin of beets". The book entitled "Rhizomania" was published by Springer in 2016 with Tamada as co-editor. Retired in 2009, he currently collaborates with the University of Padua and some Stations of USDA.

Marco de Biaggi worked at the ISCI Rovigo, Italy since 1975. Collected *Beta maritima* along the mouth of the Po River especially where Munerati sampled the populations used for his first crosses with sugar beet varieties. Part of the collections was sent to McFarlane. In the population-coded WB258, resistance to rhizomania and root-knot nematode was ultimately found. The populations from 1978 were sown into field plots, along with several CLS-resistant 2n families, in a rhizomania-infected field. There de Biaggi and Biancardi identified resistance both to CLS and rhizomania. In the same trials, good resistance to these diseases was found also in 2n multigerm strains derived from Munerati's breeding pool. In 1980, de Biaggi left the Rovigo Station to start a CLS-rhizomania selection and breeding project for the private seed company SES-Italia located at Massa Lombarda, Ravenna. Here, de Biaggi established variety trials in a severely rhizomania-infected field at San Martino,

Bologna. Based upon the Rovigo findings, a set of multigerm diploid entries derived from old families released by Rovigo were tested. The test was under rhizomania severe attack and nearly all plots were hardly damaged. It was possible to select about 100 mother roots from five multigerm entries. The most promising family was the coded 2281. Following overwintering, the selected beets from each family along with two CMS F_1 lines were transplanted into five isolation plots, producing ten experimental hybrids. In 1982, these hybrids were drilled into ITB trials under severe rhizomania conditions near Phitiviers, Loiret, France (Biancardi et al. 2002). All hybrids showed good resistance to rhizomania, but the strain 2281, recoded as ITBR1, showed a very high sugar yield as well. This high-performing hybrid was retested both in healthy and rhizomania diseased trials in 1983 and 1984 at Loiret and Erstein, Bas Rhin, France. The high sugar yield confirmed resistance to rhizomania. In 1985, this new hybrid was named "Rizor" and was commercially grown the year after. The female monogerm parent of Rizor had high yield performance as well, but it was evident that the resistance to rhizomania came from the pollinator (de Biaggi 1987). The resistance from the 2281 pollinator expressed dominance (Biancardi et al. 2002). This first important source of resistance to rhizomania then appears likely to trace back to the populations of Munerati that had *Beta maritima* germplasm introgressed. The fact that similar or identical resistance was found in *Beta maritima* collection WB258 collected at Po di Levante in 1978 supports the hypothesis.

Brian Ford-Lloyd is Emeritus Professor of Plant Genetic Conservation, School of Biosciences, University of Birmingham, UK. The research carried out by Ford-Lloyd on sea beet has included taxonomy, evolution, and domestication, assessments of molecular genetic diversity for conservation, and use of beet genetic resources and risk assessment of gene flow. This has been underpinned by collecting expeditions particularly to Turkey and the Canary Islands. The most important conclusions from his revision of Genus *Beta* Section *Beta* (Ford-Lloyd et al. 1975) were that levels of microspeciation have occurred among wild forms in the center of diversity, with hybridization between wild sea beet and cultivated forms, resulting in a difficult taxonomic situation. Because of predominant continuous variation, the taxonomy was simplified, and a new view of the origin of cultivated beets was proposed. With the development of new molecular genetic markers, the relationships among annual and perennial forms of sea beet could be revealed clarifying the status of sub-specific taxa including "*adanensis*" and "*trojana*", and subspecies *maritima* was found to be more polymorphic than either "*macrocarpa*" or "*adanensis*" at the population and subspecies levels (Shen et al. 1996). The sea beet of Section *Beta* was also used to determine genetic distance between the four sections of the genus, two major findings being the confirmation that Section *Procumbentes* should be regarded as a separate genus (*Patellifolia*) and that Sections *Nanae* and *Corollinae* are very closely related (Shen et al. 1998). The work on beet also led to the isolation of a set of SSR markers from sea beet (Cureton et al. 2006) which then enabled gene flow among populations to be indirectly estimated, and risk assessment of transgene escape to be made. An important conclusion was that the likelihood of transgene spread from crop to wild sea beets is habitat dependent and that this needs to be taken into account when estimating isolation distances for GM sugar beet (Cureton et al. 2006).

George N. Skaracis Professor Emeritus of Plant Breeding and Biometry and Director of the Biometry Science Center, Agricultural University of Athens. He served as former Dean of the School of Agriculture, Engineering and Environmental Sciences, Head of the Department of Crop Science, and Head of the Plant Breeding and Biometry Lab. Before joining the University, he has worked as an Associate Research Geneticist, USDA-ARS, Co., Director of Plant Breeding and Biotechnology and Director of Strategic Planning and Development at Hellenic Sugar Industry S.A. He earned his MSc in Cytogenetics-Chromosome Engineering and his Ph.D. in Quantitative Genetics-Plant Breeding, both from Colorado State University, USA.

A big part of his work on sugar beet breeding focused on developing lines and hybrids resistant to the cercospora leaf spot and/or rhizomania disease, and thus he was heavily involved in evaluating and exploiting germplasm of *Beta maritima* origin. This mainly was accomplished through cooperation in the framework of the IIRB, but most importantly by collaborating with the sugar beet genetics and breeding programs of the USDA research groups at Fort Collins, Co and Salinas, CA, and the respective group at Rovigo, Italy. He also participated in extensive sea beet germplasm collection in the mainland and the Greek islands.

Leonard W. Panella is a Research Geneticist and plant breeder with the USDA-ARS Sugar Beet Research Unit in Fort Collins, Colorado, and has been at the station since 1992. His field program develops sugar beet germplasm with good agronomic characteristics, and increased resistance to rhizoctonia root rot, cercospora leaf spot, the curly top virus, and other important diseases. Enhanced sugar beet germplasm developed in Fort Collins is released to the sugar beet industry. There is a history of over 50 years of continued germplasm development from this program, with most rhizoctonia resistance used in commercial cultivars derived from released sources.

Laboratory research includes programs in *Beta* genomics to explore the potential applied uses of traditional, biochemical, and molecular techniques in a sugar beet germplasm improvement program. These techniques and tools are used to (i) investigate the genetic relationships among cultivated and wild beets, to bring new sources of resistance into the cultivated gene pool, and to better manage our USDA-ARS germplasm resources; (ii) determine genetic control of pathogenicity in important sugar beet pathogens, the genetic control of resistance in the sugar beet, and genetic control of the interactions between this pathogen and sugar beet; and (iii) increase our understanding of the genetic control of sugar beet physiology, particularly the mechanisms of flowering (Reeves et al. 2007). The germplasm and knowledge developed in these research programs maintain a successful breeding program that releases enhanced germplasm to the sugar beet seed industry.

He succeeded Devon Doney as Chairman of the Sugar Beet Crop Germplasm Committee (CGC) and has continued the important work of evaluating the *Beta* collection of the USDA-ARS National Plant Germplasm System (NPGS) and incorporating the disease-resistant genetic resources that are discovered (Panella and

Lewellen 2007). The current goal of the Sugar Beet CGC is to evaluate over 500 *Beta maritima* accession currently in the collection (Chap. 8). He is active in the World Beta Network and the USDA-ARS NPGS. His breeding efforts have been in rhizoctonia root rot resistance, cercospora leaf spot resistance, beet curly top resistance, and sugar beet cyst nematode resistance. He has actively supported plant exploration missions to fill gaps in the USDA-ARS NPGS *Beta* collection, having co-authored a proposal for a collection trip to Greece for *Beta nana*, authored a proposal and participated in the collection trip for *Beta maritima* and *Beta patellaris* in Morocco.

Detlef Bartsch is plant ecologist by training. He began his studies of *Beta maritima* in 1992 from the perspective of gene introgression from genetically modified (GM) sugar beet and its consequences. Initially, the German sugar beet breeder KWS developed BNYYV virus-tolerant sugar beet and sought assistance for environmental risk assessment. Public funding enabled altogether two decades of biosafety research, including basic research in the crop-wild relative complex of *Beta vulgaris*. In 1993, he was the first performing field trials with GM beet on potential environmental impacts of this new plant breeding technology. His interest focussed immediately on the fact that gene flow will happen, and therefore any risk assessment needs to address the consequences for fitness and genetic diversity of native *Beta maritima* populations. He was interested in the full range of beet—environment interactions including plant performance, phytopathology, vegetation science, persistence, and invasiveness.

He studied—together with a number of students—various geographical areas like Germany, Italy, Ukraine, and USA (California). Major findings were that current genetic diversity of *Beta maritima* and some other relatives like *Beta macrocarpa* are to a large extent manmade, and past gene flow from cultivated or weedy forms to *Beta maritima* has more or less broadened the distribution range and genetic diversity of this species. Any environmental impact of modern breeding technologies needs to be set into context of societal/economic needs and environmental protection goals. It is important to manage and use *Beta maritima* as a plant genetic resource in a sustainable manner, taking into account the very dynamic habitats where this plant is found. Since 2002, Bartsch has been working as a technology regulator in the governmental German Authority responsible for GMO risk assessment and management, including applications of GM sugar beet. He still keeps his University of Aachen ties by lecturing and supervising Ph.D. students.

J. Mitchell McGrath is a Research Geneticist with the USDA-ARS with primary responsibility for sugar beet genetics, genomics, and germplasm enhancement, since 1996. He has released several enhanced germplasms targeting traits needed for the rain-fed Eastern US growing region, including reduced soil tare (smooth root) material with increasing levels of disease resistance and improved agronomic performance. In particular, EL54 (PI 654357) was released with Aphanomyces damping-off resistance derived from Beta maritima WB879. Molecular investigations uncovered novel mechanisms contributing to seedling vigor and methods were developed to allow for selection of improved seedling vigor using aqueous solutions, including germination in saline solutions to improve salt-tolerant germination, resulting in release of EL56 (PI 663211) derived from NPGS Accession "Ames 3015". Development of the sugar beet crop is regular and committed post-germination, and a developmental phase change was characterized at ca. 5 weeks post-emergence in the greenhouse, concomitant with the accumulation of sucrose and biomass as well as the acquisition of adult-type (chronic) root disease resistances. Genetic signatures of selection are currently being uncovered through construction of the EL10 sugar beet genome reference assembly and application of additional cultivar and germplasm short-read sequences. Such genome archaeology is proving illuminating in describing genes that define and sustain the beet sugar industry, as well as the various table, fodder, and leaf crops of Beta vulgaris.

Henri Darmency as specialist of weed biology and herbicide resistance, was questioned about the potential consequences of growing genetically modified (GM) herbicide-resistant sugar beet varieties for the agriculture and the environment. He focused the research on the behavior of weed beets. Weed beets growing in root production areas are known to be sugar beet volunteers or progeny of hybrids between sugar beet and inland wild beet occurring in seed production areas. The research topic encompassed the components of the gene flow between the crop and its weed relative and all aspects of the life cycle of the weed. Darmency examined two approaches to gene flow. One was the monitoring of multiyear farm-scale experiments where naturally occurring weed beets grew in GM sugar beet fields and set aside fields. When GM bolters occurred, it showed the production of spontaneous hybrids and the buildup of a soil seed bank containing herbicide-resistant weed seeds (Darmency et al. 2007). There was no fitness cost due to the transgene in the subsequent generations (Vigouroux and Darmency 2017). The second consisted of specific experimental designs to describe the pollen flow using male-sterile target plants. Hybrids were found more than 1 km away from the pollen source, and the pollen was dispersed in agreement on a negative power law with a fat tail, which confirmed that large amounts of pollen migrate at large distances from the field (Darmency et al. 2009). In addition, he tested the diluting effect of increasing weed population sizes with various levels of consanguinity (Vigouroux 2000, and 2019 in preparation). The fate of the domestication traits was also investigated. In order to anticipate the agronomic consequences of the gene flow and enable management options, Darmency collaborated with Colbach to model the effects of the farming systems on the demography of weed beet populations. Several key periods of the life cycle of the weed beet were experimented in order to estimate equations and parameters: staircase-shaped seed longevity and yearly variation of seed dormancy (Sester et al. 2006a), and plant growth and reproduction in agreement on hosting crops (Sester et al. 2004). The model, called GeneSys, was built (Sester et al. 2006b, 2008) and tested through sensitivity analysis (Colbach et al. 2010; Tricault et al. 2009) and run with data set from farm surveys. All these data could help predict coexistence issues between GM and non-GM varieties and recommend management procedures. The model is also helpful to anticipate the behavior of weed beets as undesirable plants in arable fields (Colbach et al. 2011).

Nina Hautekeete Her research is underpinned by rather fundamental interests in the evolution of biodiversity in its broad sense: intraspecific, specific, and ecosystem diversity, under the current context of global changes. The work focuses on the continuum of species responses to global change, ranging from adaptation, observed at a specific level, to the consequences of global change on species communities, through migrations, local extinctions, and changes in plant–pollinator interaction networks.

She started working on sea beet, and more largely on the *Beta* complex, during her Ph.D. thesis under the supervision of Henk van Dijk and Yves Piquot, at the University of Lille, and kept on working on this species as a Postdoc Researcher at the University of Leiden and then as a Lecturer in Lille in 2002. Her work focused also on the adaptation to natural and anthropogenic selective factors, and more specifically on the evolution of life history traits under the influence of natural environmental factors and of human-related disturbance.

These studies have been conducted mainly by correlative approaches in space, taken as a substitute to time, experimental experiments, and also experiments of ecological "resurrection". The general objective was to identify the potential of genetic evolution in a changing environment over time and then to verify the existence, or not, of such evolution on some climate-related traits. This work first revealed the geographical variability of many life history traits in *Beta vulgaris maritima*, i.e., life span, seed dormancy, flowering phenology, and estimated their heritability. We also revealed the ecological factors correlated to this variability. We moreover showed a genetic evolution of some traits related to flowering phenology, which can be linked to the recent climate change and that corresponds to a shift of 39 km northward in 20 years.

Henk van Dijk is Emeritus Professor of Evolutionary Ecology, Université de Lille, France. His research, together with his Ph.D. students Pierre Boudry, Benoît Desplanque, Nina Hautekeete, and Kristen Wagmann, was principally on life history variation and its geographic distribution. This included the moment of first flowering (without or with passing a winter period); the number of reproductive periods, which is connected with life span; aging effects; and germination behavior. Special attention was given to the phenomenon of "weed beets" (Boudry et al. 1993) and the possible risk they form for the introduction of transgenic herbicide-resistant sugar beets (Desplanque et al. 2002). Artificial selection experiments were successfully carried out to change the sensitivity for day length requirement (van Dijk and Hautekèete 2007) and the earliness of flowering of plants without cold requirement. The wild populations sampled in 1989 along the French and neighboring coasts, which formed the basis of most of the research, were sampled again in 2009 which enabled a direct comparison, thanks to the fact that sea beet seeds do not decline in germination rate even after 20 years. Although the number of generations was limited during 20 years, a shift in flowering time under controlled conditions, pointing to a genetic change, could be established.

Piergiorgio Stevanato has collaborated since 1999 at ISCI-CRA Experimental Station, Rovigo, Italy. Under the direction of Biancardi, his mission was the study of the natural populations of sea beet in some areas of Adriatic coastline, paying special attention to the coastal areas of Po Delta. A further objective of this study was the evaluation of the influence on the biodiversity of these populations of the presence of large areas cultivated with sugar and seed beets. In 2006, he moved to the Department of Agricultural Biotechnology, University of Padua, continuing the collaboration with ISCI-CRA.

The aims of the project can be summarized as follows:

- Identification of the different populations of sea beet along the mentioned coastal areas;
- Evaluation of the dimensions and phenotypic variability in these populations;
- Mapping the biodiversity within the *Beta vulgaris* species present in the areas, and monitoring the variation over time of the evaluated diversity;
- The relationships existing among the different sea beet populations and differences between them and the sugar beet commercial varieties are evaluated and quantified in detail;
- Identification of possible gene flow between the wild and the cultivated gene pool and *vice versa*;
- Quantification of the diversity still present in the sea beet populations is also evaluated.

The methods employed for reaching these aims are (i) in situ identification of sea beet populations, recording of their geographical coordinates, and of the number and size of each populations; (ii) molecular studies carried out on the plants belonging to the populations. The distribution, relative frequency, and the polymorphism level for each locus identified are compared within the natural populations under study, and between them and sugar beet commercial varieties; and (iii) statistical analysis of the data relative to the genetic loci examined. The information gained during this study is of great value for protection of the biodiversity of sea beet and for the correct management of the sugar beet.

Pascal Touzet I am an evolutionary geneticist and have been working on wild beets of Section *Beta* for more than two decades in the University of Lille following two axes: (i) Mating system evolution with a focus on gynodioecy in *Beta maritima*: its evolutionary dynamics in populations, its impact on mitochondrial genetic diversity, its genetic architecture; (ii) Speciation in Section *Beta*: the origin of tetraploid *Beta macrocarpa* in the Canary Islands.

Ourania I. Pavli Assistant Professor of Genetics and Plant Breeding, at the Department of Agriculture Crop Production and Rural Environment, University of Thessaly. She earned her M.Sc. in Plant Breeding and Genetic Resources and her Ph.D. in Plant Science, both from the Wageningen University, the Netherlands. Prior to that, she has been an Assistant Researcher at the Breeding Station of Hellenic Sugar Industry, a Research Associate and a Postgraduate Researcher at the Laboratory of Plant Breeding and Biometry, Dpt. of Crop Science, Agricultural University of Athens.

Her research has been concentrating on breeding for disease resistance in sugar beet, with special emphasis on rhizomania. Mostly during her research work at Hellenic Sugar, she was occupied with efforts to introgress germplasm of *Beta maritima* origin into elite lines of an applied program, utilizing populations segregating for genetic male sterility stemming out of the USDA sugar beet research program at Salinas CA.

References

Bartsch D, Cuguen J, Biancardi E, Sweet J (2003) Environmental implications of gene flow from sugar beet to wild beet–current status and future research needs. Environ Biosafety Res 2: 105–115

Beguinot A (1910) Contributo alla conoscenza della flora litoranea del Polesine. Rivista Agraria Polesana 12:232–242

Biancardi E, de Biaggi M (1979) *Beta martima* L. in the Po Delta. In: ISCI (editor) Proc Convegno Tecnico Internazionale in Commemorazione di Ottavio Munerati. Rovigo, Italy, pp 183–185

Biancardi E, Lewellen RT, de Biaggi M, Erichsen AW, Stevanato P (2002) The origin of rhizomania resistance in sugar beet. Euphytica 127:383–397

Boudry P, Mörchen M, Saumitou-Laprade P, Vernet P, van Dijk H (1993) The origin and evolution of weed beets: consequences for the breeding and release of herbicide-resistant transgenic sugar beets. Theor Appl Genet 87:471–478

Campbell LG (2010) Registration of seven sugarbeet germplasms selected from crosses between cultivated sugarbeet and wild *Beta* species. J Plant Reg 4:149–154

Colbach N, Darmency H, Tricault Y (2010) Identifying key life-traits for the dynamics and gene flow in a weedy crop relative: Sensitivity analysis of the GeneSys simulation model for weed beet (*Beta vulgaris* ssp. *vulgaris*). Ecol Model 221:225–237

Coons GH (1936) Improvement of the sugar beet. 1936 Yearbook of Agriculture. USDA, Washington, DC, pp 625–656

Coons GH (1953a) Breeding for resistance to disease. Yearbook of Agriculture 1953:174–192

Coons GH (1953b) Disease resistance breeding of sugar beets 1918–1952 Phytopathology 43:297–303

Coons GH (1954) The wild species of *Beta*. Ibidem 8:142–147

Coons GH (1975) Interspecific hybrids between *Beta vulgaris* L. and the wild species of *Beta*. Ibidem 18:281–306

Coons GH, Owen FV, Stewart D (1955) Improvement of the sugar beet in the United States. Adv Agron 7:89–139

Coons GH, Stewart D, Bockstahler HW, Deming GW, Gaskill JO, Lill JG, Schneider CL (1950) Report on 1949 tests of U.S. 216 × 225 and other varieties from sugar beet leaf spot resistance breeding investigation of the U.S. Department of Agriculture. Proc ASSBT 6:209–217

Cureton AN, Newbury HJ, Raybould AF, Ford-Lloyd BV (2006) Genetic structure and gene flow in wild beet populations: the potential influence of habitat on transgene spread and risk assessment. J Appl Ecol 43:1203–1212

Dahlberg HW (1938) Some observations on the wild beet (*Beta maritima*). Proc ASSBT 1:76–79

Darmency H, Klein E, Gestat de Garanbé T, Gouyon PH, Richard-Molard M, Muchembled C (2009) Pollen dispersal in sugar beet production fields. Theor Appl Genet 118:1083–1092

Darmency H, Vigouroux Y, Gestat de Garambé T, Richard-Molard M, Muchembled C (2007) Transgene escape in sugar beet production fields: data from six years farm scale monitoring. Environ Biosafety Res 6:197–206

de Biaggi M (1987) Mehodes de selection - un cas concret. Proc IIRB 50:157–161

Desplanque B, Hautekèete NC, van Dijk H (2002) Transgenic weed beets: possible, probable, avoidable. J Appl Ecol 39:561–57

Doney DL (1993) Broadening the genetic base of sugarbeet. J Sugar Beet Res 30:209–220

Doney DL (1995) Registration of y317, y318, y311 and y387 sugarbeet germplasms. Crop Sci 35:947

Doney DL, Whitney ED, Terry J, Frese L, Fitzgerald P (1990) The distribution and disperal of *Beta maritima* germplasm in England, Wales, and Ireland. J Sugar Beet Res 27:29–37

Ford-Lloyd BV, Williams ALS, Williams JT (1975) A revision of *Beta* section *Vulgares* (Chenopodiaceae), with new light on the origin of cultivated beets. Bot J Linn Soc 71:89–102

Jones HA, Davis G. (1944) Inbreeding and heterosis and their relation to the development of new varieties of onions. Beltsville MD. USDA Tech Bull No 874

Lewellen RT (1997) Registration of 11 sugarbeet germplasm C79 lines with resistance to rhizomania. Crop Sci 37:1026

Lewellen RT (2000) Registration of rhizomania resistant sugarbeet × *Beta vulgaris* subsp. maritima germplasms C26, C27, and C51. Crop Sci 40:1513–1515

Lewellen RT, Schrandt JK (2001) Inheritance of powdery mildew resistance in sugar beet derived from *Beta vulgaris* ssp. *maritima*. Plant Dis 85:627–631

Lewellen RT, Whitney ED (1993) Registration of germplasm lines developed from composite crosses of sugar beet x *Beta maritima*. Crop Sci 33:882–883

Marchesetti C (1897) Flora di Trieste e dè suoi dintorni. Museo Civico, Trieste, Italy

McFarlane JS (1969) Breeding for resistance to curly top. Proc. IIRB 4:73–83

McFarlane JS (1975) Naturally occurring hybrids between sugarbeet and *Beta macrocarpa* in the Imperial Valley of California. J ASSBT 18:245–251

McFarlane JS (1993a) The Savitsky Story: Part I. J Sugar Beet Res 30:1–36

McFarlane JS (1993b) The Savitsky Story, Part II. J Sugar Beet Res 30:125–141

McFarlane JS, Price C, Owen FV (1948) Strains of sugarbeets extremely resistant to bolting. Proc ASSBT 5:151–153

Munerati O (1932) Sull' incrocio della barbabietola coltivata con la beta selvaggia della costa adriatica. L'Industria Saccarifera Italiana 25:303–304

Munerati O (1946) Il problema della barbabietola. Consulta Regionale Veneta dell' Agicoltura e delle Foreste. Stamperia Editrice Zanetti, Venice, Italy

Munerati O, Mezzadroli C, Zapparoli TV (1913) Osservazioni sulla *Beta maritima* L. nel triennio 1910–1912. Staz Sper Ag Ital 46:415–445

Owen FV (1941) Asexual propagation of sugar beets. J Hered 32:187–192

Owen FV (1942) Inheritance of cross- and self-sterility in *Beta vulgaris* L. J Agr Res 64:679–698

Owen FV (1944) Variability in the species *Beta vulgaris* L. in relation to breeding possibilities with sugar beets. J Am Soc Agron 36:566–569

Owen FV (1945) Cytoplasmically inherited male-sterility in sugar beets. J Agr Res 71:423–440

Owen FV (1948) Utilization of male sterility in breeding superior-yielding sugar beets. Proc ASSBT 5:156–161

Owen FV (1950) The sugar beet breeder's problem of establishing male-sterile populations for hybridization purposes. Proc ASSBT 6:191–194

Owen FV (1952) Mendelian male sterility in sugar beets. Proc ASSBT 7:371–376

Owen FV (1954) Hybrid sugar beets made by utilizing both cytoplasmic and Mendelian male sterility. Proc ASSBT 8:64

Owen FV, Abegg FA, Murphy AM, Tolman B, Price C, Larmer FG, Carsner E. Curly-top-resistant sugar-beet varieties in 1938 (1939). Washington, D.C., United States Department of Agriculture. Circular No. 513. 10 pp.

Owen FV, Carsner E, Stout M (1940) Photothermal induction of flowering in sugar beets. J Agr Res 61:101–124

Owen FV, McFarlane JS (1958) Successive annual backcrosses to a nonbolting inbred line of sugar beets. J ASSBT 10:124–132

Owen FV, Murphy AM, Ryser GK (1946) Inbred lines from curly-top-resistant varieties of sugar beet. Proc ASSBT 4:246–252

Panella L, Lewellen RT (2007) Broadening the genetic base of sugar beet: introgression from wild relatives. Euphytica 154:382–400

Rimpau W (1891) Kreuzungproducte landwirtschaftlicher Kulturpflanzen. Landwirtschaft Jahrbuch, vol 20, Berlin, Germany

Savitsky H (1950) A method of determining self-fertility of self sterility in sugar beet based upon the stage of ovule development shortly after flowering. Proc ASSBT 6:198–201

Savitsky H (1960) Meiosis in an F_1 hybrid between a Turkish wild beet (*Beta vulgaris* ssp. *maritima*) and *Beta procumbens*. J ASSBT 11:49–67

Savitsky H (1975) Hybridization between *Beta vulgaris* and *Beta procumbens* and transmission of nematode (*Heterodera schachtii*) resistance to sugar beet. Can J Genet Cytol 17:197–209

Savitsky H, Gaskill JO (1957) A cytological study of F_1 hybrids between Swiss chard and *Beta webbiana*. J ASSBT 9:433–449

Savitsky VF (1952) Methods and results of breeding work with monogerm beets. Proc ASSBT 7:344–350

Savitsky VF, Murphy AM (1954) Study of inheritance for curly top resistance in hybrids between mono- and multigerm beets. Proc ASSBT 8:34–44

Schindler F (1891) Über die Stammpflanze der Runkel- und Zuckerrüben. Botanisches Centralblatt 15:6–16

Sester M, Delanoy M, Colbach N, Darmency H (2004) Crop and density effects on weed beet growth and reproduction. Weed Research 44:50–59

Sester M, Dürr C, Darmency H, Colbach N (2006a) Evolution of weed beet (*Beta vulgaris* L.) seed bank: Quantification of seed survival, dormancy, germination andf pre-emergence growth. Europ J Agronomy 24:19–25

Sester M, Dürr C, Darmency H, Colbach N (2006b) Modelling the effects of the cropping systems on the seed bank dynamics and the emergence of weed beet. Ecol Model 204:47–58

Sester M, Tricault Y, Darmency H, Colbach N (2008) GeneSys-Beet: A Model of the effects of croppin systems on gene flow between sugar beet and weed beet. Field Crops Research 107:245–256

Shen Y, Ford-Lloyd BV, Newbury HJ (1998) Genetic relationships within the genus *Beta* determined using both PCR-based marker and DNA sequencing techniques. Heredity 80:624–632

Shen Y, Newbury HJ, Ford-Lloyd BV (1996) The taxonomic characterisatoin of annual *Beta* germplasm in a genetic resources collection using RAPD markers. Euphytica 91:205–212

Theurer JC (1978) Registration of eight germplasm lines of sugarbeet. Crop Sci 18:1101

Tricault Y, Darmency H, Colbach N (2009) Identifying key components of weed beet management using sensitivity analyses of the GeneSys-Beet model in GM sugar beet. Weed Research 49:581–591

Vigouroux Y (2000) Betteraves transgéniques et betteraves adventices:étude des fluz de génes et de leurs conséquences. University of Bourgogne, France

von Proskowetz E (1892) Über die Stammpflanze der Runkel- und Zucker-rübe. Österreiche- Ungarische Zeitschrift für Zuckerindustrie und Landwirtschaft 29:303–317

von Proskowetz E (1894) Über die Culturversuche mit *Beta maritima* L. (und *Beta vulgaris* L.) im Jahre 1893. Österreiche-Ungarische Zeitschrift für Zuckerindustrie und Landwirtschaft 31:201–223

von Proskowetz E (1895) Über die Culturversuche mit *Beta* im Jahre 1894 und über Beobachtungen an Wildformen auf naturlichen Standorten. Österreiche-Ungarische Zeitschrift für Zuckerindustrie und Landwirtschaft 32:227–275

von Proskowetz E (1910) Über das Vorkommen der Wildformen der Zuckerrüben am Quarnero. Österreiche-Ungarische Zeitschrift für Zuckerindustrie und Landwirtschaft 47:631–640

Appendix B
Lifespan of Authors and Scientists Cited in the Text

If the case, the complete Latin names or pseudonymous is written between parentheses

Adrian (*Publius Aelius Traianus Hadrianus*) 76–138
Albertus Magnus 1113–1206
Aldrovandi Ulisse 1522–1605
Apicius (*Apuleius Barbarus*) 320–370
Arcangeli Giovanni 1840–1921
Aristophanes 446–386 BC
Aristotle 384–322 BC
Augustus (*Gaius Julius Caesar Augustus*) 63 BC–14 AD
Aven Roshdi (Averroes) 1126–1198
Averroes (Ibn Rusd) 1126–1198
Avicena (Ibn Sina) 980–1037
Bauhin Gaspard (*Bauhinus*) 1560–1624
Bauhin Johann 1541–1613
Blackwell Elizabeth 1797–1755
Boccone Paolo 1663–1704
Bock Hieronymus (*Tragus*) 1498–1554
Briem Hermann 1846–1910
Brotero Felix 1744–1828
Bruhnfels Otto (*Brunfelsius*) 1488–1534
Cato (*Marcus Porcius Cato*) 234–145 BC
Catullus (*Gaius Valerius Catullus*) 84–54 BC
Cesalpino Andrea (*Cesalpinus*) 1524–1603
Chabraeus Dominicus (*Chabraeus*) 1610–1665
Cicero (*Marcus Tullius Cicero*) 106–43 BC
Coles William 1826–1662
Columella (*Lucius Junius Moderatus Columella*) 4–70 AD

This is a U.S. government work and not under copyright protection in the U.S.; foreign copyright protection may apply 2020
E. Biancardi et al. (eds.), *Beta maritima*,
https://doi.org/10.1007/978-3-030-28748-1

Coons George H. 1885–1980
Crateuas 120–63 BC
Culpeper Nicholas 1616–1654
D'Alambert Jean Baptiste 1717–1783
Dalechamps Jacques 1513–1588
Darwin Charles 1809–1882
de Candolle Alphonse 1806–1893
de Candolle Augustin 1778–1841
de Crescenzi Pietro (*Crescentius*) 1233–1320
de Lobel Matthias (*Lobelius*) 1538–1616
de Medici Lorenzo 1449–1492
de Serres Oliver 1539–1619
de Villlanova Arnaldo (*Novavilla*) 1240–1313
de Vilmorin Jacques 1882–1939
de Vilmorin Luis 1816–1860
de Vries Hugo 1848–1935
Desfontaines Reneè Louiche 1750–1833
Diderot Denise 1713–1784
Diocles Caristos 380–420? AD
Dioscorides (*Pedanius Dioscurides Anazerbaeus*) 40–90? AD
Dodoens Rembert (*Dodoneus*) 1517–1585
Dorsten Johann (*Dorstenius*) 1643–1706
Dudok van Heel Johannes 1891–1971
Durante Castore 1529–1590
Engler Adolf 1844–1930
Fuchs Leonhart (*Fuchsius*) 1501–1556
Galilei Galileo 1554–1642
Galen (*Claudius Galenus*) 129–216
Gerard John 1545–1611
Gessner Conrad (*Gesnerius*) 1516–1565
Gunter Johan (*Gunterius*) 1505–1574
Hildegard von Bingen (*Hildegarda*) 1098–1179
Hill John 1714–1773
Hippocrates 460–370? BC
Homer 900–800? BC
Hooker William J. 1785–1865
Ibn Beith 1179–1248
Johnson Thomas 1684–?
Lamarck Jean-Baptiste 1744–1829
Linnè Carl (*Linnaeus*) 1707–1778
Malpighi Marcello (*Malpighius*) 1628–1694
Margraaf Andreas Sigismund 1709–1782

Martial (*Marcus Valerius Martialis*) 40–104 AD
Mattioli Pietro Andrea (*Matthiolus*) 1501–1597
McFarlane John S. 1915–1994
Mendel Gregor 1822–1894
Moquin-Tandon Horace 1804–1863
Morison Robert (*Morisonus*) 1620–1683
Munerati Ottavio 1875–1949
Naccari Fortunato Luigi 1793–1860
Owen Forrest V. 1899–1962
Parkinson John (*Parkinsonus*) 1567–1660
Pena Pierre (*Pena*) 1535–1605
Pitton de Tournefort Joseph (*Tournefortius*) 1656–1708
Plautus (*Titus Maccius Plautus*) 253–184 BC
Pliny the Elder (*Gaius Plinius Secundus*) 23–79
Rabano Mauro (*Hrabanus Maurus*) 780–856
Ray John (*Raius*) 1627–1705
Reichenbach Heinrich 1824–1889
Reichenbach Ludwig 1793–1879
Rimpau Wilhelm 1842–1903
Rini Benedetto (*Rinius*) 1485–1565
Ritzi Valerius (*Valerius Cordus*) 1515–1544
Roccabonella Niccolò (*Roccabonella*) 1386–1459
Savitsky Helen 1901–1986
Savitsky Viacheslav 1902–1965
Schindler Franz 1881–1920
Silvatico Matteo (*Sylvaticus*) 1285–1348?
Smith James Edvard 1759–1828
Squalermo Luigi (*Anguillara*) 1512–1570
Strabo 64 BC–24 AD
Svetonius (*Gaius Svetonius Tranquillus*) 70–126
Tanara Vincenzo ?–1644?
Theodorus Jacob (*Tabernaemontanus*) 1520–1590
Theophrastus 371–287 BC
Varro (*Marcus Terentius Varro*) 116–23 BC
Virgil (*Publius Virgilius Maro*) 70–19 BC
von Lippmann Edmund 1857–1940
von Proskowetz Emanuel 1846–1944
Weinmann Johann Wilhelm 1683–1741
Zanichelli Giovanni Geronimo 1662–1729.

Appendix C
Beta Chronology

Chronology of the more significant progresses regarding *Beta* subsp. *vulgaris* and *Beta subsp. maritima*

Before 8500 BC	Leaves of wild beets collected for food
Around 8500 BC	Domestication of wild beets likely in Middle East
After 8500 BC	Leaf beet cultivation spread across the Mediterranean basin
Around 3500 BC	Leaf and likely root beets in Egypt
Around 1200 BC	Leaf beet in Syria
Around 1000 BC	Leaf beet in Greece
Around 700 BC	First written mention regarding leaf beet
Around 600 BC	Leaf beet in China
Around 460 BC	Leaf beet (τευτλον-*teutlon*), black beet (τευτλον μελαν), and wild beet (βλιτος-*blitos*) described by Hippocrates and Theophrastus also for medicinal purposes
Around 420 BC	*Teutlon* cited by Aristophanes
Around 400 BC	*Teutlon* named "*cicla*" likely in Sicily
Around 300 BC	First written mention of τευτλον άγςια (wild beet) by Diocles from Carystos
Around BC 300	*Teutlon* (*cicla, limonium, neuroides*, etc.) in Rome
274 BC	*Teutlon* or *cicla*, etc. named "*Beta*" by Cato
After 250 BC	Diffusion of root (garden or red) beet
After 50 BC	Diffusion of *Beta* crops across the Roman Empire
78	*Beta*, including βλιτος-*blitos-Beta silvestris* (wild beet), described by Pliny
Around 80	Medicinal uses of *Beta*, including *Beta silvestris*, described by Dioscorides, Apicius, Galen, and others
1000–1300	*Beta, selga, silga*, etc. cited by Arabian authors

(continued)

(continued)

Middle Age	*Beta sylvestris* and synonyms cited by several European authors
Around 1500	Diffusion of fodder beet in Central Europe
1551	*Beta sylvestris* named "*Beta sylvestris marina*" by Aldrovandi
1623	*Beta sylvestris marina* named "*Beta sylvestris maritima*" by Bauhin
1657	*Beta sylvestris* named "sea beet" by Coles
1762	*Beta maritima* classified as species by Linnaeus
1768	*Beta maritima* transported casually in California by Spanish ships
End of 1700	Selection of sugar beet in Germany likely from fodder beet x *Beta maritima* casual hybrids
Early 1900	*Beta maritima* crossed with sugar beet for breeding purposes in Germany, Austria, and Italy
1937	Release of cercospora leaf spot-resistant[a] varieties in Italy and USA
1948	Selection of male-sterile lines in USA
1955	Selection of monogerm lines[b] in USA
1960	Release of genetic monogerm hybrids in USA
1960	Release of multigenic rhizomania-resistant[a] varieties in Italy
Around 1980	Employment of biotechnology-assisted breeding methods on *Beta maritima*
1980	Release of fodder beet monogerm varieties
1985	Release of monogenic rhizomania-resistant[a] varieties in Italy, USA, and France
1992	Release of fodder beet rhizomania-resistant[a] varieties
1993	*Beta maritima* classified as subspecies of the species *vulgaris*
1998	Release of Roundup-resistant varieties in USA
2002	Release of cyst nematode-resistant[a] varieties in USA and Germany
2006	Release of powdery mildew-resistant[a] lines in USA
2006	Release of varieties with multiple resistance to rhizomania
2015	"Rizor" and "Holly" rhizomania resistance behave the same lineage
2016	Molecular mechanism of bolting

[a]Trait surely transferred from *Beta maritima*
[b]Trait likely transferred from *Beta maritima*

Appendix D
Synonyms of *Beta maritima*^(*)

Greek: βλιτος (blitos); λειμωνιον (leimonium) (1) (2);

Latin: *lonchitis, sinapi aselli* (3); *Beta silvestris, limonium, neuroides* (4); *lapathium*
(5); *tarlus, beta, secla, sencon* (6); *blitum silvestre, polysporon anguillarae; Beta
sylvestris marina* (8); *tintinabulum terae* (9); *plumbago, molybdena Plinii* (12); *trifolium palustre, lampsana, bistorta, pyrola, mysotis, potamogaton, carduus pratense,
plantago aquatica, lapathum* (13); *Beta erythrorhiza, Beta platicaulis* (14); *Beta
maritima* (14), *Beta marina* (14), *Beta decumbens* (14), *Beta triflora* (14), *Beta carnulosa* (14), *erecta* (14), *noeana* (15); *Beta sylvestre spontanea maritima* (16), *Beta
communis viridis* (16); *Beta agrigentina, Beta atriplicifolia* (19); *Beta sylvestris maritima* (20); *Beta commune viridis* (21); *Beta sylvestris spontanea marina* (f23); *Beta
marina syl. major* (24); *Beta marina syl. minor* (24); *Beta marina decumbens* (25);
Beta marina triflora (25); *Beta marina noëana* (25); *Beta marina annua* (25); *Beta
rapacea* (27); *Beta maritima* subsp. *mediterraneum* (29), *Beta maritima subsp. danica* (29); *Beta perennis* (30); *Beta vulgaris* L. subsp. *maritima* (L.) Arcang. (36);
Beta carnulosa (38); *Beta vulgaris* L. var. *bengalensis* (37), *Beta vulgaris* L. var.
orientalis (37); *Beta vulgaris* subsp. *perennis* (L.) (38); *Beta orientalis* (38); *Beta
bengalensis* (38); *Beta orientalis* var. *bengalensis* (38); *Beta vulgaris* subsp. *perennis*
var. *maritima* (38); *Beta vulgaris* subsp. *lomatogonoides* (38); *Beta vulgaris* subsp.
maritima var. lomatogonoides (38); *Beta vulgaris* subsp. *maritima* var. *glabra* (38);
Beta vulgaris subsp. *maritima* var. *foliosa* (38); *Beta vulgaris* subsp. *maritima* var.
pilosa (38). *Beta vulgaris* subsp. *orientalis* (38); *Beta maritima* subsp. *danica* (38);
Beta maritima subsp. *meditrraneum* var. *erecta* (38); *Beta maritima* subsp. *mediterraneum* var. *prostrata* 38); *Beta maritima* subsp. *mediterraneum* var. *atriplicifolia*
(38); *Beta trojana* (38); *Beta vulgaris* subsp. *maritima* var. *grisea* (38); *Beta vulgaris*
subsp. *maritima* var. *atriplicifolia* (38); *Beta vulgaris* subsp. *provulgaris* (38); *Beta
palonga* (38); *Beta caudicantibus foliis* (39); *Beta radice buxea* (39): *Beta quinquinervia* (39); *Beta centinervia* (39);

E. Biancardi et al. (eds.), *Beta maritima*,
https://doi.org/10.1007/978-3-030-28748-1

Italian: pyrola (6); piantagine acquatica, giegola silvestre, elleboro bianco (10); caprinella, herba di Sant'Antonio, dentilaria (12); beta campestre (34), bietola o barbabietola marittima (12);

German: Vintergrun, Holtz mangold (6); Wintergrün, Holtzmangold, Waldmangold (9): Wald Mangold, Winter grün, Winter grün Pyrola, *Betula theophrasti* (11); Wildbete; wilde Rübe, wilde Runkelrübe (12); Bieza, Brittannica Bete, Scutla Beta (28); Strand-Runkelrube (35); Beisz-Izol, Romisch-Izol, Rograz, Mangolt (39);

French: limoine, beta de pre (6); bette sauvage (9); betterave maritime (12); pyrole (22); betterave sauvage (37);

English: wintergreen (6); spinach beet (7); Indian spinach; perpetual beet; savoy beet (12); sea beet (16); common green beet (21); cliff spinach (36);

Spanish: belesa (12); *Pyrola rotundifolia mayor*; acelga salvage, bleda boscana, ramolacha maritima (26);

Finnish: Rantajuurikas (12); **Danish**: strandbede (12); **Dutch**: strandbiet (12); **Iranian**: silijah, silaigah (6); **Czech**: Flepa bngalsky (12); **Syrian**: menda (6); selka (27); **Croatian**: primorska blitva (12); **Rumanian**: dacina (6); **Slovenian**: primorska pesa (12); **Swedish**: strandbeta (12); **Nepalese**: bangaali paaluugo (12); **Russian**: svekla primorskaia (12); **Portuguese**: Acelga brava (12); **Arabian**: Selq (18); Silk (27); **Hindi**: palangsag, palak, palanki (12); **Not specified languages**: cimonion; lynchitis, napi onjou, mendruta, lycosephalon, eleborosemata, scillon, cor lupi, veratrum nigrum (6) "*palung*" and "*mitha*", could have been a locally adapted sea beet (37) (Watt 1899 cited by von Lippmann 1925).

(1) Theophrastus Eresius (295 BC?) *Historia plantarum*. Reprinted in: Mancini FF (1900) La storia delle piante di Teofrasto. Ermanno Loescher & C, Rome, Italy

(2) Foës A (1657) *Magni Hippocrates medicorum omnium facile principis: Opera omnia quae extant in VIII sectiones ex Erotiani mente distributa*. Samuelis Chouët, Geneva, Switzerland

(3) Dioscorides (89 AD?) *Materia medica libro primo*. Reprinted by Ruellium (1529) *Dioscoridae phamacorum simplicium etc*. Apud Johannes Schottum, Strasbourg, Germany

(4) Pliny the Elder (75 AD?) *Historia naturalis*. Reprinted in: Storia naturale, vol 4–5. Giulio Einaudi Editore (1998), Milan, Italy

(5) Galen (190 AD?) *De almentorum facultatibus*. Reprinted in: Kühn CG (1833) *Medicorum graecorum opera*, vol 20. Officina Libraria Caroli Cnoblochii, Lipsia, Germany

(6) Roccabonella N (1457) *Liber simplicibus*. Manuscript, Church S.S. Johannes et Paulus, Venice, Italy

(7) Day HA (1917) Vageculture. Methuen & Co, London, UK

(8) Soldano A (2003) L'erbario di Ulisse Aldrovandi, vol 8–11. Istituto Veneto di Lettere, Scienze ed Arti, Venice, Italy

(9)   Fuchs L (1551) *De historia stirpium commentarii insignes*. Apud Balthazarii Arnolletum, Lyon, France

(10) Squalermo L (1561) *Liber de simplicibus etc.* Valgrisi, Venice, Italy

(11) http://www.plantnames.unimelb.edu.au/Sorting/Beta.html

(12) Durante C (1635) Herbario nuovo. Jacomo Bericchi et Jacomo Ternierij, Rome, Italy

(13) Chabray D (1666) *Stirpium sciatigraphia et icones ex musaeo Dominici Chabraei*. Colonia, Germany

(14) Dalechamps J (1587) *Historia generalis plantarum*. Gulielmum Rouillium, Lyon, France

(15) Parkinson J (1655) *Matthiae de L'Obel stirpium illustrationes*. Warren, London, UK

(16) Coles W (1657) Adam in Eden or natures paradise. Printed by F Streater, London, UK

(17) Hooper D (1937) Useful plants and drugs of Iran and Iraq. Field Museum of Natural History, Chicago MI, USA

(18) Sontheimer G (1845) Heilmittel der Araber. Frieburg, Germany

(19) Gandoger M (1910) *Novus conspectus florae Europeae etc.* Hermann et fils, Paris, France

(20) Bauhin G (1623) *Pinax theatri botanici etc.* Ludwig Regis, Basilea, Switzerland

(21) Parkinson J (1629) *Paradisi in sole paradisus terrestris*, or a garden of all sorts of pleasant flowers. Printed by Humfrey Lownes and Robert Young, London, UK

(22) Bauhin J (1651) Neu wollkommen Krauterbuch etc. JL Koenig, Basilea, Switzerland

(23) de Tournefort JP (1700) *Institutiones rei herbariae*, vol 1. Thypographia Regia, Paris, France

(24) de Lobel M (1591). *Icones stirpium seu plantarum tam exoticarum quam indigenarum etc.* Christoffel Plantyn, Anterwep, Belgium

(25) Ulbrich E (1934) *Chenopodiaceae*, pp 379–584. In: Engler A, Harms H (eds) Die natürlichen Pflanzenfamilien, vol 16c. Verlag von Wilhelm Engelmann, Lipsia, Germany

(26) Willkomm M, Lange J (1870) *Prodromus florae hispanicae*, vol 1. Sumptibus E Schweizerbart, Stuttgart, Germany

(27) http://www.jewishenciclopedia.com

(28) Fischer-Benzon R (1894) Altdeutsche Gardenflora. Untersuchungen über die Nutzpflanzen des deutschen Mittelalter. Verlag von Lipsius & Tischer, Lipsia, Germany

(29) Krashochkin VG (1959) Review of the species of the genus *Beta*. Trudy Po Prikladnoi Botanike, Genetike i Selektsii 32:3–35

(30) Reichenbach L, Reichenbach HG (1909) *Icones florae Germanicae et Helveticae*, vol 24. Sumptibus Federici de Zezschwitz, Lipsia, Germany

(31) http://www.ars-grin.gov

(32) http://www.mansfeld.ipk-gatersleben.de

(33) http://www.plantnames.unimelb.com

(34) Berti-Pichat C (1866) Corso teorico e pratico di agricoltura, vol 5. Unione Tipografico-Editrice, Turin, Italy

(35) von Lippmann EO (1925) Geschichte der Rübe (*Beta*) als Kulturpflanze. Verlag Julius Springer, Berlin, Germany

(36) **Note 1**: *Beta maritima*, now classified *Beta vulgaris* L. subsp. *maritima* (L.) Arcang (see Chap. 4), is called for the sake of brevity "*Beta maritima*". http://www.plantnamesunimelb.com.

(37) Becker-Dillingen J (1928) Die Wuerzelfructe (Rueben). In: Handbuch des Hackfruchtbaues und Handelpflanzenbaues. Paul Parey, Berlin, Germany

(38) Cordus V (1551) Cited by Letschert JPV (1993) *Beta* Section *Beta*. Dissertation. Wageningen Agricultural University. Papers 93-1, Wageningen. The Netherlands

(39) Cesalpinus A (1583) *De Plantis*. Apud Georgium Marescottum, Florence

Appendix E
Essential References

General Information about *Beta maritima* and genus *Beta* is available in the following publications:

Arnaud JF (2008) Importance de la dispersion dans la structuration génétique et l'évolution du système de reproduction chez une espèce gynodioique. Dossier de candidature. Université des Sciences et Technologies Lille 1, Lille, France

Bandlow G (1955) Die Genetik der *Beta vulgaris* Rüben. Der Züchter 4–5:104–122.

Bartsch D, Lehnen M, Clegg J, Schuphan I, Ellstrand NC (1999) Impact of gene flow from cultivated beet on genetic diversity of wild sea beet populations. Molecular Ecology 8(10):1733–1741

Biancardi E, Campbell LG, Skaracis GN, de Biaggi M (eds) (2005) Genetics and breeding of sugar beet. Science Publishers, Enfield NH, USA

Biancardi E, Panella LW, Lewellen RT (2012) *Beta maritima*: The Origin of Beets 9781461408420, pp 1–293

Bosemark NO (1993) Genetics and breeding. In: Cooke DA, Scott RK (eds) The sugar beet crop. Chapmann & Hall, London, UK, pp 67–119

Coons GH (1936) Improvement of the sugar beet. Yearbook of Agriculture, pp 625–657

de Vilmorin JL (1923) L'hérédité de la betterave cultivée. Gauthier-Villars, Paris, France

Dohm, J.C.a,b,c, Minoche, A.E.a,b,c, Holtgräwe, D.d, Capella-Gutiérrez, S.b,c, Zakrzewski, F.e, Tafer, H.f, Rupp, O.d, Sörensen, T.R.d, Stracke, R.d, Reinhardt, R.g, Goesmann, A.d, Kraft, T.h, Schulz, B.i, Stadler, P.F.f, Schmidt, T.e, Gabaldón, T.b,c,j, Lehrach, H.d, Weisshaar, B.dEmail Author, Himmelbauer, H.a,b,cEmail Author View Correspondence (jump link)

Driessen S (2003) *Beta vulgaris* subsp. *maritima* an Deutschlands Ostseeküste. Dissertation, RWTH Aachen, Germany

Frese L (2011) Conservation and access to sugar beet germplasm. Sugar Tech 12:207–219

Institut für Zuckerrübenforschung (IFZ) (1984) Geschichte der Zuckerrübe. 200 Jahre: Anbau und Züchtung. Albert Bartens, Berlin, Germany

Knapp E (1958) *Beta* Rüben. In: Roemer H, Rudorf W (eds) Handbuch der Pflanzenzüchtung, vol 3. Paul Parey, Berlin, Germany, pp 196–284

Letschert (1993) *Beta* section *Beta*, Biogeographycal pattern of variation and taxonomy. Dissertation, Wageningen University of Agriculture, Wageningen, the Netherlands

McFarlane JS (1971) Variety development. In: Johnson RT, Alexander JT, Bush GE, Hawkes GR (eds) Advances in sugar beet production. Iowa State Univ. Press, Ames IO, USA, pp 402–435

Munerati O (1979) *Opera omnia*. Reprinted in: Sulla barbabietola da zucchero. ISCI, Rovigo, Italy

Panella L, Lewellen RT (2007) Broadening the genetic base of sugar beet: Introgression from wild relatives. Euphytica 154(3):383–400

Saccomani M, Stevanato P, Trebbi D, McGrath JM, Biancardi E (2009) Molecular and morpho-physiological characterization of sea, ruderal and cultivated beets. Euphytica 169(1):19–29

The genome of the recently domesticated crop plant sugar beet (Beta vulgaris) (Article) (Open Access)

Villain S (2007) Histoire evolutive de la section *Beta*. Dissertation, Université des Sciences et Technologies de Lille, France

von Lippmann EO (1925) Geschichte der Rübe (*Beta*) als Kulturpflanze. Verlag von Julius Springer, Berlin, Germany

von Proskowetz E (1892–1910) Über die Stammpflanze der Runkel- und Zuckerrübe. Österreiche-Ungarische Zeitschrift für Zuckerindustrie und Landwirtschaft Nat 505(7484):546–549 (2014)

WEB Sites

http://www.botanicus.org
http://www.biodiversitylibrary.org
http://www.bioversityinternational.org
http://bibdigital.rjb.csic.es/ing/index.php
http://gallica.bnf.fr
http://catalogo.bnportugal.pt
http://www.archive.org
http://www.springerlink.com
http://www.sma.unibo.it/erbario/erbarioaldrovandi.aspx
http://books.google.com
http://scopus.com
http://scholar.google.it

Libraries

Accademia Veneta di Scienze Lettere ed Arti, Venice;
Biblioteca Nazionale Marciana, Venice;
Biblioteca ApostolicaVaticana, Rome;
Biblioteca Nazionale, Rome;
Biblioteca Ulisse Aldrovandi Bologna;
Biblioteca «Gabriele Goidanich» Bologna;
Biblioteca Centrale Universitaria, Bologna;
Biblioteca Ariostea, Ferrara;
Biblioteca dell'Accademia dei Concordi, Rovigo;
Biblioteca dell'Orto Botanico, Padua;
Biblioteca dell'Abbazia Benedettina, Praglia, Padua;
Biblioteca Nazionale «Vittorio Emanuele III», Naples.
Sopraintendenza per i Beni Culturali di Napoli e Pompei;